UNCONVENTIONAL NATURAL GAS

UNCONVENTIONAL
NATURAL GAS

Resources, Potential and Technology

Edited by M. Satriana

NOYES DATA CORPORATION

Park Ridge, New Jersey, U.S.A.

1980

Published in the United States of America by
Noyes Data Corporation
Noyes Building, Park Ridge, New Jersey 07656

Library of Congress Cataloging in Publication Data
Main entry under title:

Unconventional natural gas.

(Energy technology review ; no. 56)
Bibliography: p.
Includes index.
1. Gas, Natural. I. Satriana, M. J. II. Series.
TN880.U44   553.2'85'0973   80-15215
ISBN 0-8155-0808-5

Foreword

The quest for energy independence and the changing relative cost of energy exploration versus cost of importing fuel have encouraged the reexamination of energy deposits once considered impractical, such as unconventional sources of natural gas.

Unconventional natural gas deposits have been known for some time; however, the incentive to develop their potential has only recently come about, due to the Federal Energy Regulatory Commission's partial decontrol of "high-cost gas sources." While estimates of the resource vary, one source puts the figure at 65,000 trillion cubic feet, or roughly 50 times known U.S. natural gas reserves (*Chemical Week,* November 14, 1979). The main portion of unconventional natural gas occurs as geopressured methane; coal bed methane, tight sands formations and Devonian shale account for the remaining part.

This book provides, for those interested in exploring the subject, in-depth discussions of each of the four resources and its potential. Engineering and chemical personnel will find information on the methodology, equipment, and fracture fluids used for the release and recovery of unconventional natural gas. Several case studies, of ongoing projects involving tight gas basins and Devonian shale, are cited.

Because the information in this book is taken from multiple sources, it is possible that certain portions of the book may disagree or conflict with other parts of the book. This is especially true of opinions of future potential. These different points of view are included, however, in order to make the book more valuable to the reader.

Advanced composition and production methods developed by Noyes Data are employed to bring these durably bound books to you in a minimum of time. Special techniques are used to close the gap between "manuscript" and "completed book." Industrial technology is progressing so rapidly that time-honored, conventional typesetting, binding and shipping methods are no longer suitable. We have bypassed the delays in the conventional book publishing cycle and provide the user with an effective and convenient means of reviewing up-to-date information in depth.

The expanded table of contents is organized in such a way as to serve as a subject index and provides easy access to the information contained in this book which is based on various studies produced by and for government agencies. These primary sources are listed at the beginning of each chapter or section and at the end of the volume under the heading *Sources Utilized.* The titles of additional publications pertaining to topics in this book are found in the references at the end of the chapters or sections.

Some of the illustrations in this book
may be less clear than could be desired;
however, they are reproduced from the
best material available to us.

Contents and Subject Index

Introduction

As the energy shortage continues and as the supply of natural gas from conventional sources is depleted, development of the so-called unconventional natural gas resources becomes an increasingly attractive prospect. The sources of unconventional natural gas are:

(1) Tight gas basins such as the Uinta, Piceance, Green River and San Juan basins.

(2) Devonian shale of the Appalachian and other Devonian-Mississippian basins.

(3) Coal seams.

(4) The geopressured aquifers of the Gulf Coast.

Estimates of the quantity of methane occurring in these sources vary widely; this is especially true of estimates of methane in the geopressured aquifers. Estimates of the amounts of methane recoverable from these sources also vary considerably. Most experts agree, however, that with application of conventional and recently developed technology, and with additional incentive provided by partial decontrol, these sources could significantly augment the supply of natural gas.

This book describes the resources, their potential, and associated technology. The first chapter is an overview of the four resources. Each of the next four chapters gives a broader description of one of the resources. For the tight gas basins and Devonian shale, information on location, origin and extent is given, and data on physical properties such as permeability, porosity, and saturation are provided. Stimulation technology, estimates of the resource potential, recoverability and the economic feasibility of development are discussed.

The chapter on natural gas from coal seams includes data on chemical composition and methane content, discussion of differences in technology for recovery of gas from minable and unminable coal, the legal and environmental problems associated with recovery, and commercialization criteria.

For geopressured aquifers, the resource with probably the greatest quantity of methane in place, estimates are given of the total gas in place and the amount recoverable. Data are included on porosity, permeability, pressure and other physical properties. The economic feasibility of development, regulatory and permit

1

requirements, and environmental problems are discussed.

The final two chapters provide research and development studies relating to the tight gas basins and Devonian shale. The studies include reports on massive hydraulic fracturing projects, explosive fracturing experiments, production data and core analyses.

Overview of the Resources

The information in this chapter is based on *Help for Declining Natural Gas Production Seen in the Unconventional Sources of Natural Gas*, January 1980, NTIS PB80-128929, report to the Congress by the Comptroller General of the United States, prepared by the General Accounting Office, Energy and Minerals Division.

INTRODUCTION

Over the past several years the Nation has produced and consumed natural gas at a faster rate than it has discovered new reserves. The Nation has proven gas reserves estimated at 200 Tcf which will be consumed in about 10.5 years at the current rate. Yearly additions to U.S. reserves have been declining, and a long-term reversal of this trend is generally not expected.

To return to lower-48 gas production of about 20 Tcf per year would require the discovery of almost 15 Tcf per year of new gas reservoirs for about 4 or 5 years. Most consider such a level of discoveries very unlikely. Since discoveries averaged 4 Tcf between 1967 and 1976, such a discovery rate would imply finding at least 4 or 5 new reservoirs or fields over the period with reserves on the order of 10 Tcf each. The Gomez field of West Texas is the only field 10 Tcf or larger discovered since 1945 in the lower-48. Further, there have been only 2 fields between 4 and 10 Tcf discovered during the same period.

Assuming an extremely high rate of discoveries in response to higher natural gas prices, the American Gas Association estimated that lower-48 production will be about 18.6 Tcf in 1985 and 19.4 Tcf in 1990. However, the Association's projections show that additions to reserves will peak in 1983 at 20.0 Tcf and fall to 17.3 Tcf in 1990. These declines in reserve additions would eventually be reflected in production decreases.

The Conference Committee report on the Natural Gas Policy Act of 1978 (Public Law 95-621) shows that the deregulation provisions of the act are not expected to reverse the expected production decline in the long run (1); production from the lower-48 states is expected to be about 16.6 Tcf in 1985.

The recent increased gas supply situation has not changed the expectation of declining production. The current supply situation has been brought about by the curtailments and fuel switching of previous winters and the availability of added gas supplies from the intrastate markets. Because no fundamental changes in domestic natural gas production have contributed to the current level of supplies, this supply situation should be considered temporary.

Some of the Nation's natural gas customers must switch to other fuels or rely increasingly on alternative sources of natural gas in the future. This trend has started already. Previous gas production losses have been replaced by imported oil or by switching to coal, electricity, or other fuels. For example, about one-half of the oil import growth from 1974 through 1977 has been directly attributed to declining natural gas production, according to the American Gas Association. Also, the U.S. continues to import gas from Canada and has negotiated purchases of Mexican gas.

However, importing natural gas and oil contributes to the U.S. balance of payments deficit. For example, if domestic gas supplies are 1.5 Tcf greater in 1985 than they would otherwise be, U.S. oil imports could be reduced up to 750,000 barrels per day. This would represent about 12% of total oil imports and, based on the cost of oil at $20.00 per barrel, could contribute up to $5.5 billion toward improvement in the balance of payments.

Even though the U.S. may continue to purchase oil and gas from foreign sources, imports could be reduced to some extent by (a) greater conservation, (b) increased U.S. oil and gas production, and (c) conversion to alternative domestic supplies such as coal and renewable energy sources. Within that framework, the major options for maintaining domestic gas production past 1985 appear to be limited to new conventional sources, such as the Outer Continental Shelf, coal gasification, and producing gas from deposits which are usually termed unconventional natural gas sources. Continued gas production enables the Nation to take advantage of the substantial pipeline network and infrastructure of the gas industry. For the Department of Energy (DOE) to push immediate commercialization of the unconventional gas sources successfully, technical feasibility and profitability should be demonstrated.

This chapter provides an overview of the following four unconventional gas sources:

Tight gas basins—the low-permeability or tight gas sands of the Rocky Mountain Region;

The gas-bearing Devonian shales of the eastern U.S.;

The methane contained in coal seams; and

The methane contained in the geopressured aquifers of the Gulf Coast Region.

TIGHT GAS BASINS

Twenty major tight sands basins with large amounts of natural gas stretch from New Mexico northward into Canada and eastward into Arkansas and Louisiana. Many sandstones in these basins also have a high resistance to gas flow which poses recovery problems. However, industry has been active in all types of tight sand deposits and gas from one type of deposit is produced in significant quantities. DOE is attempting to stimulate the development of the tight sand deposits which present great technical challenge. Based on technical feasibility, the prospects for profitable operations, and industry interest, continuing increases in production from tight sand areas can be expected.

The Resource Base

The locations of tight sand gas deposits have been known for over 30 years, but with the exception of the most favorable areas, this resource has not been developed. The three basins with the greatest exploration history are in the Rocky Mountain area—the Piceance Basin of Colorado, the Green River Basin of Wyoming, and the Uinta Basin of Utah. These basins are commonly referred to as the Western Tight Gas Basins.

Drilling has occurred in many other tight sands gas areas. However, most were quickly described as noncommercial and little geologic data are available on these deposits in comparison to conventional oil and gas fields. Although the characteristics of these deposits vary considerably, they are generally classified in four ways:

Tight Blanket Sands, blanket shaped, relatively continuous deposits of great lateral extent often found at great depth;

Western Tight Basins, blanket and lens-shaped (lenticular) deposits;

Tight, Lenticular Basins, in contrast to the Western Tight Basins, these areas have larger lens-shaped deposits at shallower depths;

Shallow Basins, shallow deposits which vary from blanket to lenticular; and

Other, reservoirs with special development and engineering problems.

In regard to depth, it should be noted that there are no set industry standards for defining various drilling depths. However, one DOE official provided the following guidelines: shallow wells extend to 4,000 ft, medium depth to 10,000 ft, and deep wells below this depth.

Many of these basins cover large land areas, exploration data are limited and, consequently, the resource estimates vary widely. Estimates of gas in place are as high as 1,200 Tcf and estimates of recoverable gas range from 25 to over 600 Tcf. The Lewin report (2) concluded that 14 of the tight basins contained 409 Tcf of recoverable gas.

How much of this resource can be produced? Under favorable recovery and price assumptions some estimate that about half of this resource can be produced, including significant quantities by 1990. For example, the Lewin report estimated that as much as 7.7 Tcf of gas could be produced each year by 1990 at prices up to $4.50 per Mcf. However, a large portion of this production would likely be produced from the blanket sands formations where industry has already been active and has amassed a considerable amount of basic geologic and engineering data.

Industry Perspective

Substantial development efforts have been and are underway in several areas. By far, most drilling is in the blanket sands areas where the reservoir distribution is more predictable and the lateral continuity of the gas deposits improves recovery. In addition to the production activities, industry is conducting research with Government assistance in tight sand areas. Cumulative production in tight sand areas through 1975 was at least 12.5 Tcf; about 84% of this amount was produced from the blanket sands. As shown below, production data from 1974 through 1975 identified over 6,000 producing wells.

Target Formation	Number of Producing Wells	Annual Production (Tcf)
Blanket	4,371	.44
Tight lenticular	1,408	.12
Western (lenticular)	373	.06
Shallow	280	.01
Other	174	.11
Total	6,606	.74

DOE estimates current annual production is over 0.85 Tcf, which is about 4.5% of total U.S. gas production.

A common characteristic of the tight sands is their low permeability. This resistance to gas flow is 5 to 2,000 times greater than typical oil and gas producing formations. Also, the gas-bearing formations range from single, relatively thin (10 to 100 ft thick) gas-bearing beds of generally uniform thickness covering large areas to multiple lens-shaped layers of sands interbedded with clays and shales. The varying shapes of the gas-bearing sands and their high resistance to gas flow have made exploration and production from these deposits much more difficult than typical conventional gas deposits.

In order to increase the flow of gas from tight sands deposits, they are fractured, often using hydraulic pressure. Advances in fracturing effectiveness are one key to increasing production from tight sands. For the lens-shaped deposits it will be important to develop drilling and fracturing methods which will intersect multiple sand lens layers. In addition, hydraulic fracturing treatments must be designed to prevent wells from clogging with sand or clay which may lie in the well area.

Government Actions to Spur Development

The Government has taken action to promote further development of this resource. The Western Gas Sands Project has been established within DOE to inventory and characterize the tight sand resources and to test well stimulation methods. The Natural Gas Policy Act of 1978 provided that, after November 1, 1979, gas producers may charge deregulated prices following (1) submission of documentary evidence describing qualifying gas bearing deposits to the jurisdictional agency; (2) approval by the jurisdictional agency; and (3) approval by the Federal Energy Regulatory Commission (FERC). (The FERC has authorized the decontrol of tight sands gas prices as of November 1, 1979.)

The Western Gas Sands Project: Through fiscal year 1978, DOE spent $10.2 million on the Western Gas Sands Project, with $7.5 million budgeted for fiscal year 1979 and $8.8 million requested for fiscal year 1980. Project activities include: a five year resource characterization program being performed by the U.S. Geological Survey for DOE; laboratory research and development; and field projects to test methods of stimulating tight sand gas formations. The field projects are expected to cost $23.2 million with eight companies providing $13.6 million of the total. DOE is centering its research and development activities in the three western basins containing lenticular formations.

The objectives of DOE's Western Gas Sands Project have been to define the resource base more accurately, determine the reservoirs' physical and chemical properties, determine appropriate stimulation technology, and assess potential reserves and demonstrate economic productivity to encourage industrial development.

Future Production Possibilities

Future production levels are conditioned by the type of sand deposit, its characteristics, technical developments, gas prices, and in turn, the pace of industry activity. These factors will, in combination, play the major roles in determining the extent these resources will be produced. With the exception of blanket sands, many of which are considered commercial at this time; additional knowledge of the sand's characteristics and geologic structure of the basins will greatly improve the potential for recovery. Such resource knowledge, combined with technical developments which overcome the difficulties of production, can reduce the risks of exploration and production.

With or without dramatic improvements in recovery rates, blanket sand deposits are likely to continue to play an important role in future natural gas production levels. Predictions of gas production from the other tight sand deposits are speculative and dependent on technical advancements and prices which permit higher cost recovery methods to be profitable. The Lewin report (2) estimated that as much as 7.7 Tcf of gas could be produced annually from the tight sands areas by 1990 including production from the blanket sand deposits. Except for the blanket sand areas, technical advancements are needed in order to achieve the estimated production levels.

Conclusions

The blanket sand areas have been under commercial development by industry for years. For the other types of tight sand deposits, the high resistance to gas flow is a major technical problem which must be overcome by improvements in well fracturing technology. This problem is compounded in the areas with multiple, thin, lens-shaped sands deposits. For these deposits, fracturing technology must be developed to intersect the multiple-sand lenses simultaneously.

DEVONIAN SHALE GAS

The eastern Devonian gas shales are located in an area from New York to Alabama and extending westward to Ohio, Kentucky, and Tennessee. Similar deposits occur in Illinois, Indiana, and Michigan. They contain sizeable volumes of natural gas. Industry interest in Devonian shale gas areas is evident, but generally low gas recovery rates have prevented widespread development. DOE is attempting to improve the recovery rates, and the price of shale gas has recently been deregulated. Based on technical feasibility, the prospects for profitable operations, and industry interest, gradual increases in natural gas production from the eastern shales could be forthcoming.

The Resource Base

Close to urban and commercial centers in the eastern U.S. are shales containing an abundant accumulation of natural gas originating from the decomposition of plant and marine life. The gas has been used as fuel by a few resourceful individuals, gas utilities, and independent producers since 1820. Annual Devonian shale gas production is about 0.1 Tcf and cumulative production is about 3.0 Tcf.

The Devonian shales derive their name from the geologic period in which they accumulated some 350 million years ago. Some geologists believe these shales may extend over a 500,000 square mile area from the Appalachians to the Rocky Mountains and from Canada to Mexico. Major concentrations of current interest are located from New York to Alabama, and extend westward into Ohio, Kentucky, and Tennessee, with similar deposits in Illinois, Indiana and Michigan.

Available evidence indicates that virtually all Devonian shale contains some gas, and its gas-producing ability is generally indicated by color which ranges from gray to deep brown to black. The black and very dark brown shales are generally believed to be better gas producers and brown shale areas support most commercial wells today.

Estimates of the Devonian shale gas resource base vary considerably as well as the amounts which are recoverable. Before DOE's research and development program, the resource base had been estimated to range from 75 to 700 Tcf. Recently, officials of the Department of the Interior's Geological Survey, involved in mapping and characterizing this resource as part of the DOE program, made a preliminary resource estimate of 10 to 520 Tcf. The range of these resource estimates reflect that a high degree of uncertainty continues to exist.

In many areas the concentration of free gas within the shales may be too low to be produced at reasonable prices. Some experts have estimated that only 10% of the shale gas can be recovered due to low concentration. In contrast, typical oil and gas reservoirs may produce 40 to 60% of the total resource in place. Further testing of the resource is needed to obtain better estimates of the amounts of gas which are recoverable.

Industry Perspective

The gas industry has drilled over 8,300 wells in the Big Sandy Devonian shale gas field of eastern Kentucky, and about 4,600 in areas of other eastern States which contain shale deposits. Although data from some of these wells have been used for evaluating shale gas well profitability and future production, all of these wells cannot be classified as shale gas producers. Other gas-producing strata above or below the Devonian shales are important contributors to the gas supplied by these areas.

Because there are other gas-producing strata in shale gas areas, a Federal Energy Regulatory Commission official said it may be difficult for a producer to determine if wells in shale areas are producing gas only from the Devonian shale or from both shale and other producing strata at the same time. When such situations occur, it will be difficult to apply consistent regulatory treatment. A method for resolving this question has not been developed.

The shale deposits which have been analyzed show that gas is trapped tightly in the shale itself, and the shale provides only small spaces to hold gas or other substances. Thus, the ability of gas to flow through the small spaces in the shale is low. Flow through a porous medium is termed permeability, and low-permeable substances offer great resistance to the free flow of gas. The permeability of shale deposits has been shown to be 2.5 to 1,000 times lower than typical oil- and gas-producing deposits. The permeability of typical oil- and gas-bearing strata ranges from 5 to 2,000 md; Devonian shale ranges from 0.001 to 2 md. Measures of shale porosity, or the extent that pores within the shale are filled by gas, indicate that porosity levels are 4% or less. Typical producing formations are 8 to 30%.

These characteristics of gas-bearing shales vary. Some shale deposits have such low quantities of gas or potential for releasing gas that they can never be produced. In other deposits, these conditions result in slower gas recovery rates than from conventional reservoir rocks even though shale wells are fractured. Although shale gas wells have produced at slower rates than industry norms, they continue to produce gas at relatively constant rates for extended periods of time. This is not as attractive as conventional gas deposits which pay off investment costs in less than half the time required for shale gas wells. Nonetheless, some companies have found that shale gas wells can be profitable.

The Columbia Gas System Service Corporation, one of the largest corporations involved in shale gas production, provided examples of shale well economics. Its information is based on production data for about 2,000 wells and prices set before passage of the Natural Gas Policy Act of 1978.

Columbia found that most of its shale gas wells have produced at commercial rates for more than 35 years. However, the shale wells have provided a low return on investment compared to conventional wells. Also, the break-even point for investments in shale wells has been different. The company can recover the cost of establishing a conventional well in less than 2 years, in contrast to its shale wells which require about 5½ years to recover initial investment costs.

This slow rate of recovery appears to be an important consideration to the driller exploring for gas in the Appalachian region. Because these drillers are typically small independent operators, they usually lack the resources to assume risks greater than the industry norm. If an operator's financing arrangements are set with the expectation of conventional well payoff times, shale wells will not appear to be an attractive investment.

If technical improvements in shale well fracturing methods increase the initial gas recovery rates without decreasing long-term production, shale well profitability will improve. The goal of fracturing is to open fractures or joints in the shale so gas can flow more readily into the well. Two basic fracturing methods have been used; they are, applying hydraulic pressure and setting off explosions in the wells. There are several methods within these two categories and further variations within each method.

For example, hydraulic fracturing can be carried out with water, foam, or liquefied gas. In each hydraulic fracturing treatment, a variety of ingredients such as sand, water, and chemicals are injected into the shale after drilling. A basic variation of this method is massive hydraulic fracturing which includes higher injection rates and larger quantities of fluid than regular hydraulic fracturing treatments. The choice of fracturing method and mix of ingredients for Devonian shale is site-specific and there is no agreement whether hydraulic or explosive fracturing is generally the better method. Both methods are currently in use, and DOE is testing variations in fracturing methods to improve shale gas recovery rates.

Technical variations in drilling are also being considered as potential methods for improving shale gas production. For example, deviated drilling, or drilling at an angle less than perpendicular, is being attempted by DOE and private industry to intersect more sets of natural fractures and to increase the potential area of gas drainage. Although more costly than vertical drilling, it is hoped that such wells will increase gas flow. Further experimentation and production history is required before advanced drilling and stimulation practices are regarded as reliable, effective methods of improving shale well gas production.

Government Actions to Spur Production

The Government has taken two major actions to spur the commercial development of this resource. In 1976, DOE initiated the Eastern Gas Shales Project to perform research, development, and demonstration activities. The second major action was the decontrol of Devonian shale gas prices as part of the Natural Gas Policy Act of 1978 (3)(4). Several additional options also remain open for Government action including variations in DOE's research and development program and the addition of financial incentives. The rural development proposals of the Administration would also provide incentives for small rural communities to use these resources.

The Eastern Gas Shales Project: Three important issues are being addressed by DOE in this project: How much gas is contained in the shale? Where is it located? How can it be developed commercially? The project was designed as an 8 year effort to develop the information and technology needed to attract large-scale commercial production of shale gas. About 35 organizations including State geologic survey teams, universities, private research laboratories, and gas companies are participating in the project along with the U.S. Geological Survey. DOE's proposed budget for the complete project was $135 million, with industry contributing an additional $45 million. The project is scheduled for completion in 1984.

The project includes efforts to inventory and characterize the shale resources, improve and test well completion methods, and disseminate the results to industry. The first effort is designed to determine the characteristics of the resource, the resource's location, and the quantity which is in place. The second effort is designed to increase the profitability of shale fracturing methods. In addition, the project had several specific goals:

> To increase the average open-flow production rate of new shale wells from 100 to 300 thousand cubic feet or more of gas per day;
>
> To increase the average total gas reserves added per well drilled from 300 to 600 million cubic feet; and
>
> To add 3.5 to 7.0 Tcf of gas to the proven reserves in the Appalachian basin by 1985.

According to several gas industry officials, these goals are reasonable, achievable, and adequate to stimulate significant growth in gas production from the Devonian shales.

Through fiscal year 1978, the Government has spent $35.5 million and industry has contributed an additional $8 million. Officials of DOE and the U.S. Geological Survey, who are coordinating the resource characterization and mapping work, are due to complete the work during fiscal year 1980. The U.S. Geological Survey will make another estimate of the resource before completion.

As of March 1979, 34 wells have been completed and evaluated by the Eastern Gas Shales Project. Twenty of the wells showed increased gas flow rates following a fracturing treatment, and 14 wells were classified as unsuccessful. The results indicate that the preferred stimulation method is hydraulic fracturing in areas where the shales have a low level of natural fractures. In areas with a high level of natural fractures, explosive stimulation appears to be the preferred method. The cost of recovery so far has ranged from $2 to $6 per thousand cubic feet. This price range indicates that some areas may be commercial, but further improvements in flow rates are still needed to make large areas commercial.

DOE's fiscal year 1980 budget proposes a $9 million reduction from the $18 million budget of fiscal year 1979. This budget cut reflects the advanced status of resource characterization, the state of technical development compared to the other unconventional resources, and the expectation of additional production due to the Natural Gas Policy Act.

Devonian Shale Gas Prices Decontrolled: This study performed by the Office of Technology Assessment (5) (see chapter in this book entitled "Devonian Shales) and two 1978 studies performed for DOE indicated that prices between

$2 and $3 per Mcf should be sufficient to cover the costs of shale gas production in some areas. The Office of Technology Assessment study used constant 1976 dollars as a basis for economic calculations, while the other studies used constant 1977 dollars. Prices above $2 may now be charged according to the Natural Gas Policy Act of 1978 since Devonian shale gas prices have been decontrolled. Under decontrol, the incremental pricing rules require that certain production costs must be passed-through to industrial consumers; but, shale gas pricing can now be determined by market decisions. The incremental pricing rules prescribed in Title II of the Natural Gas Policy Act would operate to increase prices to certain industrial gas consumers until the price they pay equals an equivalent price of substitute fuel oil. This provision provides some protection against rapid price increases for residential gas users and others such as schools, hospitals, and agricultural facilities.

Previous analyses show that decontrol should result in prices for Devonian shale gas which are sufficient to cover production costs. However, these analyses did not discuss the effect of an incremental pricing provision on high cost gas. Several DOE officials have said that the pricing provisions of the Natural Gas Policy Act are likely to increase attention on conventional gas deposits, but not the unconventional deposits. This option is based on the theory that industry drillers will continue to be attracted to conventional deposits where payoff times are shorter due to higher gas flow rates, and risks are decreased due to familiarity with the resources.

Other Actions to Spur Development: Additional financial incentives could be provided to accelerate industry drilling for Devonian shale gas and tax incentives can be considered an effective tool for reducing payoff times for Devonian shale wells. According to the Office of Technology Assessment, a 22% depletion allowance would be as effective as a price increase of $0.50 per Mcf. This would certainly affect the payoff time gap between a conventional and a Devonian shale gas well.

The President proposed a set of special incentives for the unconventional gas resources in May and July 1979. These include:

A $0.50 per Mcf tax credit which would be gradually reduced as prices increase so that at the world oil price equivalent of $28 per barrel no tax credit would be provided; and

A $300 million grant, loan and loan guarantee program for rural communities to develop coal bed gas or Devonian shale gas projects if they would provide local benefits.

The decontrol of shale gas prices under the Natural Gas Policy Act may provide some incentive, even though the precise effect of decontrol and the incremental pricing provisions on production cannot be predicted with certainty. The principal benefit of decontrol with incremental pricing and the proposed tax credit in the near-term would likely be accelerated exploration of shale gas areas and the additional resource knowledge that would result. If exploration of shale gas areas accelerates, additional gas production could follow.

The proposed grants, loans, and loan guarantees for rural committees are dependent upon the successful completion of a demonstration program. Demonstrations are being undertaken in six communities in areas where shale deposits and coal beds exist, to determine if dependable gas supplies can be developed for community use at competitive prices. A $700,000 grant has been made to the American Public Gas Association to initiate the six recovery projects. Details of the $300 million program, including funding requests, will be finalized following successful completion of the six demonstration projects.

Future Production Possibilities

As described above, the pace of industry's activity will be conditioned largely by shale gas quality, technical improvements, and gas pricing. These factors are interrelated and will, in combination, play a major role in determining the future of Devonian shale gas. Any estimates of future production are based on preliminary assessments of these factors, and numerous assumptions are made when projecting future production rates. The pace of industry activity could also be influenced by other factors including time required to obtain drilling permits and environmental clearances, obtaining legal rights to the gas, and the available ability of workers and materials.

How much of the shale gas resource can be produced in the next 15 years? The Lewin report (2) estimates that 0.7 to 0.9 Tcf might be produced each year by 1995 based on prices up to $4.50 per Mcf. Also, the study by the Office of Technology Assessment indicated that yearly production could reach 1 Tcf about 1990, based on prices in the range of $2.00 to $3.00 per Mcf. The study concluded that

> There appears to be no practical way short of creating the economic incentives necessary to induce an extensive drilling effort, to ascertain whether the Appalachian Basin shale might actually contribute, more or less than 5% of the total U.S. natural gas supply.

Accordingly, economic and technical uncertainties must be resolved in order to realize this production potential.

Conclusions

Gas production from the eastern shales is commercial in some areas today. The principal deterrents to widespread investment in shale gas wells in the past have been traditionally low production rates and low gas prices. These conditions favored investments in conventional gas wells which recover investment costs in less than half the time required for shale gas wells. If DOE's research and development program and industry, acting both with DOE and independently, continue to demonstrate techniques which improve gas recovery rates within the range of competitive prices, investments in shale gas wells will be more favorable.

Changes in gas pricing are also affecting the profitability of shale gas wells. Decontrol of shale gas prices permitted by the Natural Gas Policy Act appears to make additional shale gas prospects commercially attractive; but the effects of decontrol coupled with incremental pricing are uncertain.

METHANE FROM COAL SEAMS

Coal beds in the eastern U.S. contain significant quantities of natural gas, but insufficient information is available to judge the overall quality and potential of methane in western coal beds. Eastern coal beds have produced small, but locally important quantities of gas and local use of coal bed gas may increase. Coal bed methane has also caused disastrous coal mine explosions. The coal industry has been primarily interested in the removal of this methane from coal mines as a safety precaution. The gas industry has shown minimal interest in the energy potential of this resource to date.

The Resource Base

In the eastern and western coal States considerable quantities of methane are trapped within and adjacent to the coal beds. In the East this resource is located near major population and industrial centers. Little use has been made of this coal gas in the U.S., but in some European countries coal bed gas production is common with coal production, although the gas is often mixed with air.

Methane in coal was generated naturally during the coal formation process and is trapped within the coal itself, in the natural fractures within the coal, and in the strata adjacent to the coal beds. Because coal is relatively impermeable, any methane which is recovered must generally flow through the natural fractures of the coal. For this reason, coal beds which are highly fractured appear to be the best sources of methane.

The content of methane in coal varies considerably, and limited data have been collected on this subject. Due to lack of data, estimates of the resource size are speculative. However, studies of eastern coals have provided some information. Based on the U.S. total coal reserves, estimates of coal bed methane resources range from 72 to 860 Tcf.

However, much less of the gas than the total resource base is recoverable. Again, estimates of the recoverable quantity are speculative due to the lack of data. Nevertheless, experts agree that some coals contain little recoverable methane and many coal beds are too thin for commercial recovery. Estimates of recoverable coal bed methane range from 2 to 487 Tcf.

Limited data on the location and characteristics of some types of coal are part of the information problems. Although resource and geologic data on mineable coal fields in the eastern U.S. are extensive, practically no data exist for western coal and unmineable coal—coal which is too deep or thin to be mined currently. The importance of such data is evident because unmineable coal, especially in the western States, is assumed to be very plentiful. Also factors such as the locations of promising sites, the permeability, fracture system, and methane content of coal influence recovery economics.

Industry Perspective

The gas industry has made few attempts to recover coal bed methane; its attention has been focused on more readily exploitable conventional gas deposits. The coal industry has been concerned primarily with diluting methane with air during mining to prevent explosions. Current practice is to ventilate coal mine entries with air to dilute the methane below explosive concentrations and to exhaust it to the atmosphere. Presently, neither industry is planning major efforts to produce coal bed gas.

The Gas Industry: There are examples of gas industry production of coal bed methane. The Equitable Gas Company of Pittsburgh has been involved in two recovery efforts in the Appalachian area. In 1892, gas flows were detected from the Pittsburgh coal bed during the closing of an old gas well which had been drilled through the coal bed. This well was recompleted to produce the coal bed methane and remained productive until 1968. In the same vicinity, an additional 23 wells were drilled into this naturally-fractured coal bed before 1950. Through 1974 these wells produced 1.7 billion cubic feet of gas.

The Equitable Gas Company is continuing to develop gas prospects in the Pittsburgh coal bed. The company believes substantial quantities of methane could be recovered from Appalachian area coals. Equitable has gas rights for about 6% of the land in several counties under lease which are estimated to contain the equivalent of 10 years' gas supply required by Equitable.

An Equitable official described the company's efforts to obtain coal bed methane. Equitable contacts coal owners, describes the recovery operations, and pays a royalty for any gas recovered. The coal beds currently being tapped are not now being mined; the methane content is high, and no fracture stimulation technology is used when completing the wells. The time required for wells in these areas to recover investment costs is 3.5 to 5.0 years. According to Equitable, the major barrier to widespread production in the East is convincing information which shows coal owners that future mining will not be affected or damaged by gas production from coal beds.

Very little interest in western coal bed methane has been expressed by the gas industry. An official of one company which operates in the western States said that there is almost a complete lack of knowledge of western coal depth, porosity, permeability, and methane content. More information on these factors must be obtained before recovery attempts can be made.

The Coal Industry: The frequency and severity of coal mine disasters caused by explosions of coal bed methane are well-known. These disasters have highlighted the need for methane removal. For many years improving ventilation or slowing the pace of mining were the only methods of reducing methane concentrations in the mines. The current practice of the U.S. coal industry is to increase the rate of mine ventilation. Between 70 and 90 billion cubic feet of methane are annually vented to the atmosphere. Unless the industry begins to capture this methane, the amount will increase with the pace of deep coal mining.

In Europe, mining practices now include capture and use of coal bed methane. In this country, the U.S. Bureau of Mines, Department of the Interior, has been developing methods to drain coal beds of methane to reduce coal mining hazards. These methods can be used to recover the methane for use or sale, but are not yet used widely by the coal industry.

The coal industry's primary interest is the coal itself; its only concern with methane is how fast it can be vented to maintain mine safety. Therefore, it has been considered a nuisance. While noting that the Bureau of Mines methane drainage methods have potential, coal companies are not yet convinced these methods are sound investment. Even so, at least two coal companies are now using the predrainage techniques to vent the gas, and other coal companies are now expressing interest in this resource.

However, the coal industry is concerned about the use of well stimulation technology to extract gas from coal beds. If stimulating coal beds to increase gas flows also causes unstable mine roof conditions, mining operations become more difficult, costly and may be foregone. Although the Bureau of Mines has successfully applied stimulation methods during several experiments, Consolidation Coal Company cited its own attempt which caused a roof fall adjacent to the well.

The Ownership Question: Coal companies believe ownership of methane in coal is inseparable from ownership rights to the coal itself. In contrast, the oil and gas industry believes that if gas rights remain with the surface owner then they have the right to produce gas from any formation including the coal beds. A problem obviously arises when the coal rights and the gas rights are held by different individuals. Very few deeds or leases for coal rights mention the methane in the coal, but courts have ruled that coal owners have the legal right to remove gas for safety reasons as part of their access right to coal.

Complicating the problem are deeds or leases that name coal as a specific substance and then include a general statement about "other minerals." In most States, the meaning of "minerals" includes oil, gas, and petroleum products unless another meaning is specified in the legal document. The definition of minerals as it applies to methane in coal may require clarification.

Several court opinions support the petroleum industry view of ownership, but a pending case in Pennsylvania again raises the issue. A coal owner is attempting to prevent a gas producer with gas rights from drilling into a coal bed. Even though such legal questions could delay development, industry officials believe that these legal questions could be resolved if there were sufficient interest in producing the resource.

Government Actions to Spur Development

The U.S. Bureau of Mines and DOE studies have indicated that the economic feasibility of coal bed methane gas recovery is highly probable. The accumulation of sufficient data to encourage industry participation is a principal goal of Government research. In addition, the price of methane from coal beds has been deregulated under the Natural Gas Policy Act of 1978 (3)(4), but its potential effect on production is unknown at this time. The main market for this gas in the future may be limited to on-site space heating or local power generation.

Responsibility for work on methane in coal beds is split between DOE and the Bureau of Mines. DOE is responsible for the methane recovery, capture, and utilization from coal beds which are too deep or thin to be mined. The Bureau of Mines is responsible for mine health and safety research which includes draining methane from coal beds in advance of mine openings. Coordination between the agencies is informal.

Department of Energy: DOE's Methane Conservation Production and Utilization Project is developing methods and systems for using gas from coal seams which are too deep or thin to be mined. The project's purpose is to promote commercialization of the resource for industry use or for communities close to potential deposits. The project was initiated in fiscal year 1978 with a $2 million appropriation. Funding for fiscal year 1979 was doubled, and $5.0 million was requested by DOE for fiscal year 1980. About 20 contracts are now funded and contract activities include 8 tests of recovery and utilization techniques. Multiple completions in thin coal beds, directional drilling techniques, hydraulic fracturing, and various utilization schemes such as space heating and power generation are being tested.

In addition, the Administration's May 1979 rural development initiatives include a $700,000 grant to the American Public Gas Association to initiate projects to recover gas from coal or shale at six selected communities. The grant, to be administered by DOE, could amount to $3.8 million if the projects are completed.

Should these pilot projects demonstrate that dependable gas supplies can be developed at competitive prices, $300 million in grants, loans and loan guarantees are proposed to be made available for widespread adoption of the proven recovery techniques. Program details are to be developed at a later date and implemented if the demonstrations are successful. Under the initial plans, the Departments of Agriculture and Commerce would administer this program and provide funding through existing legislative authority. The May 1979 White House initiative estimates that 6,500 rural communities with populations of 10,000 or less are potential beneficiaries of such a program.

The U.S. Bureau of Mines Methane Control Research Program: The Bureau's Methane Control Research Program was started in 1964 to develop technology for safe and economic mining of methane-laden coal beds. The technology is aimed at eliminating mine disasters caused by accidental methane ignition during shaft sinking and subsurface excavation. According to the Bureau's 5-year plan, beginning with fiscal year 1979, the expected cumulative funding is $11.5 million.

The Bureau's results to date have been promising and several methods of methane removal have been developed using holes bored vertically into the coal beds from the surface and holes bored horizontally from the bottom of mine shafts and mine entries. These methods are to be applied 3 to 5 years in advance of the mine opening so methane levels can be reduced. The Bureau's methods have been successfully applied in the Pittsburgh coal bed. One of these methods includes hydraulic stimulation to enhance gas flow from the vertical boreholes. The coal industry, however, remains skeptical as to the safe mineability of coal seams after such stimulation.

The Bureau cites the major benefits from methane drainage as reduction of hazards, lower ventilation costs, reduced mine development costs and increased productivity, and small capital investment with the potential to recover investments from gas sales. In addition, an economic analysis using Bureau methane drainage methods projected that $10 million could be earned from the sale of gas for an additional investment of $2 million.

Even though the Bureau believes its demonstrations are clearly effective, their methods have not yet been widely accepted by the coal industry. Less than one-third of the 2.5 Bcf of gas removed from mines during Bureau-sponsored demonstrations has been sold. The Bureau also points out that effective drainage can be achieved without stimulating coal beds through holes drilled from mine shafts and entries.

Coal Seam Methane Price Decontrolled: Sufficient examples upon which to base a reliable portrait of coal bed methane economics have not yet been developed. However, various studies have concluded that methane drained from coal beds in advance of mining, and coal bed methane wells not related to mining operations could be produced at prices ranging up to $3.00 per Mcf, depending upon the effort required for recovery. Of course, the costs could be higher or lower, depending on the specifics of each recovery operation.

The decontrol of coal bed methane prices permitted by the Natural Gas Policy Act should result in prices which are sufficient to cover production costs in some areas (3). As with shale gas production, the incremental pricing rule could affect industry's interest in producing this resource. However, local users may constitute the initial market for this product and the incremental price rules may not apply in such cases.

Other Actions to Spur Development: Federal efforts could advance the technology and demonstrations needed to show industry that this resource is economically advantageous and answer the uncertainties of fracturing techniques. Also, Federal demonstrations might prove that the use or sale of methane removed from coal beds prior to mining is an economically attractive addition to traditional mining operations.

In addition to decontrol and the research and development activities, additional incentives are proposed for encouraging coal bed methane recovery. As mentioned earlier, the Administration's May 1979 rural development initiatives include a proposal for $300 million in grants, loans and loan guarantees. These incentives would be made available to rural communities if DOE demonstrations prove the resource can be of local benefit. The President's July 1979 energy initiatives also include a proposed $0.50 per Mcf tax credit.

Future Production Possibilities

Future production from both mineable and unmineable coal beds is dependent on added research and development, but, some recovery methods are now technically feasible and potentially profitable. For these areas, such as coal bed methane drainage in advance of mining, additional demonstrations appear to be necessary if convincing proof of economic feasibility is desired. Before reliable estimates of future production can be made, the feasibility of proposed recovery methods must be tested and the resource characterized. Because of uncertainties in these areas, estimates of future production are highly speculative.

The following areas require further research: resource definition, recovery techniques, and the economics of collecting and marketing the resource. For unmineable coal, the thickness, quality, and locations which will support recovery operations must be identified and demonstrations performed.

According to an official of the Bureau of Mines, methane drainage from coal in advance of mining could contribute 0.5 Tcf annually by 1986 and 1 Tcf annually by 2000. The Lewin study (2) performed for DOE was not as optimistic; it projected production of 0.05 Tcf annually by 1990.

Conclusions

As the pace of coal mining increases, so will the venting of methane trapped in coal beds. Venting wastes a natural resource at a time when the Nation needs to use its energy resources wisely. Analysis indicates that methane production from both mineable and unmineable coal beds is economically attractive, but demonstrations of recovery methods are necessary for encouraging commercial development. Also, if mineable coal bed methane drainage techniques are used in advance of mining, they should help reduce the risks of coal mine explosions.

Because coal fracturing methods are proposed as a method of encouraging increased gas flows, the fear of mine roof instability is an issue. Due to concerns for miner safety, this issue should be addressed fully if coal companies are expected to place reliance on this drainage technique. Other methods are available for methane drainage from mineable coal so it appears that development of this resource could proceed without use of the fracturing method.

More research and development is required to assess and develop the potential of coal bed methane. Additional information on the applications of current technology, and evidence of economic feasibility from demonstration projects could attract industry or community interest and raise the production from this resource over the next 15 to 20 years. Although production appears to be feasible and economic, there have not yet been enough demonstrations to attract widespread interest.

METHANE FROM GEOPRESSURED AQUIFERS

Initial estimates of the total resource base for geopressured zones in Texas and the Louisiana Gulf Coast area have been vast. But, the important question is the amount that can be economically recovered. Some oil and gas industry experts are cautiously optimistic about the potential of this resource while others are skeptical that the energy of geopressured zones can be recovered in any sizeable quantity. Barring unexpected dramatic results from DOE's research and development program, the commercial potential of this resource is not expected to be known before the mid-1980s.

The Resource Base

Geopressured aquifers are found in various parts of the world, such as China, the Soviet Union, and the North Sea. One of the largest known geopressured areas lies under the Texas and Louisiana Gulf Coast. These deposits, water-bearing reservoirs, occur onshore and offshore at depths ranging from 5,000 to 18,000 ft. They are characterized by abnormally high pressures of 4,000 to 15,000 psi and temperatures of 200° to 400°F. Normal pressure at these depths in the Gulf Coast region ranges from about 2,325 to 8,370 psi. Normal temperature at these depths in the Gulf Coast region ranges from about 70° to 250°F.

Geopressured zones resulted from compaction of sandy sediments. Water would normally have been forced from these sands due to the buildup of strata above, but these areas were covered with rock which prevented the water from escaping and permitted pressure buildups. As a result of compaction, isolated units of sand and mud under great pressure developed, and methane formed as the sediments were buried in the hot briny waters of these deposits.

Three forms of energy are available from these geopressured zones: (1) the potential energy of fluids under pressure; (2) the heat of the briny waters; and (3) the methane or natural gas dissolved in these waters. Due to the methane content of these geopressured zones, they have been classified as an unconventional source of natural gas. However, it may be possible to extract the three energy forms simultaneously. For example, the thermal and kinetic energy could be used through conversion operations for electricity, and the methane could be extracted and sold to pipeline companies.

Estimates of the geopressured resource base show a wide variance, but recent studies indicate that the higher estimates may be optimistic.

Type of Energy	Size of Resource Base (quad)*
Thermal	44,000–176,000
Kinetic	198–693
Methane	3,000–115,000

*1 quad = 1 quadrillion Btu; also, the energy equivalent of 1 Tcf of natural gas

These differences are due to differing assumptions about the amount of the resource, the amount of dissolved natural gas per barrel of water, the extent and thickness of the deposits, and the porosity and permeability of the reservoirs. Recent studies have not been optimistic about the size of the recoverable resource; the studies are based on assumptions which can only be verified with field testing. Changes in the assumptions would alter the study results.

One such study was performed for DOE by Louisiana State University (6) and is based on an analysis of 6,000 wells previously drilled by industry. These wells were drilled in the known geopressured areas of onshore Louisiana and those offshore areas under Louisiana jurisdiction. Analysis of the well records showed that about 61%, or 3,626, of these wells were geopressured. The study estimated that 34.4 quads of energy are recoverable from this geopressured area including 13.6 Tcf of natural gas. These estimates are described in the study as optimistic.

Oak Ridge National Laboratory (7) reviewed and critiqued previous resource estimates. For example, the Louisiana State estimates would increase if the geopressured zones contain free gas in addition to gas-saturated brine. One of DOE's test wells in Louisiana, Edna Delcambre No. 1, indicated that geopressured zones may contain such free gas. In addition, the Oak Ridge study notes that a study of the Texas resource area (8) emphasized the geothermal aspect of the resource. Due to this emphasis, the Texas study did not include lower temperature aquifers which could have potential to produce methane alone, but have less potential for thermal energy recovery. The Oak Ridge study concludes that until more data are collected, the resource cannot be assessed with confidence.

Industry Perspective

Industry has encountered geopressured reservoirs along the Gulf Coast since 1936, and many thousands of wells have penetrated geopressured zones in the search for conventional oil and gas. The Humble Oil and Refining Company (now Exxon) performed a study over a 10 year period in the 1940s and 1950s and found high concentrations of hydrocarbons in the subsurface waters. There is general agreement that the prospects for commercial development, from industry's perspective, depend on uncertain economics and production risks.

Disagreement was expressed about production conditions, impediments to development, and the technology required for production. For example, one company stated that the methane constitutes the major value of the resource, and that the hot, pressured waters were of limited value due to the costs of converting this resource to energy. Another company believes that all three energy sources must be produced in combination for this resource to become a feasible economic prospect.

Others believe that the technology base is adequate and implied that building suitable equipment for producing this resource is a question of redesign. Some thought that the equipment design questions are more serious, particularly if all three resources are to be produced simultaneously. Disagreement was also expressed on production theories, regulatory questions, and environmental concerns.

There is some cautious optimism within industry that this resource could make a major energy contribution. Even among those whose views are optimistic, there is little expectation that this resource will be an important source of energy within the realm of current prices. On this basis industry spokesmen judge the resource too risky a prospect for industry investment today; therefore, they believe the Government research and development program is appropriate for defining the resource potential and demonstrating the economics of production.

Government Actions to Spur Development

The Government has taken action in four major ways to spur the commercial development of this resource. In 1975, DOE initiated a research and development program as part of its overall research on geothermal resources. As part of the Energy Tax Act of 1978 (Public Law 95-618), a tax incentive and a depletion allowance are now provided to stimulate interest in the resource, and the price of methane obtained from geopressured zones is now decontrolled under the Natural Gas Policy Act of 1978 (3)(4).

DOE Research and Development Efforts: Three important issues are being addressed by DOE in its research: How much energy is contained in geopressured zones? Where is it located? How can it be developed commercially? Two production methods are being studied by DOE. First, the methane recovery potential of shallow geopressured reservoirs is being assessed. Second, DOE is assessing the potential of deep, high-temperature reservoirs for producing methane, electric power, and heat. The research program includes efforts to inventory and characterize the geopressured resources, design and test recovery methods and technology, and study the environmental and institutional problems.

Including the amount budgeted for fiscal year 1979, DOE has spent $56.2 million for geopressured resource activities since 1975. DOE's fiscal year 1979 budget was $27.7 million, and DOE requested a budget of $36.0 million for fiscal year 1980. Industry is not yet making large contributions to this program. Program funds have largely been spent for resource characterization and background studies of environmental, institutional, and production difficulties. These studies have also identified prospects for geopressured reservoir confirmation drilling and testing.

DOE is also testing geopressured areas to further the characterization of the resource. The reservoir confirmation program was initiated in 1977 with the test of an oil well drilled through a geopressured area. This abandoned well in Vermilion Parish, Louisiana, produced at rates up to 12,000 barrels of brine per day. Tests are now underway on a second well and several other potential candidate wells have been identified.

Another test in Brazoria County, Texas, which was a widely publicized effort to complete a geopressured well, was not completed. Shale strata collapsed during drilling and the well was abandoned. A second attempt is being made to complete a well near the original Brazoria County, Texas, site. As of June 1979, the well had been drilled and preparations were underway to begin the first production tests. If successful, the well will be tested for 2 years. Between fiscal year 1980 and 1984, an additional 20 wells are now planned along with engineering and economic feasibility studies and construction of a pilot power plant and methane separation facility. Until these activities have produced results, the commercial potential of the resource cannot be assessed with confidence according to DOE officials.

Price and Tax Incentives: A variety of incentives are now offered for geopressured resource development. The price of gas produced from geopressured deposits is not controlled, and, an investment tax credit for equipment used to extract the resource and a depletion allowance for geopressured wells drilled are available to encourage production. In addition, the Administration proposes that geopressured methane production should receive a tax credit.

Economic incentives address only one aspect of the production question. Technical problems and risks remain and, until more is known about the production of geopressured methane, the economics of production remain speculative. Also, industry officials do not expect commercial production unless DOE's research and development program produces successful results. As a result, providing an incentive such as the tax credit may not result in commercial production at this time.

Other Actions to Spur Development: Based on the uncertainties of this resource, DOE's research and development program will be the key to future production. Based on the current schedule, DOE believes that the commercial potential of the resource will not be known until the mid-1980s. Industry believes that the proper role of Government at the present time is to continue research on this resource because it would not otherwise be accomplished.

Future Production Possibilities

Future production levels will be determined by a variety of factors which are just beginning to be defined and explored in DOE's research program. These factors fall in three general areas including the potential of the resource itself, the cost to extract and convert the resource to useable forms, and the requirement to protect areas which might be developed from environmental damage. Should favorable answers result from the research and lead to commercial interest, ownership issues must be resolved before large-scale energy production can occur. Due to the large number of uncertainties, experts do not expect any large-scale commercial production of geopressured methane before 1990.

Potential of the Resource: Efforts to estimate the potential of the geopressured resource have been concentrated on the overall size of the resource base and the characteristics which an individual reservoir must possess to produce the three available sources of energy. Several experts generally agree upon the following minimum characteristics for a potentially commercial geopressured reservoir: temperature should be over 250°F; the gas content of the water should be near saturation levels; the reservoir must be 300 ft thick and extend for 3 cubic miles; and the reservoir must produce 40,000 barrels of water per day for 20 years.

Candidate areas for testing have been identified by DOE's research and development program. Previous geologic studies indicate about 3 to 4% of Louisiana's sub-surface area may contain such economically sized prospects. However, other areas may also prove economic under different production assumptions—particularly if geopressured zones contain more gas than can be dissolved in brine.

Extraction and Conversion Costs: Once candidate reservoirs have been identified they must be drilled and produced. Since the reservoirs of greatest potential are expected to be found below 7,000 ft, drilling costs will be high. Also, poorly consolidated sand in these reservoirs may cause wells to clog or reduce the flow rates. Industry has a number of well completion techniques to control sand problems, but these have not been tested in wells where the pressures and required production rates are so high.

Also, long-term production must be proven and this will require equipment designed to withstand the corrosive effects of processing the hot brines. While experts believe such technology is available today, the equipment must be designed and tested. High efficiency methane extraction facilities and conversion units must also be designed. The design goal for a methane extraction unit, for example, would be to recover at least 85% of the methane in the well waters.

Since geopressured wells would produce great quantities of brine, an acceptable method must be found to dispose of this fluid at reasonable cost. Discharging the brine into the Gulf of Mexico is a possibility if any necessary cooling is performed. Disposal into subsurface aquifers is also possible if the aquifers will accept the large quantities of brine and if the brine does not contaminate fresh water or producible oil and gas reservoirs.

Reinjection of the brines into producing geopressured reservoirs has also been considered as a method of maintaining well pressure. While reinjection is theoretrically possible, equipment must be specifically designed to test the reinjection theories. However, some industry officials believe the costs of disposal in this manner may be prohibitive.

Environmental Concerns: Besides disposing of hot brines in an environmentally acceptable way, the possibility of surface subsidence is being studied. Subsidence might be caused by compaction of deep sediments as large quantities of pressurized reservoir fluid are produced. The amount of compaction will depend on pressure decline, reservoir thickness, and reservoir compressibility. If compaction occurs across a fault, fault reactivation or a lateral shift in land level may occur in addition to subsidence. DOE plans extensive monitoring to track subsidence activity around its test wells.

The impact of subsidence would vary depending upon the location and use of the land. In areas of low elevation along the coast, the effects of even a small amount of subsidence could be highly destructive. The effects might include flooding or large areas, loss of important animal habitats, and damage to roads and buildings.

There is considerable uncertainty that subsidence will occur or how much may occur with full scale development. Some authorities do not expect any subsidence although they agree the question cannot be answered until actual production occurs (9). Those who do not expect subsidence believe that it would not occur with production from great depths where the geopressured zones are sealed with overlying caprock. However, subsidence from natural compaction of sediments in the Gulf of Mexico is a continuing process resulting in flooding of coastal marshes. Additional subsidence from geopressured operations might compound these existing problems.

During the development phase of this resource, the Environmental Protection Agency plans to control discharges through the issuance of permits on a case-by-case basis. It plans to issue periodic guidance on known or expected environmental effects, state-of-the-art control technology, and environmental impact reviews.

Legal Questions: Although legalities are not now as critical to development as resource characterization and economic feasibility, there are legal questions to be resolved before commercial production can expand. The major questions concern the definition of the resource and ownership rights. Presently, it is uncertain whether geopressured aquifers are legally classified as a mineral, water, or another substance and if extraction rights belong to the surface owner or the mineral owner.

Conclusions

The potential of geopressured methane is too speculative to depend on as a major contributor to the Nation's energy supplies at this time. Such substantial research and development is required that the commercial potential of the resource is not expected to be known until the mid-1980s. Regardless of whether the recoverable resource may be lower than some of the high initial estimates, the resource continues to merit attention. However, a $0.50 per Mcf tax credit may not result in commercial production at this time because the economics of production are speculative and industry believes that research and development is appropriate given the high potential costs and risks associated with production.

REFERENCES

(1)   Natural Gas Policy Act of 1978 (Public Law 95-621), Joint Explanatory Statement of the Committee on Conference, p. A-14.
(2)   *Enhanced Recovery of Oil and Gas,* Lewin and Associates, Inc., prepared for DOE (February 1978).
(3)   Natural Gas Policy Act of 1978, sections 107(c), 121(b), and 503.
(4)   Federal Energy Regulatory Commission's Interim Rules of October 29, 1979, "Defining and Deregulating Certain High-Cost Natural Gas."
(5)   *Status Report on the Gas Potential from Devonian Shales of the Appalachian Basin,* Office of Technology Assessment (November 1977).
(6)   *Investigations on the Geopressure Energy Resource of Southern Louisiana,* Louisiana State University (April 1977).
(7)   Samuels, G., *Geopressure Energy Resource Evaluation,* ORNL (May 1979).
(8)   Bebout, D.G., et al, *Frio Sandstone Reservoirs in the Deep Subsurface Along the Texas Gulf Coast—Their Potential for Production of Geopressured Geothermal Energy,* Bureau of Economic Geology Investigation Report No. 91, University of Texas, Austin (1978).
(9)   Wilson, J.S., Shepherd, B.P., and Kaufman, S. of Dow Chemical, *An Analysis of the Potential Use of Geothermal Energy for Power Generation Along the Texas Gulf Coast,* p. 49 (October 15, 1975).

Tight Gas Basins

TYPES AND PROPERTIES OF TIGHT FORMATIONS

The information in this and the following three sections is based on *National Gas Survey Report to the Federal Energy Regulatory Commission by the Supply-Technical Advisory Task Force on Non-conventional Natural Gas Resources,* DOE/FERC-0010, June, 1978, "Gas in Tight Formations," prepared by Sub-Task Force IV, F. Stead of the U.S. Geological Survey, T. Jennings of National Gas Survey, P. Brown of Columbia Gas Transmission Corp., J.M. Dennison of University of North Carolina, W. deWitt, Jr. of the U.S. Geological Survey, W.K. Overby, Jr. of Energy Research and Development Administration and A.B. Waters of Halliburton Services for the U.S. Department of Energy. References for these sections are on p 48.

The so-called tight formations occur at one extreme as single, relatively thin (10 to 100 ft) gas-bearing zones of generally uniform thickness over a large area. At the other extreme would be relatively thick (possibly 1,000 or more ft) sections, containing multiple, lenticular gas-sand zones scattered throughout the section as in nonmarine formations of the Rocky Mountain basins.

Data for the mostly nonmarine tight sandstone formations in the western United States show the following characteristics:

	Western sandstones
Depth (ft)	Moderate to deep 4,000–20,000
Permeability (μd)	Low, 0.5–50
Porosity (%)	Low, 8–12
Water saturation (%)	Moderate to high 30–70
Gas-filled porosity (%)	Medium to low 3–6
Pressure	Normal to high

Permeability

Other factors being constant, an increase in confining pressure leads to a marked decrease in gas permeability, particularly for those rock types where the initial unconfined permeability is relatively low, at about 1 md (millidarcy) or less. This reduction is at least an order of magnitude greater than would be predicted from the reduction in pore volume due to rock compressibility (1). The strong influence of confining pressure on permeability suggests that, under compression, the interconnecting pathways, microfractures or pore throats, between pores are contracted to isolate pore space which would otherwise sustain gas flow, and that the reduction in pore volume is not a controlling factor.

Summarized in Figure 2.1 are data for permeability versus net confining pressure for sandstone core samples (2), (3) from: (1) Project Gasbuggy—in the San Juan Basin, New Mexico; samples from the Picture Cliffs sandstone at a depth of 4,000 feet; (2) Project Rio Blanco—in the Piceance Basin, Colorado; samples from the Fort Union sandstones, at a depth of 6,000 ft; and (3) Project Wagon Wheel—Green River Basin, Wyoming; samples from the Fort Union sandstones (8,000 ft) and the Mesaverde sandstones (10,000 ft).

Figure 2.1: Permeability as a Function of Confining Pressure

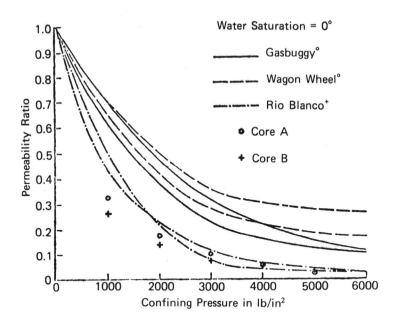

Source: DOE/FERC-0010

The reduction from unconfined or initial permeability, to the permeability at net confining pressure, equivalent to the overburden pressure less the internal pore pressure is: (1) factor of three for Gasbuggy samples at 2,700 lb/in² confining pressure; (2) factor of 10 for Rio Blanco samples at 3,200 lb/in² confining pressure; and (3) factor of five for Wagon Wheel samples at 5,000 lb/in² confining pressure. It is obvious that, given essentially the same porosities and water saturations: (1) the reduction in permeability is not a simple linear function of present overburden pressure or net confining pressure; i.e., the samples from the Piceance Basin at intermediate depth of 6,000 ft, between 4,000 ft for the San Juan Basin and 8,000 to 10,000 ft for the Green River Basin, show a much larger reduction in permeability; and (2) the reduction in permeability reflects the pressure loading-unloading history of the rock, including herein the variability introduced by tectonic stresses—folding, faulting, and thrusting, as opposed to simple gravitational loading in relatively undeformed basins.

Illustrated in Figures 2.2, 2.3, 2.4 and 2.5, and Table 2.1, for sandstone core samples from the Fort Union Formation in the Piceance Basin, the Rio Blanco site, are measured values of gas (nitrogen) permeability at initial unconfined conditions and also at various net confining pressures (2). For the samples studied, absolute permeability is greatly reduced by increased confining pressure, with the reduction at least an order of magnitude greater than would be predicted from reduction of pore volume due to rock compressibility.

Figure 2.2: Effect of Confining Pressure on Permeability at 45% Water Saturation

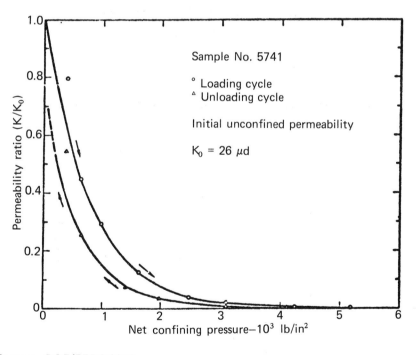

Source: DOE/FERC-0010

Data on the permeability of shales and closely related rocks such as silty shale, and shaly siltstone, are sparse, and are usually based on measurements made with water as the working fluid rather than air or other gases (4)(5).

Figure 2.3: Effect of Confining Pressure on Permeability at 16.4% Water Saturation

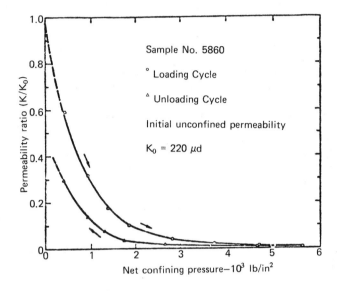

Source: DOE/FERC-0010

Figure 2.4: Effect of Confining Pressure on Permeability at 4.1% Water Saturation

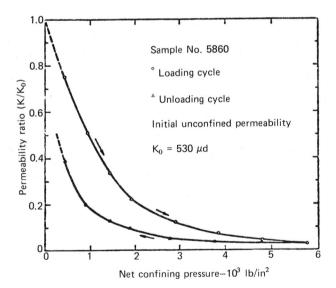

Source: DOE/FERC-0010

Figure 2.5: Effect of Confining Pressure on Permeability at Various Levels of Water Saturation

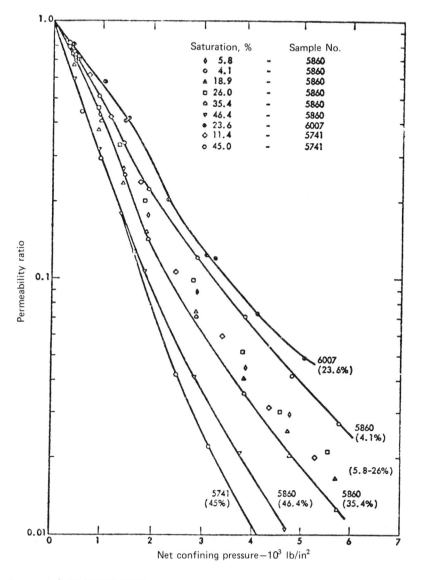

Saturation, %		Sample No.
◆ 5.8	-	5860
○ 4.1	-	5860
▲ 18.9	-	5860
□ 26.0	-	5860
⬠ 35.4	-	5860
▽ 46.4	-	5860
● 23.6	-	6007
◇ 11.4	-	5741
○ 45.0	-	5741

Source: DOE/FERC-0010

Such permeability measurements, although not precisely applicable to gas-bearing shaly rocks, do indicate the range in situ permeability to be expected—from 1.0 to 10^{-4} μd or possibly less, with a few reservoir pressure drawdown and buildup measurements suggesting as high as 0.7 μd for a silty shale. Arbitrarily, in later calculations of flow capacity of shales, a lower limit of 0.1 μd (0.0001 md) for in situ matrix permeability has been used, even though it is obvious that most true shaly rocks would be below this limit.

Table 2.1: Permeability, Porosity, and Percentage Water Saturation of Fort Union Formation Sandstone Samples

Sample Depth Below Surface (ft)	Unconfined Permeability	Permeability at Simulated In Situ Confining Pressure	Interconnected Porosity	Water Saturation
	. . . . (μd)		. . . .(%).	
5,741	31	2.2	8.76	11.4
5,741	26	0.6	8.76	45.0
5,860*	580	44	8.70	5.8
5,860	530	58	8.68	4.1
5,860	330	21	8.70	18.9
5,860	250	20	8.74	26.0
5,860	370	23	8.63	35.4
5,860	220	7.5	8.86	46.4
6,007	3.6	0.46	4.0	23.6

*Air cored.

Source: DOE/FERC-0010

Water Saturation

As shown in Figures 2.5 and 2.6, and in Table 2.1, the in situ permeability for the flowing gas phase decreases rather sharply with increasing water saturation, with permeability decreasing to essentially zero at 60 to 80% water saturation. Again, these data are for sandstone core samples from the three major industry-government, gas-stimulation projects: (1) Rio Blanco site, Piceance Basin, in Figure 2.5 and Table 2.1; (2) Gasbuggy site, San Juan Basin, in Figure 2.6, and (3) Wagon Wheel site, Green River Basin, in Figure 2.6. Similar data are lacking for the Uinta Basin.

In Table 2.1, for the Rio Blanco site, data for sample 5860 (number is sample depth in ft) show a reduction in gas permeability: (1) at 35.4% water saturation, from 370 μd at unconfined pressure to 23 μd at simulated reservoir confining pressure; and (2) at 46.4% water saturation, from 220 μd to 7.5 μd. Other data for this sample interval at 5,860 ft show: (1) by conventional core analyses, 910 μd at dry and unconfined conditions, 9.8% porosity, and 36% water saturation; and (2) by log evaluation, 10.5% porosity, and from 33 to 50% water saturation.

Although the foregoing data derived by three dissimilar methods of evaluation are reasonably compatible, it should be noted that misjudgment of in situ water saturation could lead to much larger misjudgment of in situ permeability, i.e., at 35.4% water saturation the permeability is 23 μd, and at 50% water saturation the permeability is about 5 μd, or an increase in water saturation from 35.4 to 50%, a ratio of 1:1.4 leads to a decrease in permeability from 23 μd to 5 μd, a ratio of 4.6:1.

Figure 2.6: Permeability as Function of Water Saturation

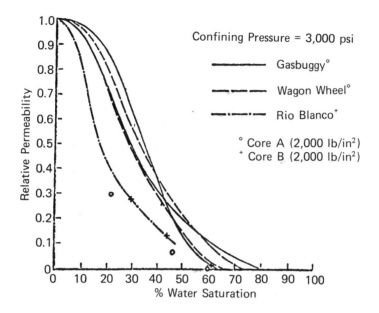

Source: DOE/FERC-0010

The combined effect of increasing net confining pressure and increasing water saturation sharply decreases the initial dry permeability. For example, the Wagon Wheel cores show an average initial permeability of 68 μd when dried and unconfined; they show an average water saturation of 50%. According to Figure 2.2, at a net confining pressure of 3,000 lb/in², the reduction from initial permeability is a factor of 0.28; and according to Figure 2.6, the reduction at 50% water saturation is a factor of 0.18 (4). The combined effect, or total reduction in permeability, is then a factor of 0.05, or an in situ value of 3.4 μd. This value of 3.4 μd has been used to predict gas well production rates in the Green River Basin using nuclear stimulation and massive hydraulic fracturing (6).

The combined effect of net confining pressure and water saturation for core samples from the Rio Blanco site, as shown in Figures 2.1, 2.5 and 2.6, is larger than the effect on cores from Gasbuggy and Wagon Wheel, although Rio Blanco is intermediate in depth at 6,000 ft, compared to Gasbuggy at 4,000 ft and to Wagon Wheel at 8,000 to 10,000 ft. The Rio Blanco cores show an initial unconfined permeability of 530 μd at 4.1% water saturation (Table 2.1). At 55% water saturation, the reduction from the initial permeability is a factor of 0.05; and at 3,600 lb/in² confining pressure the reduction is a factor of 0.08; or a combined reduction by a factor of 0.004, leading to an in situ permeability of 2.1 μd.

This compares with: (1) 14 μd based on logging data for the interval adjusted by pressure-buildup measurements in an adjacent well, Fawn Creek No. 1;

(2) 7 and 15 μd used to predict gas well production in the Piceance Basin (6); and (3) about 2 μd derived from production tests both in the original well RE-E-01 from which the core samples were obtained, and in an adjacent formation evaluation well RB-U-4, 600 feet to the northwest (7). It should be noted that, if the 930 μd permeability (air) measured on dried core at atmospheric pressure by conventional core analysis methods is accepted, then the reduction from initial to in situ permeability at simulated reservoir pressure is 931 μd to 2 μd, a reduction factor of 0.0021, or reciprocally 465:1.

Size Distribution of Sands

The gas-bearing sandstone zones in the Cretaceous and Tertiary nonmarine tight formations in the Rocky Mountains are predominantly fluvial channel-fill deposits with some point-bar sandstones. Well logs and core samples from the Mesaverde and the Fort Union Formations obtained in the Piceance, Uinta, and Green River Basins show sand lenses varying from less than a foot up to 50 feet thick, interbedded with shales, siltstones, and sandstones having no effective gas permeability. The percentage of gas-bearing sand lenses (with less than 65 to 70% water saturation) varies from about 15% to about 30% of the gross section thickness.

For example, in the massive hydraulic fracture experimental well RB-MHF-3, using limits of 65% or less water saturation and of 5% or more porosity, 45 sand lenses averaging 13 ft thickness, or 589 ft total sand, were selected in a gross section interval of 2,200 ft; lenticular sands are 26% of the gross section interval (8).

The expected sizes of channel-fill sandstone in the northern Piceance Basin would be: thicknesses from 20 to 30 ft; widths from 280 to 420 ft; and lengths from 2,800 to more than 4,200 ft. Point-bar sandstones, occurring mostly in the Tertiary Fort Union Formation and its equivalents in other basins, tend to be somewhat larger than channel-fill sandstones. Other data (9) show average length-L/width-W/thickness-H (in feet) ratios as follows:

Formation	Sandstone Type	L/W/H
Fort Union	point-bar	190/90/1
(Typical case)		(7,600/3,600/40)
Mesaverde	channel-fill	140/14/1
(Typical case)		(3,500/350/25)

Rather obviously, any reservoir evaluation, using a homogeneous flow model with constant gas-bearing sand thickness, would overestimate the resources, production rates, and ultimate recovery.

Based on these studies (9), in a sequence of channel-fill sandstones in the Piceance Creek Basin, an average conventional well will be connected to about 18% of the in-place reservoir volume in a 320-acre area, including therein allowance for erosional contact or interflow connection between sand lenses. For a well treated by massive hydraulic fracturing, with fracture wing dimensions of 2,000 ft, the well will be connected with 70% of the reservoir volume in a 320-acre area, assuming the fracture remains within the designated limits.

In 1973, as part of the National Gas Survey, the National Gas Technology Task Force carefully evaluated the effect of sand lensing on estimated productivity of stimulated wells; this evaluation is still valid and is a valuable reference. However, it should be noted that this evaluation was completed during the early developmental stage of massive hydraulic fracturing and used a 500-foot fracture-wing length for flow calculations, whereas today a fracture-wing length of 2,000 feet or more is used both experimentally and in application.

Spatial Variation of Physical Properties

To illustrate the spatial variations of physical properties within tight formations, data are taken from two major projects in the Piceance Basin, Colorado, and one in the Green River Basin, Wyoming.

Piceance Basin: The two projects in the Piceance Basin are: first, Project Rio Blanco, a joint government-industry experiment in stimulating flow of tightly held gas from deep formations (circa 6,000 ft) using nuclear-explosive fracturing; and second, the Rio Blanco Massive Hydraulic Fracturing (MHF) Experiment, a joint government-industry experiment in stimulating gas flow from the same formations using hydraulic fracturing (7) (10). These two projects are spaced less than 1 mile apart.

Independent analyses of the reservoir properties in the Project Rio Blanco emplacement well, in terms of net pay (sand) thickness and of permeability-thickness, based on core sample measurements and geophysical log interpretation, were made by VP (H.K. Van Poolen and Associates), CER (CER-Geonuclear and CONOCO), and LLL (Lawrence Livermore Laboratory). Although pressure drawdown/buildup tests were not run in the emplacement well prior to the major fracture stimulation, a nearby well at about 1,300 ft to the south had been flow tested over a limited interval; these data were extrapolated to the nuclear-emplacement well.

421 ft of net-pay sand, with individual sands ranging from 6 to 110 ft, in a total interval of 1,350 ft, were estimated using a cutoff of 30% for gas saturation and 5% for porosity. Log analyses indicate an average gas saturation of 57%, an average porosity of 9.4%, and a net pay of 421 ft (50). For the reservoir interval stimulated by the upper explosion cavity, the reservoir capacity or permeability-height, k_gh, ranged from the low VP estimate of 4.13 md-ft in four separate sands to the high CER estimate of 7.62 md-ft in three separate sands.

These values should be compared to 0.45 md-ft, derived from postdetonation drawdown and buildup tests of the top explosion cavity/chimney region, and to 0.73 md-ft from a best-fit chimney/reservoir simulation model. Where comparisons are made, in situ permeability for given water saturation and confining pressure is used in all cases (7)(8)(10). As the quantity of gas which might be produced by the top chimney/reservoir interval is directly proportional to k_gh, the observed gas production from the top chimney in the Fort Union gas sands is a factor of about 10 less than predicted from estimated reservoir properties. This lack of fit between prediction and actual drawdown/buildup tests indicates the substantial uncertainty regarding the actual physical properties of the reservoir, and the difficulty of determining physical properties such as water saturation, permeability, and individual sand thickness from exploratory well core and logging data.

Primarily to permit additional evaluation of the predetonation or undisturbed reservoir properties, a formation evaluation well was drilled 624 ft from the Project Rio Blanco emplacement well (10). After well completion, the casing was perforated from 5,836 to 5,892 ft, a 56-ft sand zone in the Fort Union; this sand zone is the equivalent of a 51- to 58-ft sand in the emplacement well. Gas production could not be obtained, and a limited hydraulic fracture treatment (approximately 16,000 gal, and 4,250 lb of sand) was performed. After the well returned approximately all of the injected fracture fluid, natural flow ceased and a pumping unit was installed.

An initial gas production rate of 53 Mscf per day was then obtained, but this decreased to 7 Mscf per day in about 3 weeks. The calculated gas permeability for the 56 ft interval was 0.0005 md; this value should be compared to: (1) the value estimated for the equivalent sand zone in the emplacement well, from a high estimate of 0.025 md for 57 ft, to a low estimate of 0.015 md for 58 ft; and (2) the value for the upper chimney/reservoir zone measured by drawdown/buildup production tests, at about 0.002 md for the 56 ft sand. Of course, these values are not directly comparable, as the first value is based on formation drawdown/buildup measurements following a limited fracture treatment in the formation evaluation well, the second value is calculated from core and log data for the emplacement well, and the third value is based on drawdown/buildup measurements following a massive fracture treatment (the upper nuclear explosion) in the emplacement well.

Nevertheless, the marked discrepancy among these values, an order of magnitude or more, indicates the difficulty of obtaining adequate data on reservoir properties.

Bearing on the lateral extent of lenticular sands, in the Project Rio Blanco bottom explosion region in well RB-E-01 and as measured in the reentry well RB-AR-2, the pressure history suggests a reservoir of limited extent, i.e., the effective mean radius of drainage must be reduced to about 400 ft in order to model the observed pressure history (7).

Based on geophysical log interpretation, at least three sands at 12, 19, and 25 ft thick are interconnected to the bottom explosion region; these sands do not correlate simply with sands at an equivalent depth in the adjacent formation evaluation well RB-U-4 at a distance of 600 ft, thus confirming the probability that the effective drainage radius is less than 600 ft. It should be noted that the estimated average channel-fill sandstone in the Piceance Basin has dimensions of 3,500 ft length, 350 ft width, and 25 ft thick, and that with adjustments for interconnection between sand lenses should have an effective drainage radius of about 900 ft (9).

The Rio Blanco Massive Hydraulic Fracturing (MHF) Experiment, a joint government-industry undertaking started in 1974, is planned to test the relative effectiveness of MHF and nuclear explosion fracturing in the same gas-producing formations; this is in accordance with the identification by the Natural Gas Technology Task Force of two emerging technologies, nuclear stimulation and massive hydraulic fracturing, that should be explored to determine their potential for developing gas resources in tight formations. The experimental well, RB-MHF-3, is located about 5,000 ft northeast of the nuclear stimulated well RB-E-01; the formation evaluation well RB-U-4 is located between these two wells, at a distance of 600 ft from well RB-E-01. Several separate fracture treatments have been executed in well RB-MHF-3 (11).

Fracturing treatment No. 1 took place on October 23, 1974, in a single Mesaverde sand at a depth of 8,048 to 8,073 ft (25 ft thick); static bottom-hole pressure at 3,450 lb/in^2; bottom-hole temperature at 242°F; gas-filled porosity at about 4%; gas flow after breakdown at 60 Mcfd.

Fracture treatment was 117,500 gal of polyemulsion fluid ($\frac{2}{3}$ naphtha-diesel oil mixture and $\frac{1}{3}$ a 2% KCl brine), with 400,000 lb of sand. Postfracture data show: flow rate at about 60 Mcfd, with the fracture treatment not increasing the productive capacity; very poor lateral propagation of the fracture compared to design length of 2,500 ft; productive capacity at about 0.15 md-ft, with an in situ permeability of 6 μd; postfracture flow rates were below predicted values by a factor of 5 to 8. Additional downhole surveys suggest that the fracture probably propagated upward to 8,000 ft and downward below the sand zone, rather than outward from the wellbore.

Fracturing treatment No. 2 was conducted on May 2, 1975, in three separate Mesaverde sands over the depth interval of 7,760 to 7,864 ft; fracture treatment was 285,000 gal of polyemulsion fluid (single-phase refined naphtha and KCl brine); postfracture gas flow averaged 137 Mcfd, a 2.5-fold increase over the pretreatment rate of 57 Mcfd, but declined steadily without reaching a stabilized flow during a 30 day test; again, the postfracture flow was well below predicted flow of 500 to 1,000 Mcfd for a fracture of 500 to 2,500 feet in length.

Fracturing treatment No. 3 was conducted on May 4, 1976, in three separate sands in the middle Fort Union Formation, corresponding to the Fort Union II sand in the nuclear stimulation well RB-E-01 and in the formation evaluation well RB-U-4, although these sands are not known to be the same in a depositional sense. Fracture treatment was 344,000 gal of gelled water-base fluid with 809,000 lb of sand; postfracture flow stabilized at 160 Mcfd or an indicated factor of four over the pretreatment rate.

Green River Basin: Data for the Green River Basin are drawn from Project Wagon Wheel, nuclear-explosion gas-stimulation experiment (12), and from three massive hydraulic fracturing experiments conducted by El Paso Natural Gas Company with partial support by the Energy Research and Development Administration (13).

The sand lenses in the Fort Union and Mesaverde Formations seem to be somewhat smaller and fewer than in the Piceance Basin, i.e., 4,000 ft gross section containing more than 100 potentially productive sands with an average thickness of 7 ft, or 17.5% sands in the total section, for the Green River Basin, compared to 2,200 ft gross section containing 45 sands averaging 13 ft thick, or 26% sand in the total section, for the Piceance Basin. Although the sand lenses in the Green River Basin exhibit the same range in porosity, water saturation, and permeability as do sands in the Piceance and Uinta Basins, the Green River sands are geopressured, i.e., gas pressure varies from 3,900 lb/in^2 (1.13 times hydrostatic pressure) at 8,000 ft, to about 8,100 lb/in^2 (1.56 times hydrostatic pressure) at a depth of 12,000 ft.

The three MHF experiments, in the El Paso Natural Gas Pinedale Unit near Pinedale, WY, are summarized below, with the number and depth of lenticular sands shown in Table 2.2.

Table 2.2: Number of Sandstone Strata Included in Pinedale Experiments

MHF Fracture Date	Well	Depth Interval (ft)	Number of Sandstone Strata	Estimate of Gas-in-Place (Bcf/mi^2)
September 11, 1974	Pinedale Unit No. 7	8,990–9,190	3	19.7
July 2, 1975	Pinedale Unit No. 5	10,950–11,180	2	40.5
October 20, 1975	Pinedale Unit No. 5	10,120–10,790	6	54.3

Source: DOE/FERC-0010

The Pinedale Unit No. 7 Well experiment took place on September 12, 1974, in three sands at 12, 19, and 20 ft thick at a depth of 8,990 to 9,190 ft containing an estimated 19.7 x 10^9 scf/mi^2; fracture treatment was 257,000 gal of polyemulsion fluid with 775,000 lb of sand; no prefracture flow measurements were made; postfracture flow decreased to 100,000 Mscfd in 1 year without reaching stabilized flow; modeling suggests an effective in situ permeability of less than 1.0 μd, and a flow capacity (permeability x thickness) of roughly 0.5 md-ft.

The Pinedale Unit No. 5 First Stage experiment took place on July 2, 1975, in two sands at 51 and 70 feet thick at a depth of 10,950 to 11,180 feet containing an estimated 40.5 x 10^9 scf/mi^2; prefracture with 46,000 gal permitted gas production in the range of 100 to 200 Mscfd at a pressure insufficient to lift liquid to maintain a bottom-hole pressure below 2,500 lb/in^2 or to obtain sufficient stability for accurate measurements; fracture treatment was 191,000 gal of polyemulsion fluid with 518,000 lb of sand; from 485 Mscfd on the 15th day to 340 Mscfd on the 37th day, presumably without reaching stabilized flow; modeling suggests an effective in situ permeability of less than 1.0 μd and a flow capacity of roughly 0.1 md-ft.

The Pinedale Unit No. 5 Well Second Stage experiment took place on October 20, 1975, in six sands at 235 ft thick in a total interval of 670 ft, at a depth of 10,120 to 10,790 ft, estimated to contain 54.3 x 10^9 scf/mi^2; prefracture breakdown flow was time limited and showed 1,100 Mscf of gas produced in a 21 hour period as wellhead pressure decreased from 2,200 lb/in^2 to 500 lb/in^2; fracture treatment was 458,000 gal of gelled water with 1,422,000 lb of sand; postfracture flow showed a peak on the 6th day at 850 Mscfd, decreasing on the 15th day to 250 Mscfd, and on the 43rd day to 150 Mscfd; calculations suggest an effective in situ permeability of about 0.2 μd, with a flow capacity of 0.04 md-ft.

Denver Basin: The Wattenberg gas field of about 980 sq mi is located in the western portion of the Denver Basin, and typifies a tight gas reservoir considered noncommercial prior to stimulation by massive hydraulic fracturing. The major gas-producing zone in this field is the Muddy J sandstone of Cretaceous age, which is marine, blanket-type sand with relatively uniform thickness over a wide area (14). The Muddy J sandstone is found at a depth of 7,600 to 8,400 ft; gross sand thickness is 50 to 100 ft; net-pay sand thickness is 10 to 50 ft; porosity is 8 to 12%; in situ permeability is 5 to 50 md; bottom-hole temperature is 260°F; bottom-hole pressure is 2,900 lb/in^2; water saturation at 30 to 50%.

Flow rates, from conventional wells, stimulated by hydraulic fracturing treatment of limited size (30,000 to 50,000 gal), were in the range of 30 to 50 Mscfd. Following massive hydraulic treatments in the range of 133,000 to 180,000 gal of polyemulsion fracturing fluid, flow rates increased three-to fourfold as compared to conventional wells (15).

This field can now be considered commercial, and serves as an example of a tight formation (i.e., less than 50 μd in situ permeability) which could be stimulated successfully. The recoverable reserves are estimated at 1.3 trillion ft^3.

In general, gas-bearing blanket-type sands, even when characterized by very low in situ permeability, can probably be exploited by MHF as currently developed; gas in such sands would then be an undiscovered recoverable reserve, rather than a nonconventional resource.

STIMULATION TECHNOLOGY

Massive Hydraulic Fracturing

Massive hydraulic fracturing, in contrast to the long-established hydraulic fracturing, is designed to create a vertical fracture extending at least 500 ft away from the wellbore in two directions (a total length of 1,000 ft or more); i.e.,

MHF is a newly developing, large-scale application of fracturing techniques, where the length of a fracture (one wing) may be in thousands of feet and the height of the fracture may be in hundreds of feet. The 1973 Natural Gas Technology Task Force report has summarized adequately: (1) the development and status of hydraulic fracturing; (2) the fracture geometry created by hydraulic fracturing under specified conditions appropriate for tight formations; and (3) predictions of flow rates for selected MHF treatments (6). More recent advances in the MHF technology have been described in various meetings (16)(17).

Shown in Figure 2.7 is a plan view of a MHF treatment with a fracture length (one wing) of 2,000 ft, or a total length of 4,000 ft; arbitrarily, the drainage area is a 160 acre unit, 1,320 x 5,280 ft, which is the basic reservoir unit for later calculations.

Figure 2.7: Plan View of MHF Treatment

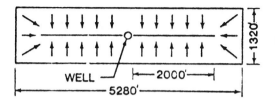

Plan view of 160 acre-unit drainage area showing flow paths (arrows) into a symmetrical, double-winged fracture. Fracture length by definition is measured from the well outward along one wing; as shown, fracture length is 2,000 ft (one wing); the total fracture length (two wings) for gas flow calculations would be 4,000 ft.

Source: DOE/FERC-0010

Data on production rates and relative costs for various combinations of fracture length, depth to pay zones, thickness of pay zones, porosity, permeability, and pressure provided by L.E. Elkins and C.R. Fast, Amoco Production Company.

It should be noted that early gas flow into a fractured well, as indicated by arrows in the figure, may be largely linear, and gradually changes from linear-elliptical to a rather radial flow configuration, a complex flow system difficult to calculate and to model. To obtain an approximation of flow to be anticipated from a MHF treatment, an effective well radius, about one-half of the designed fracture length, may be substituted in radial flow equations (18). It follows that a well stimulated by a MHF treatment designed to produce a 2,000 foot fracture length should show about a 20-fold increase in flow rate compared to the unstimulated well.

A variety of cases have been studied to determine the reservoir parameters that would permit application of MHF in stimulating gas-production from tight formations (19). These studies, backed by a moderate number of case histories, suggest that, if MHF is to be economically viable, certain minimal physical properties in a pay zone are needed as shown in Table 2.3. Additionally, a major requirement for evaluating tight formations, particularly individual pay zones, is a reliable determination of the in situ permeability (k_g) and the thickness (h), which provides the permeability x thickness factor, the k_gh, usually expressed as permeability-height, permeability-feet, or reservoir capacity.

Unfortunately, given the present state of the art, neither the in situ permeability nor the effective thickness of a pay zone is easy to obtain by geophysical log evaluation or by core analysis; in fact, as described above for carefully controlled stimulation experiments in the Piceance and Green River Basins, the flow capacity, k_gh, tends to be overestimated by a factor of 10 or thereabouts.

Table 2.3: Minimal Physical Properties in Pay Zones

Thickness, h	400 ft or greater	
Gas permeability, k_g	0.0001 md or greater	k_gh = 0.04+ md-ft
Gas-filled porosity	1.0% or greater	
Thickness, h	50 ft or greater	
Gas permeability, k_g	0.003 md or greater	k_gh = 0.15+ md-ft
Gas-filled porosity	3.0% or greater	
Thickness, h	25 ft	
Gas permeability, k_g	0.010 md or greater	k_gh = 0.25+ md-ft
Gas-filled porosity	4.0% or greater	
Thickness, h	20 ft	
Gas permeability, k_g	0.025 md	k_gh = 0.50+ md-ft
Gas-filled porosity	5.0% or greater	

Source: DOE/FERC-0010

One possible option for obtaining adequate flow-capacity data is to conduct an actual production rate test for isolated pay zones, hopefully without the necessity of running full strings of casing in the exploratory well.

MHF Gas-Well Simulator

The MHF gas-well simulator is a computer program which simulates the Darcy flow performance of a well completed in a low-permeability reservoir and stimulated with a massive hydraulic fracture treatment. Specifically, two-dimensional unsteady-state flow of gas in the horizontal plane is computed for the system consisting of a rectangular region bounded by no-flow boundaries, having a fracture and well located symmetrically within the flow region and parallel to one of the sides.

The fracture is simply a very narrow strip of the reservoir having an extremely high permeability. In addition, the following effects can be accounted for: (1) turbulent flow in the fracture, (2) variation of reservoir permeability as a function of confining pressure, (3) variation of fracture permeability as a function of confining pressure, and (4) specification of as many as three permeability zones within the fracture. A standard finite-difference solution technique is employed (provided by Amoco Production Company).

In the following figures bearing on flow performance, the terminology is: (1) normal (N) pressure is a hydraulic-pressure gradient of 0.5 lb/in^2/ft; (2) reservoir drainage area is 160 acres (Figure 2.7), with uniform properties; (3) porosity in all cases means gas-filled porosity; and (4) pay zones are centered at the stated reservoir depth, i.e., a 2,000-ft pay zone at a depth of 5,000 ft extends 1,000 ft above and below the 5,000-foot reservoir depth.

The effect of fracture length on production, for a pay zone 100 ft thick at a depth of 12,000 ft, with 5% gas-filled porosity and at 1.0 N pressure, is shown for two in situ permeabilities: (1) 0.001 md in Figure 2.8 (daily production), and in Figure 2.9 (cumulative production); and (2) 0.05 md in Figure 2.10 (daily production), and in Figure 2.11 (cumulative production). These data show that in situ gas permeability controls the shape of the production curve, and that doubling the fracture length almost doubles the production rate. In general, these data pertain to the tight sandstone formations, and should permit appraisal of potential production wherever the physical properties of pay zone are adequately known.

The effects of varying pressure, permeability, porosity, pay zone thickness, and pay zone depth on cumulative production, for a constant fracture length of 2,000 ft, are shown in Figures 2.12, 2.13, 2.14, and 2.15. In Figure 2.12, the production varies almost linearly with gas-filled porosity, but nonlinearly with permeability. In Figure 2.13, production varies almost linearly with pay zone thickness. In Figure 2.14, the production varies exponentially with depth, roughly by the square-root function. In Figure 2.15, production varies exponentially with pressure.

Figure 2.8: Effect of Fracture Length on Daily Production (k_g = 0.001 md)

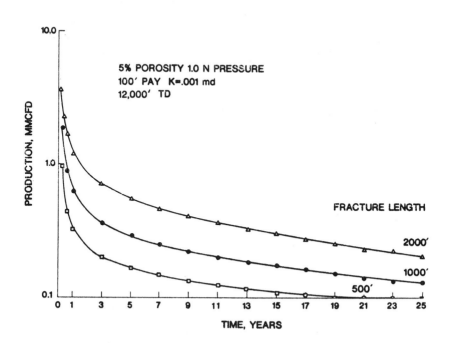

Source: DOE/FERC-0010

Figure 2.9: Effect of Fracture Length on Cumulative Production (k_g = 0.001 md)

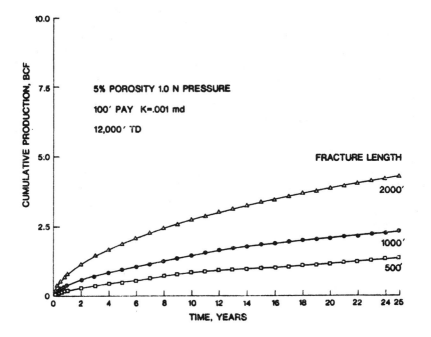

Figure 2.10: Effect of Fracture Length on Daily Production (k_g = 0.05 md)

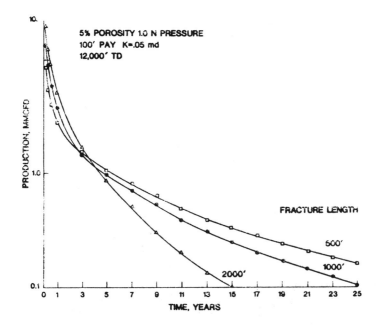

Source: DOE/FERC-0010

Figure 2.11:   Effect of Fracture Length on Cumulative Production (k_g = 0.05 md)

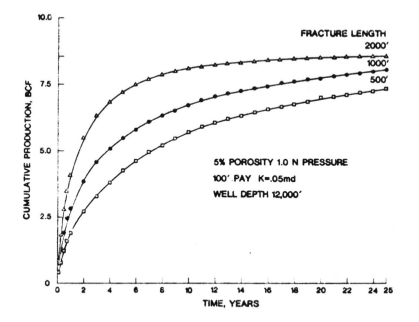

Figure 2.12:   Effect of Porosity and Permeability on Cumulative Production

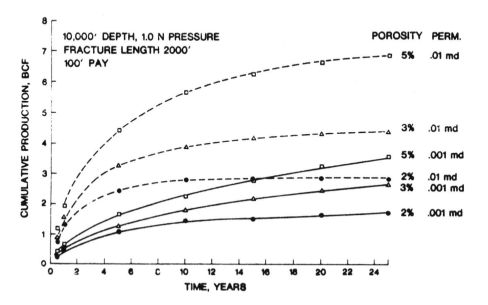

Source: DOE/FERC-0010

Figure 2.13: Effect of Pay Thickness on Cumulative Production

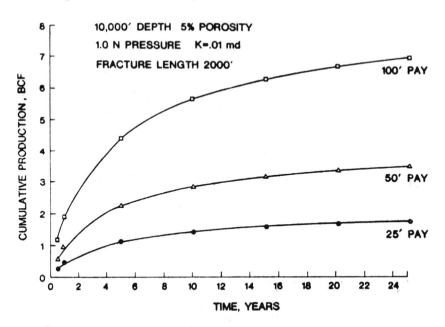

Figure 2.14: Effect of Depth of Pay Zone on Cumulative Production

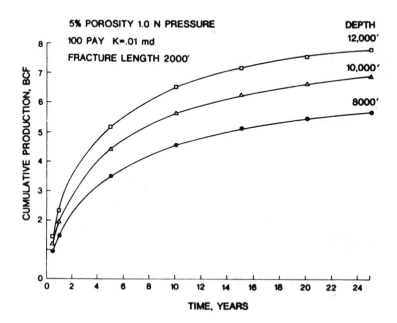

Source: DOE/FERC-0010

Figure 2.15: Effect of Pressure on Cumulative Production

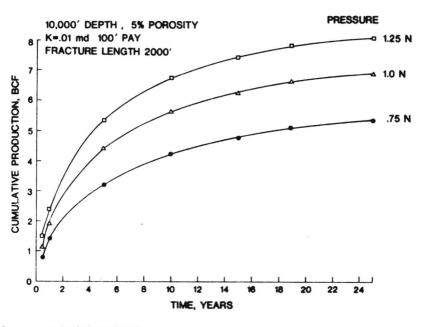

Source: DOE/FERC-0010

EFFECT OF RESERVOIR PARAMETERS ON ECONOMICS

The economic evaluations were made through a computer system named PLANS, which includes a standard discounted cash flow analysis with the additional capabilities to save selected operational data such as gas production, well drilling, fracturing, and workover activity by years. The system is modified to incorporate economic criteria and to offer ease of data input and analysis.

The basic unit of economic evaluation is the individual reservoir and includes consideration of projected production, operating expenses, and investment. Each reservoir is evaluated by computing the present worth of the net cash produced at stipulated gas prices of $0.83, $1.66, $2.50, $3.33, and $4.15 per Mcf; these prices are equivalent on a Btu energy basis to oil at $5.00, $10.00, $15.00, $20.00 and $25.00 per barrel.

In the following figures, the economic evaluations are expressed in terms of the profitability index (P.I.), where profitability index, or present worth rate of return, is that interest rate which will equate the present worth of cash income to the present worth of cash outflow; alternatively, it is that interest rate which will discount the net cash-flow series to a present worth of zero.

The effect on economics, for a constant fracture length of 2,000 ft, is shown in Figure 2.16 for varying in situ permeability and pay thickness, in Figure 2.17 for varying pressure and pay thickness, in Figure 2.18 for varying gas-filled porosity and pay thickness, and in Figure 2.19 for varying pay depth and pay thickness. It is difficult to summarize in simple terms the complex information here presented; it is best used by interpolation based on specific reservoir conditions as are either known or reasonably indicated at any given time.

Figure 2.16: Economic Effect of Permeability and Pay Thickness

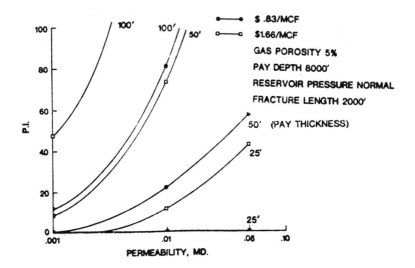

Figure 2.17: Economic Effect of Pressure and Pay Thickness

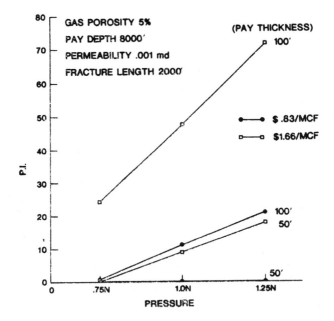

Source: DOE/FERC-0010

Figure 2.18: Economic Effect of Gas-Filled Porosity and Pay Thickness

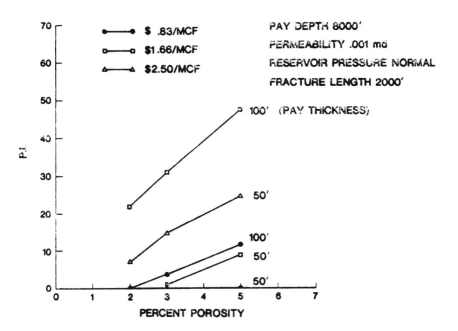

Figure 2.19: Economic Effect of Pay Depth and Pay Thickness

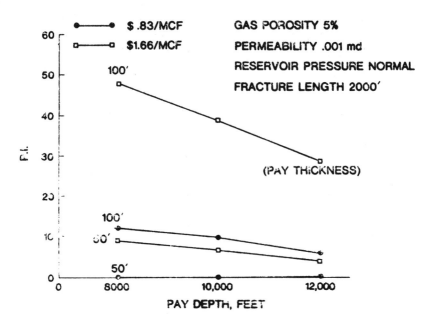

Source: DOE/FERC-0010

Rather arbitrarily in the following discussion, a P.I. of 20 (20%) is used as a cutoff limit, below which an attempt to produce gas would be uneconomical; obviously, other cutoff limits, either higher or lower, could be used dependent on specific conditions such as the total amount of gas in place within a given area, the proximity of a major market, and the availability of gas pipelines.

In Figure 2.16, by inspection and using a P.I. cutoff of 20, pay zones in the range of a few μd in situ permeability could not be stimulated by MHF treatments, unless the value of gas was in excess of $0.83/Mcf. A rather crude generalization, in terms of reservoir properties rather than economics, is that the flow capacity, k_gh (md-ft), must be above 0.1 to 0.2 md-ft to warrant consideration for MHF treatment.

In Figure 2.17, assuming a low permeability of 0.001 md, it is obvious that low reservoir pressures lead to higher wellhead costs.

In Figure 2.18, again based on a low permeability of 0.001 md, as the gas-filled porosity decreases (or conversely as the water saturation increases), the value or cost of gas production would increase to about $2.50/Mcf.

In Figure 2.19, again for a low permeability of 0.001 md, shown is the effect of reservoir or pay depth on economics, suggesting that wellhead prices in excess of $0.83/Mcf would be needed to warrant MHF stimulation of gas production.

The foregoing data on the anticipated effects of MHF treatments, although somewhat circumscribed by boundary conditions such as the 160-acre uniform reservoir drainage area and by assumptions on the lateral and vertical reach of a fracture, do provide limits within which tight formations or gas-bearing zones should fall to warrant consideration as potential resources.

ESTIMATION OF RESOURCE POTENTIAL

In the 1973 Natural Gas Technology Task Force report (47), potential gas resources were established only for the thick sequences of Upper Cretaceous and Lower Tertiary fluvial sandstones in the Piceance Basin of northwestern Colorado, in the Green River Basin in southwestern Wyoming, and in the Uinta Basin in eastern Utah. The criteria used to establish reservoirs acceptable in the resource base were:

1. Low permeability reservoir rock containing gas, not commercially recoverable with existing technology.

2. At least 100 feet of net pay, which is defined as sand having 65% or less water saturation and porosity from 5 to 15%.

3. At least 15% of the gross productive interval is pay sand.

4. The objective interval is between about 5,000 and 15,000 ft below the surface.

5. The prospective reservoir underlies at least 12 square miles.

6. The reservoirs are in remote areas.

7. Pay sands are not interbedded with high-permeability aquifers.

Some of the foregoing criteria, in particular 6 and 7, were included because of the possible use of nuclear-explosive fracturing; thus, other areas known to contain nonconventional gas resources, but not sufficiently remote from population centers for large explosion stimulation experiments, were briefly reviewed and then excluded such as: Atoka-Morrow (Pennsylvanian) sands of the Arkoma Basin, OK, and the nearby Stanley-Jack Fork (Mississippian) sands of the Ouachita Mountain province and downdip Wilcox (Eocene) and Houston (Cretaceous) sands of the Western Gulf Basin, TX. These other areas were not then considered to contain significant large resources in comparison to the three major basins; because of lack of pertinent reservoir data, these areas cannot be evaluated at this time for inclusion in the resource base.

The foregoing criteria have been modified slightly to include pay zones where: (1) thickness is as little as 20 ft; (2) gas-filled porosity is as low as 1%; (3) depth of pay zone is as shallow as 1,500 ft; and (4) remoteness from population centers and proximity to aquifers are not limiting factors. With these modifications, resources such as in the Upper Cretaceous siltstones and sandstones in the Northern Great Plains provinces, and in the San Juan Basin, can be included.

It should be clearly noted that the criteria used herein for estimating nonconventional gas resources are based fundamentally on two reservoir parameters, the gas-filled porosity and the reservoir pore pressure, which define the amount of gas contained within a given reservoir volume. Thus, these estimated resources do not imply in any way whatsoever a recoverability of a small or large fraction of the total; they state the amounts of gas in place under the stipulated reservoir conditions.

To avoid ambiguity in resource terms, using definitions established by the Department of Interior and the U.S. Geological Survey (20), the following terms may be applied to the nonconventional gas resources: (1) subeconomic—identified and undiscovered resources not presently recoverable because of technological and economic factors, but which may be recoverable in the future; (2) identified—specific accumulations whose location, quality, and quantity are estimated from geological evidence supported by engineering measurements; (3) undiscovered—unspecified accumulations surmised to exist on the basis of broad geologic knowledge and theory.

The term, reserves, is defined as that portion of the identified resource which is economically recoverable at the time of determination (using existing technology); reserves cannot be applied literally to any part of the gas resources in tight formations, which are subeconomic.

In a somewhat negative sense, as shown in Table 2.3, the flow capacity, $k_g h$, which is the cross product of the in situ permeability times the thickness of the interval at that permeability, has been used to set a minimal or lower limit below which gas in place would not be included in a resource estimate. Although this, in part, violates the concept of a resource as an accumulation of a commodity in such form that economic extraction is currently or potentially feasible, this lower boundary, together with the upper limit of 50 μd in situ permeability, was used to bracket the subeconomic resources in tight formations.

This bracketed range was selected to fit the assumed limitations of present and currently developing technologies such as massive hydraulic fracturing, chemical explosive fracturing, and deviated wellbores; presumably, if these technologies advance as rapidly as hoped, some presently unknown portion of the subeconomic resources would be developed and converted to economic reserves.

It is generally agreed that the occasional and naturally high rate of gas production from the tight formations reflects an unusual reservoir condition, where a joint-fracture system, intensively developed within a limited area and with highflow capacity for gas, has been intercepted by wellbores. Because such areas of fracturing are relatively small compared to the thousands of square miles underlain by tight formations, and because adequate data are lacking to define such fracturing at reservoir depth, no attempt has been made to adjust the following resource estimates for such an effect.

The gas-in-place resources for the tight sandstone formations in the three major basins were estimated by the 1973 Task Force (66), and are summarized as follows:

	Tcf	Bcf/mi^2
Green River Basin, Wyoming	240	120–145
Piceance Basin, Colorado	210	145–240
Uinta Basin, Utah	150	240–340
	600	

There are no compelling reasons, based on additional but still scanty data from controlled experiments over the last few years, the three nuclear and the numerous MHF experiments, to change these estimates by any significant amount. The volume of gas in place is determined by the gas-filled porosity and the reservoir gas pressure, and basic reservoir properties have not been changed except in minor degree by recent data.

Changes of some importance have occurred in the earlier "Predicted Gas Well Production Rates for the Rocky Mountain Basin" (6), because the reservoir flow capacity, $k_g h$, has been demonstrated to be significantly less than originally estimated by a factor of 5 to 10, as discussed above for experiments in the Piceance and Green River Basins. These changes in flow capacity lead to a decrease in gas production, both daily and cumulative, and to an increase in costs (6). Thus, although the resources of gas in place have not changed significantly, the resources are much less attractive as targets for currently developing techniques of gas production stimulation.

Two additional areas containing tight sandstone formations have been added to the resource base: (1) the San Juan Basin, northern New Mexico, containing an estimated gas-in-place of 63 cubic feet; and (2) the Northern Great Plains province, Montana and North Dakota, containing an estimated 130 cubic feet.

The Northern Great Plains was not previously included in natural gas resource estimates because of inadequate data, but it is now known that the area contains significantly large resources entrapped at shallow depths (less than 4,000 ft) in thin, discontinuous, low permeability Upper Cretaceous offshore siltstones and sandstones. These finegrained clastics, enclosed in a thick sequence of marine shale, were deposited on the western side of a north-south trending Interior Cretaceous seaway. Investigations by D.D. Rice of the U.S. Geological Survey using carbon isotope ratios indicate that the contained gas was generated by anaerobic bacteria at shallow depths in the accumulating sediments.

These shallow accumulations have generally been overlooked in the past; however, recent exploration and evaluation in western Canada along the trend of accumulations extended from eastern Montana indicate that major resources are present in this type of accumulation. The Suffield Evaluation Committee in 1974 (21) assigned an in-place gas reserve of 3.7 Tcf to an area of 1,000 square miles in southeastern Alberta (Canada) where the area was evaluated by a 77 well program, of which 76 wells were completed as economically producible wells. This gas-bearing facies extends into Montana and is present over approximately 35,000 square miles in the United States portion of the northern Great Plains. Using the Suffield Block reserve data, the United States portion of this province should contain 130 Tcf of gas in place and a potentially recoverable gas resource of approximately 95 Tcf. General characteristics of the area are: (1) depth of gas-bearing zones, 1,200 to 4,000 ft; (2) multiple pay zones from 20 to 100 ft individual thickness; (3) porosity from 8 to 15%; (4) water saturation at 50 to 60%; and (5) gas in place at 3 to 4 Bcf per sq mi.

Data for the San Juan Basin, for which the type locality was the Project Gasbuggy site, has been previously described: in general, the characteristics are very similar to the Piceance, Green River, and Uinta Basins, with experimental data again suggesting a flow capacity less than originally estimated.

Estimates of Nonconventional Gas Resources: Estimates of nonconventional gas resources, for these areas considered to contain the predominance of such resources, are summarized as follows:

	Tcf	Bcf/mi^2
Green River Basin	240	120-145
Piceance Basin, Colorado	210	145-240
Uinta Basin, Utah	150	240-340
San Juan Basin, New Mexico	63	30-40
Northern Great Plains, Montana	130	3-5

By inspection, the energy density of the foregoing resources, expressed here as the amount of gas in place per square mile, is significantly higher by one or two orders of magnitude for resources in tight sandstones of the four major basins in the western United States, in contrast to the energy density in the siltstone/sandstone of the Northern Great Plains. It follows that the development of reserves within the resource base, and the future gas production from such resources, will reflect in major part the initial energy distribution; i.e., for a given production level, the high energy density resources in sandstones will require proportionately fewer developmental wells, interconnecting pipelines, and access roads in a small land area, than would the low energy density resources in the Northern Great Plains.

It should be clearly noted and emphasized that these resources cannot be directly compared to present-day conventional reserves of natural gas, where the production rate per well is relatively high and where the percentage recovery of gas in place, frequently quoted and demonstrated by production records, is about 80% of the gas in place. These nonconventional resources, characterized by a low in situ gas permeability and a related low flow capacity, will provide a relatively low production rate, over a longer production interval, with a much lower percentage recovery of the gas in place.

It is not the purpose of this discussion to evaluate wellhead production costs of gas from tight formations; such cost estimates were made in the 1973 Task Force Report (6) and can be modified to fit the changing reservoir conditions. Some data have been provided on the effect of reservoir parameters on economics (Figures 2.16 through 2.19), which should permit an approximation of wellhead costs given the necessary reservoir properties. It should be noted that the Pinedale Unit MHF Experiments (12)(13) provide data which show that the production capacity in the type locality used to characterize the Green River Basin is about one-fifth of that projected by the 1973 Task Force Report (6), and that combination of larger fracture treatments and higher than estimated inflation leads to twice the cost in 1972 dollars used in the 1973 report; combining the lower production with higher costs leads to a wellhead development cost of about ten times the 47 cents per Mcf reflected in the 1973 Task Force Report. Data from the Rio Blanco site in the Piceance Basin, again the type locality used to characterize the basin, suggest an even higher increase in projected wellhead costs, because the flow capacity is about a factor of 10 lower than forecast.

Clearly, the major problems in estimating resources in tight formations are: (1) in situ permeability, (2) the thickness of pay zones, and (3) water saturation. Establishing reserves of gas within the large resource base, producible under various stimulation technologies and under various wellhead prices, will require similar data on particular reservoir volumes; such data can become available only through extensive developmental effort. Current research, both by industry and as sponsored by government, may provide answers to these problems.

REFERENCES

(1) Schock, R.N., Heard, H.C., and Stepheus, D.R. *Mechanical Properties of Rocks from Site of the Rio Blanco Gas Stimulation Experiment.* Univ. Calif. Lawrence Livermore Lab., Rept. UCRL-51280, 1972.

(2) Thomas, R.D., and Ward, D.C. "Effect of Overburden Pressure and Water Saturation on Gas Permeability of Tight Sandstone Cores." *Journal Petro. Tech.,* February, 1972.

(3) Quong, R. *Permeability of Fort Union Formation Sandstone Samples: Project Rio Blanco.* Univ. Calif. Rept. UCID-16182, 1973.

(4) Young, A., Low, P.F., and McLatchie, A.S. "Permeability Studies of Argillaceous Rocks." *Jour. Geophy. Research,* v. 69, no. 20, 1964.

(5) Bredehoeft, J.D., and Hanshaw, B.B. "On the Maintenance of Anomalous Fluid Pressures: I. Thick Sedimentary Sequences." *Geol. Soc. Amer. Bull.,* v. 79, Sept. 1968.

(6) Federal Power Commission. *National Gas Survey—Supply, Task Force Reports.* Vol. II, April, 1973.

(7) Ballou, L.B. "Project Rio Blanco—Additional Production Testing and Reservoir Analysis." Intern. Atomic Energy Agency, Vienna, Austria, *Proc. of Tech. Comm. PNE-V,* 1976.

(8) van Poolen, H.K. & Associates. *Predicted Reservoir Performance, Project Rio Blanco.* USAEC Rept. NVO-38-33, 1972.

(9) Knutson, C.F. "Modeling of Noncontinuous Fort Union and Mesaverde Sandstone Reservoirs, Piceance Basin, Northwestern Colorado." *Soc. Petrol. Engrs. Jour.,* August, 1976.

(10) Toman, John. "Project Rio Blanco: Production Test Data and Preliminary Analysis of Top Chimney/Cavity." Intern. Atomic Energy Agency, Vienna, Austria, *Proc. of Tech. Comm. PNE-IV,* 1975.

(11) Appledorn, C.R., and Mann, R.L. "Massive Hydraulic Fracturing, Rio Blanco Unit, Piceance Basin, Colorado." *ERDA Symposium on Enhanced Oil & Gas Recovery Proceedings,* v. 2, paper E-3, Petrol. Publish. Co., Tulsa, 1976.

(12) El Paso Natural Gas Company. *Project Wagon Wheel—Technical Studies Report.* USAEC Open File Report PNE-WW-1, Dec. 31, 1971.

(13) El Paso Natural Gas. *Pinedale Unit MHF Experiments—Final Report.* ERDA Rept. BERC/RI-76/19, December, 1976.

(14) Matuszczak, R.A. "Wattenburg Field, Denver Basin, Colorado." *The Mountain Geologist,* v. 10, no. 3, 1973.

(15) Covlin, R.J., Fast, C.R., and Holman, G.B. "Performance and Cost of Massive Hydraulic Fracturing in the Wattenburg Field." *Symposium on Stimulation of Low Permeability Reservoirs, Proceedings,* Colo. School Mines, 1976.

(16) Colorado School of Mines and American Gas Association. *Symposium on Stimulation of Low Permeability Reservoirs.* Colo. School Mines Petrol. Engr. Dept., 1976.

(17) Energy Research and Development Administration. *Symposium on Enhanced Oil and Gas Recovery, Proceedings, v. I & II.* Petroleum Publishing Co., Sept., 1976.

(18) Howard, G.C., and Fast, C.R. *Hydraulic Fracturing.* Soc. Petrol. Engrs. Monograph 2, 1970.

(19) Elkins, L.E. "Role of Massive Hydraulic Fracturing in Exploiting Very Tight Gas Deposits." *Natural Gas from Unconventional Geologic Sources.* National Academy of Science, Board of Mineral Resources, 1976; also, Energy Research and Development Admin. reprint, Rept. FE-2271-1, 1976.

(20) Miller, B.M., et al. *Geological Estimates of Undiscovered Recoverable Oil and Gas Resources in the United States.* U.S. Geol. Survey Circular 725, 1975.

(21) Last-Kloepfer, Ltd. Suffield Evaluation Drilling Program: Report Submitted to Suffield Evaluation Committee. For Province of Alberta (Canada), 1974.

THE PROBLEM AND THE POTENTIAL

The information in this and the following three sections is based on *Enhanced Recovery of Unconventional Gas: The Program— Volume II (of 3 Volumes),* October 1978, prepared by V.A. Kuuskraa and J.P. Brashear of Lewin and Associates, Inc., T.M. Doscher of University of Southern California and L.E. Elkins for the U.S. Department of Energy under DOE Contract No. EF-77-C-01-2705.

Despite promising developments, the potential of the Western Tight Gas Basins, the large land area that originally attracted attention, remains undeveloped and unproved. The challenges posed by the difficult geological setting—the deep, tight, lenticular gas pays—have yet to be overcome. Even in the more geologically favorable basins, in the less tight or blanket type deposits, fundamental improvements in recovery technology need to be pursued. Finally, major opportunities exist for optimizing the recovery technology and accelerating its application for recovering additional gas supplies between now and 1990.

Overall, the low permeability formations included in the study described in the rest of this chapter cover 13 major basins. The major geological and technological problems that currently impede the development of these Tight Gas Basins and that need to be overcome by R&D are summarized below.

Geological Problems: While the most popular characterization of these basins is their "tightness", low permeability is only one of several geological problems that have limited commercial development for tight gas basins. All are marked by generally low quality net pays which are often highly discontinuous or lenticular.

Permeability — Low in situ gas permeability is the defining characteristic of the tight formations. [Permeability is the resistance of reservoir rock to the flow of gas under reservoir conditions of water saturation and confining pressure, measured in millidarcies or microdarcies (one thousandth of a millidarcy)]. For this study, formations with in situ permeabilities of less than 1.0 md were considered as tight gas sands. The formations discussed range in permeability from 1.0 md down to 0.001 md (one microdarcy or μd), with the vast majority of the formations being below 0.05 md. [These are measured as in situ permeability. Under reservoir conditions, overburden pressure and high water saturations can reduce the relative gas permeability within the formation to values ranging down to 6% or less of the permeabilities measured in the laboratory (1)].

As the formation permeability drops below about 0.1 md, recovery efficiency becomes highly sensitive to small changes in permeability as shown by Elkins (2).

Permeability of the pay, however, appears to be an important impediment rather than the central problem. The Wattenberg Field (Denver Basin) with a broad, continuous net sand pay has permeabilities ranging from 5 to 50 μd and is being commercially produced. By contrast, a Uinta Basin, Chevron well (No. 212) completed in the lower Mesaverde lenticular pay encountered permeabilities of 90 μd, based on computer simulation history match of actual production data, almost twice as high as the better portions of the Wattenberg, but failed to produce at even one-half the Wattenberg rates. It appears that low permeability combines with the lenticular nature and low quality of the pay to pose the major challenge.

Lenticularity — The principal commercial successes in tight formations have been in relatively continuous, "blanket" type sands. The sands in many of the tight basins, however, are notably discontinuous—the gas-bearing pays consisting of uncorrelatable lenses within sometimes massive cross sections. The effect is to limit well drainage area and to preclude recovery from sand lenses not contacted by the wellbore.

To illustrate, a square mile (640 acres) with typical reservoir characteristics (assuming following average characteristics: 100 ft of net pay, 10% porosity, 45% water saturation, and 9,000 ft of depth) would contain approximately 32 Bcf of gas in place. With lenticular, discontinuous pay (having lens dimensions typical of the Mesaverde of 20 ft x 400 ft x 6,000 ft), and current field development practices (one well per section), a wellbore would be in contact with only 3 Bcf, or 9% of the sand lenses in the section. Even if field spacing is reduced by half, two wells per section, these two wells would be in contact with only 6 Bcf, or about 18% of the gas-bearing sand lens.

To date, MHF technology has been successful in lenticular formations only where the individual lenses are large relative to the normal well drainage area or where the individual lenses are developed in conjunction with vertically adjacent blanket formations.

Pay Quality — In addition to low permeability and frequent lenticularity, the gas-bearing portions of the tight basins are of low quality relative to conventional gas formations. First, although the gross sections can be extremely thick, ranging from two to over five thousand feet, the gas-bearing portions of such segments, the net pay, may be only a few hundred feet. Further, the net pay may be dispersed in relatively small (tens of feet or less) strata interbedded with clays and shales.

Second, the permeable sand segments often contain high levels of connate water that impede the gas flow in the fracture system. The presence of water frequently confuses initial testing and interpretation of fractured wells. Water from both net pay and adjacent strata inhibits production, until removed. Except for extremely poor quality formations, the gas will ultimately push this water out of the well, but as long as three to six months may be required before the well's productivity can be accurately ascertained.

Third, the porosities of the net pay are low, generally in the range of five to fifteen percent. Low porosity combined with relatively high water saturations of 40 to 70% reduce the gas-filled porosities from levels from less than 3% to seldom over 9%.

Finally, the net pays in tight basins often contain clays that swell when contacted by drilling or fracturing fluids (unless these fluids contain chemicals to inhibit swelling).

Technological Challenges and Goals: The geology and reservoir characteristics of the tight gas basins impose some absolute limits on the amount of commercial recovery. The limited gas in place in a given areal/vertical section, the
lenticular, discontinuous sands, and the fundamental reservoir characteristics
are immutable. The technological challenge is to exploit the limited opportunities
that lie within these constraints.

Meeting this challenge will require that fracturing technology evolve toward
four major goals:

- To stimulate all gas pay intervals exposed to the wellbore
  by using multiple fractures from the same well.

- To intersect, in lenticular formations, sand lenses not initially in contact with the wellbore.

- To maintain an effectively propped fracture, thus providing adequate fracture conductivity.

- To optimize the process and make economic the currently
  marginal and submarginal gas resources.

These technological goals appear to be within reach of a concerted research
and development effort. However, achieving them will require substantial R&D
investments in resource characterization, testing, and demonstration. Joint federal-industry collaboration could accelerate this R&D and demonstrate its commercial application (6).

Structure of the Analysis: The objectives of the analysis were to define
the required research and development that would assist in commercializing the
tight gas basins. The analysis followed six basic steps:

(1) Basic Data—Collect and analyze detailed geological and engineering data on each of the identified basins.

(2) Base Case Technology—Define the major problems and constraints that limit full exploitation of these resources by industry; define the current and near-term (next 5 years) technology and its application without federal stimulus, the Base
Case.

(3) Advanced Case Technology—Develop R&D strategies for overcoming current technical limitations and that would accelerate
the development of the target basins, the Advanced Case Technology.

(4) Economic Analysis—Simulate the production and economics
of Base Case and Advanced Case technologies to establish
the economic potential of the tight gas basins.

(5) Sensitivity Analysis—Repeat the economic analysis under alternative gas prices and technology assumptions to assess price
and technology sensitivity.

(6) Cost-Effectiveness—Compare the costs of the R&D programs
to their production benefits to establish cost-effectiveness.

The Potential of the Tight Gas Basins—Overview: *The Resource Targets* —
The thirteen tight basins included in the analysis differ geologically and in the
technological problems they pose. To facilitate the analysis, the individual
basins were grouped into five categories (called resource targets) having common
geologic features:

- Western Tight Gas Basins, the Piceance, the Greater Green River and the Uinta—that were the original focus on the FPC study are deep, generally very tight, and lenticular.

- Shallow Gas Basins—the large land area of the Northern Great Plains Province, characterized by low productivity, is shallow and has a range of deposits that vary from blanket to lenticular.

- Other Tight, Lenticular Basins—fully as tight as the Western Basins, these basins tend to be somewhat less deep and contain larger individual lenses.

- Tight, Blanket Sands—deep and very tight, these basins are favored by highly continuous, blanket-type gas deposits.

- Other Low-Permeability Reservoirs—these near-conventional formations have special development and engineering problems.

The Gas in Place — These five resource targets contain considerable gas in place—over 400 Tcf (excludes gas in place in proved and speculative areas) distributed as follows:

Target	Gas in Place (Tcf)
Western tight	176
Shallow gas	74
Other tight, lenticular	51
Tight, blanket sands	94
Other low-permeability	14
Total	409

Current Activity — Some development, in the more favorable formations and segments, is underway in each resource target. The largest amount of drilling has taken place in the tight, blanket sands, where the continuity of the gas pays improves recovery. It is estimated that nearly one trillion cubic feet (Tcf) of gas were produced from the tight gas basins in 1976. Continuing advances in the technology and improved economics will foster further development in the more favorable areas, as projected in the Base Case estimates.

Base Case Technology — Under base case technology—current and near-term advances without federal R&D supplementation—the following outcomes are projected: The tight gas formations could provide from 70 to 110 Tcf of additional recovery, for gas prices of $1.75 to $4.50 per Mcf. Price is stated in 1977 dollars and assumed maintained in constant dollars through the period of analysis, (Figure 2.20). The annual production rate could range from about 2 to over 3 Tcf per year by 1990 (Figure 2.21).

The contribution of the individual resource targets is shown below:

Base Case, Recovery, and Production—Tight Gas Basins

| | . . Ultimate Recovery . . | | . 1990 Production Rate. | |
	$1.75/Mcf	$4.50/Mcf	$1.75/Mcf	$4.50/Mcf
Western tight	4	11	0.2	0.4
Shallow gas	21	22	0.6	0.6
Other tight, lenticular	8	15	0.2	0.5
Tight, blanket	32	51	1.0	1.6
Other low-permeability	5	8	0.2	0.3
Total	70	107	2.2	3.4

Figure 2.20:   Base Case Ultimate Recovery from the Tight Gas Basins
(at Three Prices)

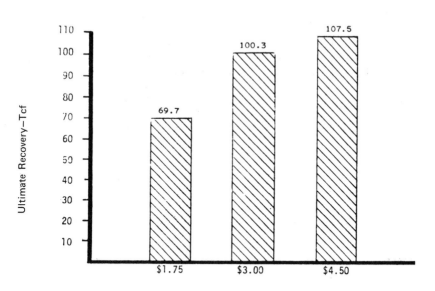

Figure 2.21:   Base Case Annual Production from the Tight Gas Basins
(at Three Prices)

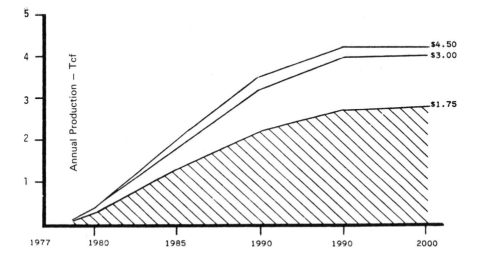

Source: DOE EF-77-C-01-2705

Under base case technology, only about 15 to 25% of the total tight gas resource in place is recovered, even at gas prices of up to $4.50/Mcf. Relative to individual targets, the proportion of gas in place recovered varies from 2 to 6% in the Western Tight Gas Basins, 34 to 54% in the Tight Blankets, and nearly 60% in the Other Low-Permeability Basins.

Advanced Case Technology — Intensive federal-industry research and development could improve the recovery efficiency and accelerate the development of the tight gas basins. These advances, going beyond industry's plans in the absence of federal sponsorship, could add substantial amounts of recovery from these resource targets. Overall, under advanced case technology: The tight gas formations could provide from 150 to 190 Tcf ultimate recovery, about 80 Tcf more than under base case technology; and the annual production rate by 1990 could range from 6 to nearly 8 Tcf per year, about 4 Tcf per year more than under the base case.

While the largest increases are in the geologically most challenging basins, federal-industry R&D substantially improves and accelerates recovery from all the resource targets, as shown below.

Advanced Case Recovery and Production—Tight Gas Basins

	. . Ultimate Recovery . .		. 1990 Production Rate.	
	$1.75/Mcf	$4.50/Mcf	$1.75/Mcf	$4.50/Mcf
Western tight	38	53	2.0	2.7
Shallow gas	23	35	0.8	1.0
Other tight, lenticular	23	24	0.7	0.7
Tight, blanket	59	66	2.5	2.8
Other low-permeability	6	10	0.3	0.5
Total	149	188	6.3	7.7

Under advanced case technology, from 35 to 45% of the resource in place can be recovered, at prices of $1.75 to $4.50 per Mcf. Substantial gains in recovery are possible in all resource targets. For example, at $1.75/Mcf, the recovery in the tight, blanket formations essentially doubles, from 32 Tcf in the base case to nearly 60 Tcf in the advanced case. In the more difficult targets, the proportion is still larger. For the Western Tight Gas Basins, the advanced case increases recovery from a base of 4 to 11 Tcf to a range of 38 to 53 Tcf, depending on gas price.

Summary: The largest total production from the tight gas basins would accrue from a combination of higher gas prices and advanced technology. At gas prices of about $3.00 per Mcf, 180 Tcf of natural gas could ultimately be recovered with substantial quantities, over 40 Tcf, available between now and 1990. The term "price" serves to summarize any combination of economic incentives such as market price, tax provisions, public subsidies, etc., that can be expressed in "price to the public" equivalent terms. The study of enhanced gas recovery examined only three prices, $1.75, $3.00, and $4.50 per Mcf, and an optimum research program for a given price. It did not seek to establish the optimum price or optimum combination of public R&D and price.

THE RESOURCE BASE

The tight gas resource base consists of twenty basins, grouped into five resource targets. The Federal Power Commission's analyses in the early 1970s focused attention on three—the Greater Green River, the Piceance, and the Uinta—referred to below as the Western Tight Gas Basins. Subsequent federal-industry collaboration has focused on these, with the somewhat later addition of the shallow, tight gas basins, in the Northern Great Plains Province. In addition to the four basins, sixteen additional basins (3) have been identified as having permeabilities too low to permit economic recovery by existing, conventional technology. Of these twenty basins, thirteen were included in the study; seven basins were eliminated from the analysis due to insufficient data (Figure 2.22).

Approach to Describing the Resource Base

Data on the tight gas resource base were collected by teams of reservoir engineers and geologists who had previously worked in these basins. These included the following organizations and individuals: Sandia, Inc.; CER Geonuclear, Inc.; C.K. Geoenergy, Inc.; Gruy Federal, Inc.; Mitchell Energy, Inc.; the Godsey-Earlougher Division of Williams Brothers Engineering, Inc.; the Bartlesville Energy Research Center; and independent geologists Gene Foss, Tom Beard, and Alan Hansen. The highest data collection priority was assigned to the four basins in which joint federal-industry research is underway.

The analysis rests on detailed reservoir characteristics and well performance data sufficient to support reservoir engineering, and computer-based simulation of gas in place, production, and economics.

All available public data, in published documents and in computer data banks, pertaining to the tight gas basins and the target geologic ages were accumulated. These were supplemented by well-location maps and well test data from state and industry sources. Detailed analysis of logs, core data, outcrop studies, and production histories served to complete the basic data used in the analysis.

These data provided the basis for dividing each basin into homogeneous sub-basin areas. These reservoir characteristics and geologic interpretations were reviewed with local geologists and with gas production firms active in the basins. The disaggregated units were placed into a reservoir data file and formed the basic units of analysis for assessing technical and economic recovery. In all, 622 areal/vertical units (reservoirs) in 13 basins were defined and analyzed.

Classification of Low Permeability Basins

Tight gas basins vary substantially in their inherent geological and technological problems. To facilitate the economic analysis and the planning of R&D programs, the basins were classified into homogeneous resource targets. Three basic geologic features were used to develop the classification:

- Permeability—Basins were classified according to the average initial in situ permeability to gas of the major deposit. Formations with permeabilities greater than 1.0 md were regarded as commercial and were excluded from the analysis of tight gas formations. Formations having permeabilities between 1.0 md and 0.05 md were considered "relatively tight" and those with average permeabilities less than 0.05 md as "very tight." The preponderance of the major gas pays analyzed in this study were found to be very tight.

- Depth—Due to the substantially different engineering problems associated with shallow versus deep formations, formations less than 2,500 ft deep were categorized as shallow, those over 2,500 ft as deep.

Figure 2.22: Location of Major Tight Gas Basins

From U.S. ERDA's Western Gas Sands, Project Plan, 8/1/77

ERDA's Primary Study Areas	Geological Area
(A) Greater Green River Basin	Tertiary and Cretaceous
(B) Piceance Basin	Tertiary and Cretaceous
(C) Uinta Basin	Tertiary and Cretaceous
(D) Northern Great Plains Province	Cretaceous
(E) Williston Basin	Cretaceous

Additional Low-Permeability Areas in the Study

(1) Big Horn Basin	Tertiary and Cretaceous
(2) Cotton Valley Trend	Jurassic
(3) Denver Basin	Cretaceous
(4) Douglas Creek Arch	Cretaceous
(5) Ouachita Mts Province	Mississippian
(6) San Juan Basin	Cretaceous
(7) Sondra Basin	Pennsylvanian
(8) Wind River Basin	Tertiary and Cretaceous

Other Low-Permeability Areas not Included in Study

(a) Anadarko Basin	Pennsylvanian
(b) Arkoma Basin	Pennsylvanian
(c) Fort Worth Basin	Pennsylvanian
(d) Raton Basin	Tertiary and Cretaceous
(e) Snake River Downwarp	Tertiary and Cretaceous
(f) Wasatch Plateau	Cretaceous
(g) Western Gulf Basin	Tertiary and Cretaceous

Source: DOE EF-77-C-01-2705

- Lenticularity—Sand discontinuity appears to be the single most severe geologic problem that must be overcome. Thus, the basins were classified as lenticular or "blanket-type" according to the nature of the major formations.

Using these three geologic conditions, five classes of basins, referred to as resource "targets," were established:

(1) Western Tight Gas Basins—the three basins, Greater Green River, Piceance, and Uinta that originally attracted the attention of the FPC task force.

(2) Shallow Gas Deposits—the shallow, low production formations of the Northern Great Plains Province and the Williston Basin.

(3) Other Tight, Lenticular Gas Sands—Sonora, Douglas Creek Arch, and the Big Horn basins.

(4) Tight, Blanket Gas Sands—the Denver, San Juan, Wind River, Cotton Valley basins, and the Ouachita Mountain Province.

(5) Other Low Permeability Reservoirs—the Bruckner-Smackover formation of the Cotton Valley Trend.

The tight gas basins initially identified but not included in this analysis due to insufficient data were the Anadarko, Arkoma, Forth Worth, Raton, Snake River, Wasatch, and Western Gulf basins.

Industry Activity in the Target Basins

Nearly all the tight basins have more favorable areas or formations where substantial development has been or is underway. While data on proved reserves at the basin level are not available, production data are reported by various state agencies and provide a valuable index of current industry activity in tight gas sands. This historical development, as well as industry's tests in the more difficult formations, provide a useful baseline for assessing industry's future activity.

The production data shows that the tight gas formations already produce substantial amounts of gas. Table 2.4 shows the number of wells and annual and cumulative production in 1974 or 1975 for the target basins. These figures apply only to nonassociated gas production from the target formations.

Of a total of over 700 Bcf in annual production from the target basins, the Tight Blanket and Other, Low Permeability formations contributed over 70%. (These data are now several years old. Given the considerable drilling that has taken place in the tight gas basins since 1975, annual production in 1977 was estimated at about 1 Tcf.) Their generally favorable geology would have suggested this outcome. Four of these basins, the Denver, San Juan, Sonora Basins, and Cotton Valley Trend are where fracturing technology has had the most extensive application.

Together, these account for over 500 Bcf per year. The geologically and technologically most difficult areas, the Western Tight Gas Basins and the Shallow Basins, together contribute less than 100 Bcf per year despite their vast land areas and considerable gas in place.

The review of current industry activity suggests that breakthrough R&D is most essential in the Western Tight, Shallow and Other Tight, Lenticular Basins.

Table 2.4: Production from the Tight Gas Basins

Target/Basin	End of Year	Producing Wells	Annual Production	Cumulative Production
			. . . . (Bcf)	
Western tight gas sands				
Greater Green River	1974	155	43	410
Piceance	1975	46	3	68
Uinta	1974	172	17	127
Subtotal		373	63	605
Shallow gas basins				
Williston and Northern Great Plains	1974	280	12	135
Other tight, lenticular				
Big Horn	1974	27	2	40
Douglas Creek	1975	127	14	126
Sonora	1975	1,254	108	462
Subtotal		1,408	123	629
Tight, blanket gas				
Cotton Valley Trend	1975	1,266	189	7,663
Denver	1975	640	51	113
Ouachita	1974	0	0	0
San Juan	1974	2,413	187	2,604
Wind River	1974	52	17	214
Subtotal		4,371	444	10,594
Other low-permeability reservoirs				
Cotton Valley Sour	1975	174	105	577
Total		6,606	747	12,540*

*Texas cumulative production from Oil Scouts 1974 Yearbook and Texas RRC Annual Production of Oil and Gas, 1975.

Source: DOE EF-77-C-01-2705

In the Tight, Blanket and Other, Low Permeability Basins, the R&D would be first directed toward stimulation technology and accelerated field development. As field development moves beyond the more favorable areas and formations, toward the less favorable margins of these basins, the technological problems will increase and will require the R&D breakthrough gained from the more difficult basins.

Scope and Areal Extent of the Resource Base

Table 2.5 lists the thirteen basins and the formations within them that were analyzed. The discussion of the resource base and the analysis of its potential relate explicitly to these basins and formations. The scope of the analysis is further limited to undeveloped areas of these basins, yet where exploration and gas shows indicate likely future potential. The gross areal extent of each basin was categorized in a scheme designed to be consistent with the reserve classification system of the Potential Gas Committee (4) and the well classifications of the American Association of Petroleum Geologists (5). Four classes of acreage were defined for each basin:

(1) Proved acreage—within the defined perimeter of existing fields.

(2) Probable acreage—extensions of existing fields and obvious corridors between fields where the direction of recent drilling suggests the fields will ultimately merge.

(3) Possible acreage—areas outside proved or probable acreage where the subject formations have had gas "shows" or production but no multiwell fields have developed.

(4) Speculative acreage—areas in which drilling through the subject formations has yielded no show of gas and areas where no drilling has taken place.

Table 2.5: Target Formations in the Tight Gas Basins

1. **Western Tights**

    (a) Green River

    Ft. Union
    Almond A
    Almond B
    Erickson
    Rock Springs/Biair
    Other Mesaverde

    (b) Piceance

    Ft. Union
    Corcoran-Cozette
    Other Mesaverde

    (c) Uinta

    Wasatch
    Barren
    Coaly
    Castlegate

2. **Shallow Gas**

    (a) Northern Great Plains

    Judith River
    Eagle
    Carlisle
    Greenhorn/Frontier

    (b) Williston

    Judith River
    Eagle
    Greenhorn

(continued)

Table 2.5: (continued)

3. **Other Tight Lenticular**
   (a) Big Horn
      Mesaverde
   (b) Douglas Creek
      Mancos
      Dakota
   (c) Sonora
      Canyon

4. **Tight Blanket**
   (a) Cotton Valley "Sweet"
      Cotton Valley Sand Trend
      Gilmer Lime

   (b) Denver
      Sussex
      Niobrara
      Dakota
   (c) Ouachita
      Stanley
   (d) San Juan
      Dakota
   (e) Wind River
      Frontier
      Muddy

5. **Other Low Permeability**
   (a) Cotton Valley "Sour"
      Bruckner/Smackover

Source: DOE EF-77-C-01-2705

To avoid "double-counting," proved acreage was excluded from the analysis. These proved areas were presumed to be included in the current AGA reserve estimates.

Speculative acreage was also excluded from the analysis. Such acreage is
analogous to the basins and formations that were excluded for reasons of insufficient data. The absence of wells with gas shows argues that too little is
known about these areas to support detailed analysis.

With these two exclusions, the area analyzed having at least definable gas
shows becomes incremental to present proved areas—the Probable and Possible
categories. To account for the fact that fast areas are seldom fully successful
when drilled, each acreage classification was weighted by the historical drilling
success ratio for the states in which the basins lie to yield "anticipated" acreage. Anticipated acreage is roughly analogous to "expected" acreage except
that only point estimates of the means (no distributions) are available. Probable areas were weighted by the success rates for developmental drilling. Possible areas were weighted by the average of within-field exploratory wells and
rank wildcats. Table 2.6 shows the anticipated productive area of each of the
basins in the analysis.

Thus, the scope of the analysis includes only those areas (having adequate
data and containing gas incremental to current proved reserves) that are anticipated to be developed given appropriate technology and economics.

Reservoir Characteristics

The reservoir characteristics of the basins and formations included in the
analysis are displayed in Table 2.7. They suggest several conclusions. The
first is that the geological problems described above are clearly demonstrated:
(1) The in situ gas permeabilities of the major formations rarely exceed 100 μd
(0.1 md). Outside of the shallow basins, the few higher permeabilities represent
the relatively small, most favorable geographic areas; (2) In the area where the
net pay is greatest, the pay is also highly discontinuous, or lenticular; and (3) All
the basins are marked by low gas-filled porosities (high water saturation with
low porosities).

Table 2.6: Anticipated Areal Extent of Tight Gas Basins

Target/Basin	Probable Area* Area (mi²)	Weight** (%)	Possible Area* Area (mi²)	Weight** (%)	Total Anticipated Area* (mi²)
Western tight				
Green River	454	78	416	25	870
Piceance	395	78	460	25	855
Uinta	698	78	297	25	995
Subtotal	1,547		1,173		2,720
Shallow gas				
Northern Great Plains	3,660	60	13,900	21	17,560
Williston	0	60	6,520	21	6,520
Subtotal	3,660		20,420		24,080
Other tight lenticular				
Big Horn	81	78	680	25	761
Douglas Creek	252	78	117	25	369
Sonora	890	85	1,070	31	1,960
Subtotal	1,223		1,867		3,090
Tight blanket				
Cotton Valley (sweet)	1,225	78	1,026	29	2,251
Denver	1,966	74	625	20	2,591
Ouachita	0	70	113	–	113
San Juan	266	93	564	41	830
Wind River	82	78	383	25	465
Subtotal	3,539		2,711		6,250
Other low-permeability				
Cotton Valley (sour)	569	78	642	29	1,211
Total	10,538		26,813		37,351

*After application of anticipated success ratios.
**Success ratios developed from American Association of Petroleum Geologists data.

Source: DOE EF-77-C-01-2705

Table 2.7: Reservoir Characteristics of Tight Gas Formations

Target/Basin	Areal Units	Formation	Depth (ft)	Gross Interval (ft)	Net Pay (ft)	Nature of Pay	In Situ Gas Permeability (μd)	Gas Filled Porosity (%)	Reservoir Pressure (psi)	Reservoir Temp (°F)
Western tight gas sands										
Greater Green River	36	Ft. Union	5,700-9,000	500-2,680	21-625	lenticular	1-50	3.4-5.0	3,150-6,334	135-194
		Almond A	8,000-10,700	400-500	9-20	blanket	9-50	4.1-4.5	4,200-6,200	180-215
		Almond B	8,000-10,700	400-500	18-45	lenticular	9-50	4.5-5.4	4,200-6,200	180-215
		Erickson	8,400-11,400	350-400	35-68	lenticular	7-20	4.1-5.4	4,400-6,500	186-231
		Rock Springs/Blair	9,700-12,500	1,500-2,500	19-80	lenticular	7-8	4.1-5.4	5,000-7,200	206-248
		Other Mesaverde	9,000-12,700	2,150-5,000	28-164	lenticular	1-9	3.4-4.5	5,850-8,250	194-220
Piceance	25	Ft. Union	5,000	600	18-44	lenticular	3-27	4.0-5.2	2,100	135
		Corcoran-Cozette	6,000	50	10-38	blanket	8-75	4.2-6.1	2,600	145
		Other Mesaverde	6,900-9,100	800-2,200	40-275	lenticular	3-60	3.6-5.4	3,000-3,400	160-170
Uinta	32	Wasatch	6,500	500	43-156	lenticular	66-600	4.4-5.8	2,795	175
		Barren	7,500	500	43-156	lenticular	30-270	3.8-5.0	3,225	195
		Coaly	8,500	500	43-156	lenticular	10-90	3.2-4.2	3,655	214
		Castlegate	9,500	250	25-75	blanket	3-30	2.6-3.4	4,275	233
Shallow gas basins										
Northern Great Plains and Williston	27	Judity River	600-1,600	30-50	8-20	blanket	17-1,000	5.2-13.7	270-680	80-85
		Eagle	1,800-2,000	30-60	3-25	blanket	17-10,000	7.4-12.2	800-900	90-100
		Carlisle	1,500	30-50	4-10	blanket	10-900	5.4-7.1	670	85
		Greenhorn/ Frontier	2,000-2,600	30-50	3-29	blanket	17-2,700	5.4-7.8	900-1,130	100
Other tight lenticular gas sands										
Big Horn	5	Mesaverde	2,285*	645	110-275	lenticular	13-120	6.6-8.7	1,100	95
Douglas Creek Arch	5	Mancos	2,845-4,045	2,400	120-300	lenticular	7-60	4.8-7.5	437	120
		Dakota	7,545	72	4-9	lenticular	10-90	3.6-4.7	1,100	240
Sonora	10	Canyon	6,000-7,000	600	30-103	lenticular**	8-84	4.4-6.3	2,100-2,700	145
Tight blanket gas formations										
Cotton Valley (sweet)	10	Cotton Valley Sand	9,000	1,100	35-88	blanket	3-30	4.0-5.3	6,000	250
Denver	5	Gilmer Lime	11,000	350	20-50	blanket	3-30	5.6-7.4	5,400	280
		Niobrara	2,300	67	11-28	blanket	3-30	2.6-3.5	950	110
		Sussex	4,460	50	11-26	blanket	3-30	3.6-4.7	1,500	185
		Dakota	8,000	50	14-34	blanket	5-50	4.0-5.3	2,900	260
Ouachita	15	Stanley	4,600-9,000	6,000-7,200	186-465	blanket	1-5	3.7-5.1	1,700-2,200	148-160
San Juan	5	Dakota	7,180	173	35-88	blanket	10-90	5.8-7.6	3,090	222
Wind River	5	Frontier	1,441*	153	20-50	blanket	33-300	6.5-8.5	550	99
		Muddy	2,529*	100	10-25	blanket	1-9	8.8-11.6	1,000	109
Other low permeability gas formations										
Cotton Valley (sour)	5	Bruckner-Smackover	12,000	900	18-44	blanket	44-400	8.0-10.5	5,600	290

*Data as reported – considerable portions of these formations are much deeper, e.g., 4,000-6,000 feet.
**Canyon lenses are very large relative to the drainage area and substantially broader than the other lenticular formations.

Source: DOE EF-77-C-01-2705

Given these conditions, it is not surprising that industry has only developed the most favorable portions of these basins and directed its activity inversely in relation to the severity of the geology.

The Western Tight Gas Basins, where there is little current activity, are beset by the most severe combination of geologic constraints—lenticularity coupled with great depth, very low permeability, and low gas-filled porosities. The Tight Blanket Sands, in which there is substantial current activity, have blanket-type sands and somewhat better pay sections (higher gas-filled porosity and lower clay content).

The single lenticular basin that has experienced appreciable development, the Sonora Basin, is favored by large lenses that approach the dimensions required to support economic productivity from a single well. Knutsen (6) has estimated that a typical lens in the Tertiary and Mesaverde sections of the Western Basins might have areal dimensions of 400 ft wide x 6,000 ft long, or a total area of about 55 acres. The typical dimensions of a lens in the Sonora Basin might be about 1,300 ft x 3,800 ft, approximately 110 acres.

Doubling of the expected lens size (and hence drainage area) mitigates the lenticularity constraint. As discussed later, the length to width ratio is also important in the effectiveness of the MHF treatment. Typically, this ratio is about 15:1 for the Western Basins, but only about 3:1 in the Sonora Basin. Thus, the severity of the lenticularity problem is largely a function of the geometry of the lenses and of the degree to which lenticularity is associated with other geologic problems. Basins with broad lenses would be nearly as commercially attractive as blanket sands, other factors being equal.

The data show that the Western Tight Gas Basins contain at least one blanket-type gas formation. The occurrence of a "stack" of sands or formations presents the opportunity for multiple MHF treatments, each of which needs to cover only slightly more than marginal fracturing costs to justify economic development.

The final observation is that the reservoir parameters, even within given formations, vary significantly. The ranges shown represent the sections of the formations that are not "dry," but have at least shows of gas. Permeability varies by an order of magnitude or more. Gas-filled porosity (porosity times gas saturation) varies by a factor of two or more. Net pay thickness varies by a factor of two to thirty.

Such variability can be observed even within the relatively narrow confines of adjacent townships of a proved field with a blanket formation, as has been shown in the Wattenberg Field (Denver Basin) (7). Even greater variations can be expected in lenticular sands.

At present, the quality of the pay in a particular area can only be known after drilling and testing the well. With the current limitations on the accuracy of available measurement, testing procedures, and devices as applied to very tight formations, even testing the well fails to remove all uncertainty. Producers in many cases must fully complete the well, including costly MHF stimulations, apply all available measurements and tests, and produce the well for considerable time to evaluate the reservoir characteristics in a particular segment of the basin. Even then, they can seldom generalize this evaluation to nearby areas.

These geological variabilities and uncertainties have deterred many producers from attempting commercial development. Federal-industry R&D involving improved measurement and resource characterization could reduce these uncertainties.

Expected Gas in Place

Table 2.8 shows the expected gas in place for each of the basins and targets derived from combining the 622 analytic units (reservoirs). Together these units cover forty thousand square miles and contain over 400 Tcf in place. The Western Tight Gas Basins as a group have the largest amount of gas in place, 176 Tcf, or 43% of the total. The Tight Blanket Formations are the next largest, at 94 Tcf, or 23% of the total.

Table 2.8: Areal Extent and Gas in Place—Tight Gas Basins

Target Basin	Analytic Units	Total Anticipated Area (mi²)	Expected Gas in Place* (Tcf)
Western tight			
Green River	216	870	91
Piceance	75	855	36
Uinta	128	995	50
Subtotal	419	2,720	176
Shallow gas			
Northern Great Plains	68	17,560	53
Williston	40	6,520	21
Subtotal	108	24,080	74
Other tight, lenticular			
Big Horn	5	761	24
Douglas Creek	10	369	3
Sonora	10	1,960	24
Subtotal	25	3,090	51
Tight, blanket gas			
Cotton Valley (sweet)	20	5,127	53
Denver	15	2,591	19
Ouachita	15	113	5
San Juan	5	830	15
Wind River	10	465	3
Subtotal	65	9,126	94
Other low-permeability			
Cotton Valley (sour)	5	1,211	14
Total	622	40,227	409

*Totals may not add due to rounding.

Source: DOE EF-77-C-01-2705

The gas in place estimates from this study contrasted with the FPC study conducted in 1972 (8) for the Western Tight Gas Basins. (The other basins were not analyzed by the FPC task force.) Although the relationship between the categories used by the FPC study and the ones employed in the present study are not exact (and a direct translation from the FPC categories and the Probable and Possible categories used here cannot be made), the overall area studied is comparable. The FPC report classified the acreage into three categories: "Essentially proved because of good well control;" "Inferred to be productive from geological interpretation;" and "Has a geological basis for being productive but is untested and must be considered speculative."

The areas and gas in place estimates for the three Western Tight Gas Basins are shown in Table 2.9. The total acreage included in the two analyses is comparable, but the present study places the gas in place estimate at less than 30% of the FPC's initial estimate. Two major factors account for this:

- Recent drilling and analysis show that the net pay of the tight formations is substantially smaller than estimated by the FPC task force. The task force estimated net pays for the various formations ranging from 500 to 1,000 ft. By contrast, as shown in Table 2.7, recent drilling has found these formations to have net pay thickness ranging from 200 to 500 ft, less than one half that initially estimated by the FPC.

- Recent well data show that the gas saturations and porosities (gas-filled porosities) are lower than estimated in the FPC study. While direct comparison is not possible, the task force estimated only one formation with the gas-filled porosity less than 4.2%, while, as shown in Table 2.7, the present study found many at levels lower than this.

Table 2.9: Areal Extent and Gas in Place for Western Tight Gas Basins (Comparison with FPC Estimates)

	. . . . . . FPC*.		. . . .Present Study . . .	
	Area (mi²)	Gas in Place (Tcf)	Area (mi²)	Gas in Place (Tcf)
Greater Green River				
Category 1	140	37.1	–	–
Category 2	500	108.4	–	–
Category 3	500	94.5	–	–
Total	1,140	240.0	870	90.5
Piceance				
Category 1	550	103.2	–	–
Category 2	650	103.9	–	–
Total	1,200	207.1	855	35.5
Uinta				
Category 1	300	101.6	–	–
Category 2	200	47.5	–	–
Total	500	149.1	996	50.2
Grand Total	2,840	596.2	2,721	176.2

*FPC, *National Gas Survey,* 1973, Volume II, p. 95.

Source: DOE EF-77-C-01-2705

Thus, this study's lower gas in place estimate arises from thinner net pays and lower gas-filled porosities. The exclusion of speculative acreage from the present study caused the total areas analyzed in the Greater Green River and Piceance Basins to be considerably smaller in the present study than in the FPC study (76% and 71%, respectively), while additional drilling and research in the Uinta Basin between 1972 and 1977 have resulted in acreage increases to almost double that used in the FPC analysis. Thus, the total area considered by the two studies is comparable.

The expected gas in place estimates by themselves, however, can only suggest the size of the potential for recovery. The amount of this gas actually produced is a function of recovery technology and economics.

TECHNOLOGICAL CHALLENGES AND GOALS

The Available Tools

Measurement and Characterization of the Resource: Effective design of field development and well stimulation programs for low permeability basins requires a level of understanding of the resource vastly greater than for conventional basins. Especially pressing is the need for improved ability to differentiate net pay within huge gross sections, greater precision in evaluating the quality of the net pay, measurements of rock strength and stress characteristics, and understanding of the geometry and orientation of gas-bearing lenses in lenticular formations.

The importance of the resource characterization goal is directly related to the severity of the geological problems besetting each basin. In the most difficult basins, e.g., the Western Tight and the Shallow Basins, progress in resource characterization is a critical prerequisite to improving and applying the technology. In basins with more favorable geology, improved recovery technology becomes dominant, as demonstrated by the trial and error approach used in the Tight, Blanket Sands. In the Wattenberg field, for example, the major operator incrementally increased the size of the fractures used until larger fractures no longer improved well performance. Once this point was reached, the majority of the future stimulations were standardized at this fracture size.

Stimulation Technology: The objective of the gas recovery technology in tight gas basins is to create an effectively propped, vertical fracture that intersects the net pay for considerable distances (sometimes up to 1,500 to 2,500 ft) in each direction from the wellbore. This fracture is created by high pressure injection of fluids—water, gels, foams, or combinations—through the well perforations into the formation rock. The fracture fluid carries and deposits solid particles (the proppant) to keep the fracture from closing when the injection pressure decreases. The fracture creates an enlarged pressure sink to all exposed gas pays, possibly cutting across the less permeable horizontal bedding planes, and provides a relatively direct, high permeability channel to the wellbore.

The effectiveness of the fracture depends on the extent to which it intersects the gas-bearing sections and remains open to the gas flow. Figure 2.23 shows isometric and plan views of typical fractures and the flow path of gas from the matrix, to the fracture, and into the wellbore.

Three key variables need to be considered as part of stimulation technology:

- Fracture height and length—Effective MHF stimulation depends on the distance the fracture remains in gas-bearing strata. Economic stimulation depends on expending minimum funds in fracturing nonproductive zones adjacent to the net pay. Fractures with excessive ratios of height to length or which tend to rise or fall out of the gas zone will be ineffective and uneconomic. Thus, improved ability to design the fracture treatments and control their height and shape are essential for full commercial application.

- Conductivity—Not only must the fracture expose the net pay, it must remain open to gas flow. The ability of the gas to flow depends on the permeability contrast between the rock matrix and the fracture. To the extent proppants crush, compact, or imbed in the fracture, or fail to reach the full extent of the fracture, this contrast is lost and the effectiveness of the fracture is reduced. Although generally not a problem in shallow formations, the greater overburden pressures of formations deeper than about 8,000 ft necessitate improved proppants and procedures for placing them in the fracture.

- Multiple Fractures—Economic exploitation of many of the basins will require that several intervals of a massive gross section be fractured from the same well, particularly when a single interval or formation must share drilling and operating costs with other formations to be commercial. On-going field research has already advanced the technology to enable some such multiple completions, but as the number, size, depth, and vertical dispersion of the treatments increase, additional improvements in well completion equipment and stimulation techniques may be required.

Figure 2.23: Isometric and Plan Views of Typical Fractures

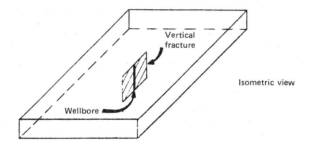

From Ron G. Agarwall, *Evaluation of Fracturing Results in Convential and MHF Applications,* SPE, Feb. 28, 1977.

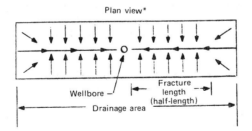

*Arrow denotes flow path.

From Lloyd E. Elkins, "The Role of Massive Hydraulic Fracturing in Exploiting Very Tight Gas Deposits, *Natural Gas for Unconventional Geologic Sources,* National Academy of Sciences, Washington, D.C. 1976.

Source: DOE EF-77-C-01-2705

The current stimulation tools are adequate for producing the near commercial Other Low Permeability Basins and the more favorable portions of the Tight, Blanket Sands. In these two areas, the R&D program should focus on optimizing field development and well stimulation technologies to yield higher recovery efficiencies, to obtain economic recovery from the less favorable (now uneconomic) segments, and to accelerate the rate of development. In the basins with more severe geological problems, substantial improvements over current gas recovery technology are required.

The required technological improvements appear to be within reach of a concerted program of research, development, and demonstration. However, the large costs of such a program and the considerable uncertainties surrounding the resource base argue that producers or service companies, acting singly, may under-invest in the R&D required to advance the technology (1). A collaborative federal-industry program appears to be required for fully commercializing these basins.

Analysis of the Technological Challenges

The requirements of commercializing the tight gas basins form the research and development goals for enhanced gas recovery. To provide a plan for the immediate, first steps, the R&D must address four key questions:

- What are the exact reservoir properties of the tight gas basins?

- To what extent can the entire net pay in the massive gross intervals of the tight gas sands be stimulated and produced from the same wellbore?

- Is it possible in lenticular formations for massive hydraulic fractures to intersect sand lenses not initially in contact with the wellbore?

- What other significant improvement in tight gas recovery would make the resource commercial in the near-term?

The importance of these four basic questions is discussed below. The analysis uses an exemplary tight gas reservoir and a single phase, finite difference reservoir simulator to ascertain the impact of research outcomes on economic exploitation of the resource.

Reservoir Characteristics of the Tight Gas Formations Used in the Sensitivity Analyses

Depth of Well, ft	9,000
Net Pay Thickness, ft	100
Fracture Height, ft	400
Drainage Area, acres	160 and 320
Flowing Bottom Hole Pressure, psi	1,000
Reservoir Temperature, °F	200
Gas Gravity	0.6 (air = 1.0)
Original Fracture Conductivity, md-ft	2,400
Final Fracture Conductivity, md-ft	300
Propping Agent, mesh sand	20–40
Producing Life, years	30
Fracture Gradient, psi/ft	0.7

(continued)

Porosity, %	10
Water Saturation, %	55
Initial in situ Permeability (to gas), md	0.001–0.10
Initial Pressure, psi	4,500

What are the exact reservoir properties of these tight gas basins? In the tight gas sands, the geology and reservoir properties dictate the limited technological interventions that can be applied. This makes it essential to know how these key properties affect economic recovery and in which range they become most restrictive. Briefly, assuming adequate gas pay thickness is available, the deposition of the pay, its permeability, and its gas-filled porosity dominate all other properties.

Sand Deposition — The most dominant feature is the sand deposition, either blanket type or lenticular, and if lenticular, the dimensions of the sand lenses. The following examples illustrate these concerns:

- Using the reservoir properties of the example formation (at 5 μd, 320 acre spacing, and a 1,000 ft fracture), a blanket sand of 100 ft would produce about 8 Bcf per well, but a lenticular sand body (100 ft net and dimensions of 400 ft wide by 6,000 ft long) would produce less than 2 Bcf per well; this is because the drainage area exposed to wellbore in a lenticular pay is less than 20% of that available to a wellbore in a blanket pay, as summarized below.

	Blanket Sand	Lenticular Sand
Single well drainage area, acres	320	55
Gas in place, Bcf		
Total per 320 acres	16	16
In contact with wellbore	16	3
30 year recovery per well	8	<2

- The critical dimension of the lenticular pay is its width rather than its length. The simulation analyses show that increasing lens width by 50% (to 600 ft) increases recovery by nearly 50%, but increasing lens length by 50% (to 9,000 feet) adds only about 5% to recovery in the economically most critical first ten years.

The second feature of sand deposition is the dominant orientation of the sand lens and the extent to which this parallels expected fracture azimuth. The impact of their relative orientations can be striking:

- If the relationship between the fracture orientation and the azimuth is random, a fracture designed for 1,000 ft of penetration (half-length) will remain in the example lens for only about 420 ft (Figure 2.24).

- When regional tectonic forces are essentially perpendicular to lens direction, effective fracture half-length is limited to 200 ft (one half of the width of the lenses), and unless other lenses can be intersected, fractures designed larger than this are wasted.

- However, should the fracture azimuth parallel the sand lens, effective fracture length could reach the full 1,000 ft of design,

and in tight (5 μd) sands, gas recovery would be twice
that in the random orientation case and three times that
in the perpendicular case.

Figure 2.24: Analysis of Effective Fracture Half-Length

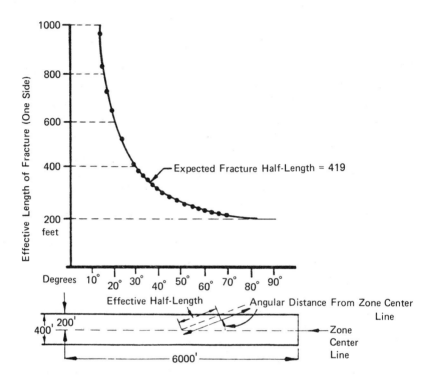

Source: DOE EF-77-C-01-2705

Permeability — The critical point of permeability appears to be about 10 μd
(or 0.01 md). At 10 μd, a 1,000 ft fracture on 160 acre spacing in a blanket-
type sand, would yield 30 year recovery efficiency of about 68% of the gas in
place. As permeability increases by an order of magnitude, to 100 μd, recovery
increases by less than one-fifth, to about 80% of gas in place. As permeability
decreases by an order of magnitude, to 1 μd, recovery decreases by nearly two-
thirds, to about 25% of gas in place.

Since the work by Thomas and Ward (9), it has been recognized that conventional laboratory permeability analysis overstates the in situ conditions by
an order of magnitude or more in tight formations. The distortion is greater at
lower permeabilities, precisely the condition that shows the greatest effect on recovery efficiency. Only a small error in the lower range could spell the difference
between a highly promising formation and one that is economically infeasible.

Gas-Filled Porosity — The amount of gas-filled porosity has direct and compounding effects: it is linearly related to gas in place and it directly affects recovery efficiency. Using the typical tight formation discussed earlier, recoveries were compared for gas-filled porosities of 4.5% (total porosity of 10% with 55% water saturation) and 2% (total porosity of 8% with 75% water saturation). The effect of the lower gas-filled porosity was to reduce total gas recovery in the lower-quality section to less than 20% of that achieved in the higher quality section, from 8 Bcf to about 1.5 Bcf per well.

Lower gas in place accounts for about one-half of the reduction and lower recovery efficiency compounds the problem. Thus, even slight overestimation of gas-filled porosity could render an apparently promising reservoir uneconomic. Moreover, since in the tight gas sands there appears to be a direct correlation between porosity and permeability, the effects of lower porosity in actual practice could be even more dramatic.

To what extent can the entire net pay in the massive intervals of the tight gas sands be stimulated and produced from the same wellbore? Many of the tight basins are characterized by massive sections containing numerous gas-bearing intervals or by the occurrence of numerous, discrete gas formations "stacked" one over another over a span of thousands of feet. Under existing practices, often less than one-third of the sand is completed and stimulated. In the Western Tight and Shallow Gas Sands, no single interval may be productive enough to be commercial on its own, yet several in combination could be economic. This requires multiple massive fracturing treatments through a common wellbore. As the number, size, and vertical dispersion of the treatments grow, it may require:

- Improved casing, cementing, and well completion practices.
- Cost-effective means of stimulating multiple intervals without damaging the production string.
- Advanced stimulation techniques to maintain the massive induced fractures in their intended pay intervals.

Should multiple completions with numerous MHF prove to be technically ineffective, much of the potential of the Western Tight, Shallow Gas, and Other Tight, Lenticular formations would become economic only at the higher ($3.00 to $4.50 per Mcf) gas prices.

Assuming that MHFs in multiple intervals can be successfully placed through the same wellbore, there still remains the challenge of stimulating as much of the quality gas pay in the formation as possible. The objective of the R&D program would be to stimulate 80% of the quality pay in the formation.

Is it possible in lenticular formations for a massive fracture to intersect sand lenses not initially in contact with the wellbore? While considerable argument can be marshalled on each side of the question, the field tests to date provide little evidence either way to this vital question.

A comparison of performance between 60 small fractures (100 to 300 ft) and 6 larger (500 to 1,000 ft) fractures showed only an insignificant improvement in anticipated gas recovery for the larger fractures. However, given the relatively small length and the low sand shale ratio (from 20 to 40%), the larger fractures had only limited probability of intersecting additional lenses.

If a fracture will not enter lenses other than those initially encountered by the wellbore, numerous small fractures appear to provide an optimum approach. However, should a large (1,500 to 2,000 ft) fracture be able to intersect additional lenses, the effect can be dramatic.

- At a rate of 1 additional lens (for each lens seen at the wellbore), the initial 10 year recovery would be about 70% higher than for the single lens; the larger fracture might add nearly 1 Bcf of additional recovery during this time, in the example formation.

- At the probable estimate of two additional lenses (for each lens seen at the wellbore) initial 10 year recovery would be more than double that for the single lens and might add 1.5 to 2 Bcf.

What other significant improvements in tight gas recovery should be pursued for accelerating commercialization of the tight gas basins? Beyond the above research questions, numerous additional opportunities should be pursued for optimum, economic exploitation of these tight sands. As in any scientific discipline, major advancements in measurement and reservoir analysis capability must parallel all technological advances. Of these, five stand out:

- Designing the optimum size fracture with respect to any given set of geological conditions. Figure 2.25 shows that long fractures (1,500 ft) are effective in low, 1 μd to 10 μd, blanket sands but contribute little to recovery efficiency over short fractures (500 ft) in the higher permeability sands (50 μd and higher).

- Ensuring adequate fracture conductivity, particularly through the use of higher sand concentrations and new proppant materials (e.g., bauxite).

- Engineering optimum fracture height, particularly in relation to the available net pay and desired length. Figure 2.26 shows the relationship of fracture height and length (when using a high quality fracture fluid) and how unnecessary height impedes effective fracture length at any given volume of fluid.

- In blanket sands, well placement, given fracture azimuth, determines the shape of the effective drainage area. At higher permeabilities, 0.1 md and greater, square drainage patterns with relatively short fractures were most efficient, while at low permeabilities rectangular patterns with long fractures are most efficiently drained (Figure 2.27).

- Establishing optimum field development in relation to sand deposit. Since only the lenses connected to the well (or to the fracture) can be drained, it may be necessary to use substantially closer spacing in formations marked by long, thin lenses. Closer spacing, combined with multilens fractures, could be the key to recovering substantial portions of the gas in place in these basins.

Summary: In summary, several important technological questions must be answered if massive hydraulic fracturing is to become commercial in any but the most geologically favorable areas. The solutions to any of these problems, however, await improved resource characterization and evaluation and a fuller understanding of how the new stimulation technology will actually perform.

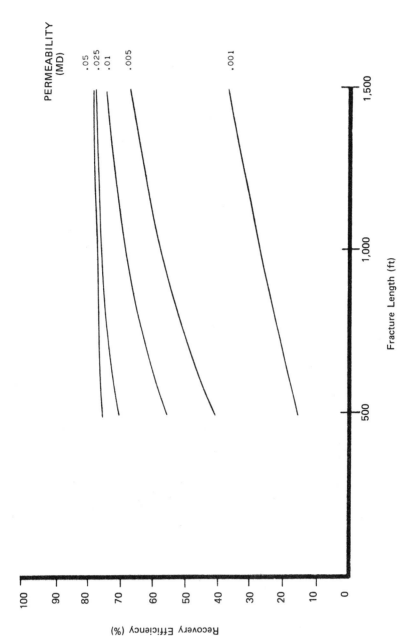

Figure 2.25: Recovery Efficiency as a Function of Fracture Length and Permeability (Assuming 30 Year Production Life)

Source: DOE EF-77-C-01-2705

Figure 2.26: Relationship of Fracture Volume, Fracture Height, and Fracture
Half-Length Using a High Quality Fracture Fluid

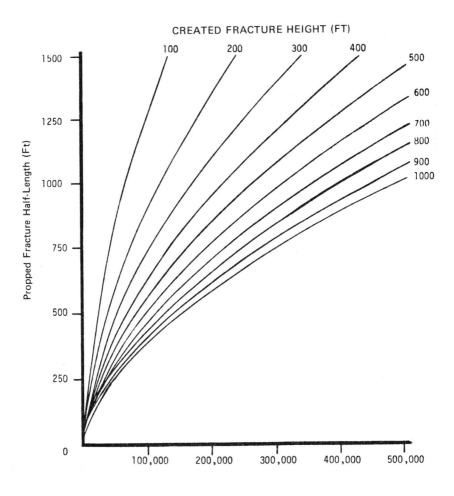

Source: DOE EF-77-C-01-2705

Figure 2.27: Simulated Twenty Year Recovery for Three Drainage Shapes

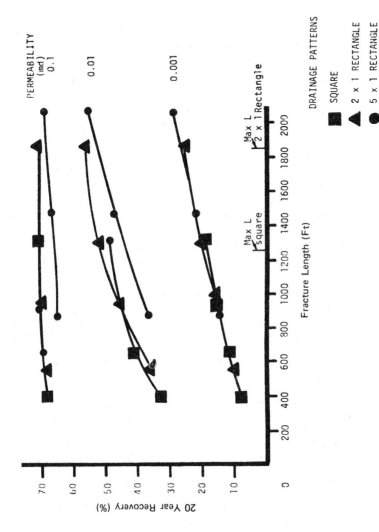

Source: DOE EF-77-C-01-2705

For all but the "cream" of the tight basins, resource characterization, technology improvements, and demonstrations of these new capabilities must occur in sequence, each establishing a knowledge base on which the next is designed and tested.

THE POTENTIAL OF THE TIGHT GAS BASINS

Approach

This section estimates the amount of gas that could be economically recoverable under alternative technological and economic assumptions. Three elements are essential for estimating this potential:

- Technology—Two sets of technological assumptions were projected:

  Base Case: the level of the technology expected to be attained by industry during the next five years without active federal involvement.

  Advanced Case: the level of the technology expected to be attained by virtue of active federal-industry collaboration.

- Gas Production—The analytic units in the resource base were developed and produced using a reservoir simulator and development model.

- Economics—Each analytic unit was evaluated using actual field development costs, return on capital requirements, and at three gas prices.

Base and Advanced Technology Cases: The Base Case reflects the current state of the art of MHF technology plus anticipated advances in the next five years. The assessment of current technology was based on ongoing field tests and discussions with industry's leaders in massive fracturing. The conclusions drawn from these steps were consistent with the results of a survey of key financial and technical managers of 92 companies in all phases of the gas industry (1).

The Advanced Case represents evolutionary technological advances that hold reasonable promise of being achieved through focused R&D. The projection of technological advances draws on the theoretical analysis and laboratory tests of the leading gas recovery experts and on preliminary pilot tests by the most prominent MHF practitioners.

Table 2.10 displays the major technological elements and how they differ between the Base and Advanced Cases. The salient parts are summarized below.

- Resource Characterization. The Advanced Case assumes all formations are eligible for economic testing; the Base Case assumes industry's interest would be confined to areas and formations that have been demonstrated to have favorable geologic characteristics. The Advanced Case assumes lower dry hole rates stemming from accelerated resource definition and improved geologic and reservoir measurement technology. Numerous additional advances are achieved through the resource characterization efforts, but these are prerequisite to the other differences shown, e.g., ability to make and interpret measurements required for more effective and efficient field development and well stimulation, acceleration of development in less defined areas, and reduced risks. These differences are reflected in items listed under technology, economics, and development in the table.

- Technology Advancement. The Advanced Case assumes improved ability to design and control the fractures and to increase their intersection of widely disseminated gas pay. In lenticular formations, spacing is reduced and fractures are assumed capable of intersecting net pay lenses not encountered at the wellbore. This combination of advances would increase the proportion of net pay in a section in contact with a wellbore from 17% to about 80%.

- Cost and Economic Criteria. In the Advanced Case, the fracture is assumed to achieve the engineered level of performance, up from 80% effectiveness in the Base Case. Improved capacity for predicting the technology, as demonstrated by successful pilot tests, would reduce industry's risk premium from the present 26% ROI (Return on Investment) (after tax) to a more conventional level of 16% ROI (after tax).

- Acceleration. The Advanced Case assumes accelerated application of the technology and timing of field development. The initial technological advances are estimated to become effective in 1981.

Simulated Expected Production: A single phase, finite difference reservoir simulator, was used to estimate gas production and recovery from the individual formations and areal units. It was developed at Texas A&M University by Doctors Steven A. Holditch and Richard Morse (10) and was validated, during the study, against field data and other simulation models used in the production industry. Dr. Holditch and his staff at Sovereign Engineering, Inc. conducted the numerous analyses.

The grid pattern of the simulator consists of 300 cells (20 x 15) and uses a fine breakup near the wellbore and along the fracture and small time steps, as small as 0.001 days, to provide accurate simulation of the early transient flow periods. The model uses a direct solution technique, alternating diagonal, matrix inversion, for solving the equations that describe a fractured reservoir.

The reservoir simulator was modified to include two important phenomena that occur in tight gas reservoirs. First, the effects of non-Darcy flow on the well performance were incorporated using published correlations to simulate pressure gradients under non-Darcy flow conditions. The second modification was to simulate increasing closure pressure on fracture conductivity. If sand is used as a propping agent, fracture conductivity may decrease by an order of magnitude during the life of a well. Adjustments were made that explicitly account for proppant type, proppant concentration, the formation embedment pressure, the value of the least principal stress, and the flowing bottom hole pressure.

This model was used to simulate the performance of each formation and each areal unit in the resource base. Where more than one formation was vertically adjacent, and could be produced by multiple completions, the gas production was individually calculated and then combined.

Table 2.10: Summary of Major Differences Between the Base and Advanced Cases—Tight Gas Basins

Strategy/item	Base Case	Advanced Case
Eligible formations	Limited to those demonstrated to be geologically favorable	All
Dry hole rate		
Lenticular	30%	20%
Blanket	20%	10%
Technology		
Fracture height	4 x net pay limit 600 (200' minimum)	3 x net pay limit 400' (150' minimum)
Fracture length (one way)		
Shallow gas sands	200'	500'
Near tight gas sands	500'	500'
Tight gas sands	1,000'	1,500'
Fracture conductivity	Decreases with depths using current proppants and methods	(With improved proppants and methods maintaining adequate conductivity)
Field development		
Lenticular	320 acres/well (2 wells/section)	107 acres/well (6 wells/section)
Blanket	160 acres/well (4 wells/section)	160 acres/well (4 wells/section)
Net pay contacted		
Lenticular gas sands		
320 acres drainage	17%	—
107 acres drainage	—	80%
Blanket	100%	100%
Economics		
Cost of delivered fracture	120%	100%
Risks, reflected in discount rate of	26%	16%
Development		
Start year for drilling		
Probable acres	1978	1981 (RD&D effect begins)
Possible acres	1987	1987
Development pace		
Probable acres	17 years to completion	13 years to completion
Possible acres	17 years to completion	15 years to completion

Source: DOE EF-77-C-01-2705

Economics: A net present value (discounted cash flow) model was used to simulate the economics of production. The unit of analysis was a well with its drainage area representing a specific areal unit. The areal units for which the discounted net cash flow exceeded zero were deemed economic and developed, according to the timing model.

State-level drilling and completion costs were drawn from the API/AGA Joint Association Survey of Costs (11). Well stimulation costs were provided by major service companies. Surface equipment and operating costs were developed from studies by Gruy Federal, Inc. Dry hole and exploration costs were functions of drilling and completion costs and the areas involved. Taxes, royalties, burden rates, and accounting procedures were drawn from actual applications in each region. All costs were varied as a function of depth and geographic region. Constant 1977 dollars were used throughout the analysis.

The analysis was conducted for three gas prices, $1.75, $3.00, and $4.50 per Mcf. Where the well, representing a specific areal unit, was found to be economic, it was extrapolated to the full unit. Each area was assumed to be developed according to a fixed schedule. The better defined (Probable) areas were assumed to be developed first, followed by the less well defined (Possible) areas.

Technically Recoverable Gas

Setting the price constraint aside and analyzing the formations under the Advanced Case technology provides an estimate of technically recoverable gas, a useful benchmark for subsequent analyses.

Table 2.11 displays the technically recoverable gas from each of the five resource targets. Figure 2.28 shows the same information in graphical form:

- Overall, slightly less than half of the gas in place is technically recoverable in 30 years under Advanced Case technology.

- The 30 year recovery efficiencies differ substantially among the resource targets, ranging from 38 to 82%:

    The geologically more favorable targets, the Tight Blanket and Other Low Permeability targets, have technological recovery efficiencies of 70 to 80%.

    The geologically more difficult target, the Western Tight Gas Basins, the Shallow Basins, and the Other Tight, Lenticular Basins, have technological recovery efficiencies of about one-half these levels, at 40 to 50%.

The technically recoverable gas provides a useful benchmark for assessing the benefits of lifting economic constraints and improving technological performance. For the Advanced Case, it represents a limit of what can be realistically expected.

The Base Case Estimates: Base Case production estimates, at the three gas prices of $1.75, $3.00, and $4.50 per Mcf, represent the amount of recovery that can be anticipated from the tight gas basins without advanced technology and without a federally sponsored research and development program.

Ultimate Recovery — In terms of ultimate recovery (estimate based on thirty-year well life), the Base Case for the tight gas basins shows:

- About 70 Tcf is recoverable at a gas price of $1.75 per Mcf.

Table 2.11: Gas in Place, Technically Recoverable Gas, and Technical Recovery Efficiency—Tight Gas Basins

Target/Basin	Gas in Place (Tcf)	Technically Recoverable (Tcf)	Technical Recovery Efficiency (%)
Western Tight			
Greater Green River	90.5	35.5	39.3
Piceance	35.5	12.1	34.1
Uinta	50.2	18.0	35.8
Subtotal	176.2	65.6	37.2
Shallow Gas			
Northern Great Plains	53.4	18.4	34.4
Williston	20.9	16.5	79.2
Subtotal	74.3	34.9	47.0
Other Tight, Lenticular			
Sonora	23.9	15.8	66.3
Douglas Creek	3.3	0.3	8.5
Big Horn	23.4	7.8	33.5
Subtotal	50.6	23.9	47.2
Tight, Blanket Gas			
Cotton Valley (Sweet)	67.1	49.7	74.1
Denver	18.5	13.0	70.5
Ouachita	4.9	1.4	28.6
San Juan	15.0	12.0	79.9
Wind River	2.7	1.0	36.5
Subtotal	108.2	77.1	71.3
Other low permeability reservoirs			
Cotton Valley (Sour)	13.8	9.9	71.7
Total	423.1	211.4	50.0

Source: DOE EF-77-C-01-2705

- An additional 30 Tcf, for a total of 100 Tcf, can be re-covered at $3.00 per Mcf.
- Raising the price from $3.00 to $4.50 increases ultimate recovery by 8 Tcf, to a total of 108 Tcf.
- At $3.00 and under Base Case technology, industry will recover about one-fourth of the gas in place.

Industry efforts are expected to be concentrated in the geologically more favorable basins (Figure 2.29).

- At $3.00 per Mcf, the Tight, Blanket formations are anti-cipated to produce more than 50 Tcf, or about three-fourths of the technologically recoverable gas.
- By contrast, at this gas price, the Western Tight Basins pro-duce less than 10 Tcf, or only about 14% of the technologi-cally recoverable gas.
- The remaining targets vary between these extremes.

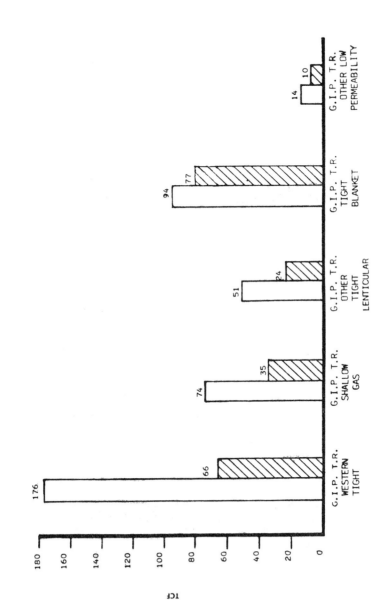

Figure 2.28: Comparison of Gas in Place and Technically Recoverable Gas–by Tight Gas Resource Target

Source: DOE EF-77-C-01-2705

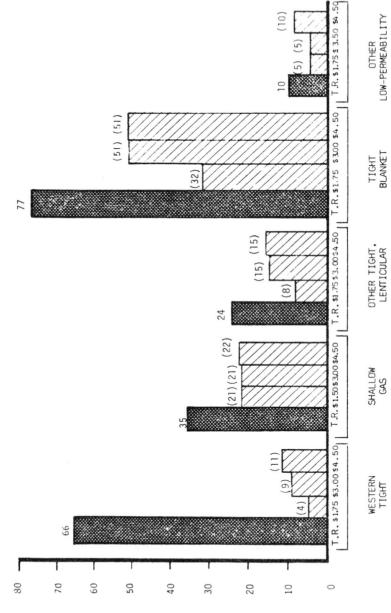

Figure 2.29: Base Case Ultimate Recovery—by Tight Gas Resource Target

Source: DOE EF-77-C-01-2705

These results point up three important conclusions. First, only a very small portion of the ultimate potential of the geologically difficult, lenticular basins will be developed under Base Case technology. Development of such targets requires substantial resource characterization and technological advance, well beyond the levels assumed in the Base Case.

Second, even in the geologically favored targets where considerable industry activity is projected, the amount of recovery can be improved through Advanced Case technology.

Third, ultimate recovery in the Base Case shows considerable sensitivity to gas price. The estimates for the Western Tight Basins, the Other Tight, Lenticular Sands, and the Tight Blanket Formations show sizeable price/supply sensitivity between $1.75 and $3.00 per Mcf, but little sensitivity beyond $3.00. The Other Low Permeability Sands show sensitivity between $3.00 and $4.50 per Mcf; the Shallow Basins show practically no price sensitivity in the Base Case. The overall price/supply sensitivity is greatest between $1.75 and $3.00 per Mcf.

Production Rate — While ultimate recovery is a valuable indicator from a total resource perspective, many of today's concerns center on: "How much additional gas can we produce in 1985 or 1990?"

Figures 2.30 and 2.31 show that the total Base Case production rate from the tight gas basins rises gradually from 1979 to a peak of 4.0 Tcf per year by 2000. These estimates are incremental to current tight gas production from proved reserves.

- The Tight, Blanket basins provide over one-half of the total production rate, nearly 1.6 Tcf per year in 1990 and 2.0 Tcf per year at their peak in 1995, at $3.00 per Mcf.

- The contribution of the other four targets becomes significant after 1990, equaling the contribution of the Tight, Blanket basins.

Figure 2.30: Base Case Annual Production to the Year 2000 (at $3.00/Mcf)— by Tight Gas Resource Target

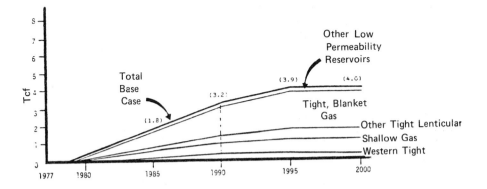

Source: DOE EF-77-C-01-2705

Figure 2.31: Base Case Cumulative Production to the Year 2000 (at $3.00/Mcf) by Tight Gas Resource Target

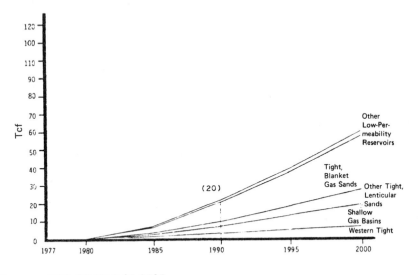

Source: DOE EF-77-C-01-2705

The cumulative recovery curve (at $3.00/Mcf) shows that only about 20 Tcf of the 100 Tcf of ultimate recovery could be produced by 1990. By 1995 and 2000, cumulative production is about 40 and 60% of the ultimate, respectively. This argues that acceleration as well as increased recovery should be the goals of federally sponsored research and development.

Improving on the Base Case — While important quantities of gas can be recovered from the tight gas basins, even under Base Case technology and without a federally sponsored research and development program, the analysis of technical recovery efficiency and the status of the technology show that major improvements are feasible and should be pursued through advanced technology.

The Advanced Case Estimates: The impact of the Advanced Case technology, achieving the resource characterization and technology goals of the joint federal-industry R&D strategies, was analyzed at the same three prices. The difference between the Advanced Case and Base Case is the incremental production stimulated by a successful R&D program.

Ultimate Recovery — Under Advanced Case technology, almost 150 Tcf is projected to be ultimately recoverable at $1.75/Mcf. Raising the gas price to $3.00/Mcf for all the basins would yield another 32 Tcf, raising total recovery to above 180 Tcf (Figure 2.32). Further increasing the price to $4.50/Mcf adds only about 6 Tcf.

Under Advanced Case technology and at $3.00/Mcf, about 43% of the gas in place could be recovered. This contrasts with the Base Case where at the same price, about 24% of the gas in place is economically recoverable.

As shown in Figure 2.33, the increase in recovery is not proportional across all targets, although significant benefits are available from all:

- The largest benefits are those in the geologically difficult Western Tight Gas Basins. At $1.75/Mcf, the Advanced Case ultimate recovery is almost nine times the Base Case estimate. At $3.00/Mcf, the Advanced Case ultimate recovery is over five times the Base Case estimate and yields recovery of about three-fourths of the technically recoverable gas.

Figure 2.32: Ultimate Recovery from the Tight Gas Basins (at Three Prices)

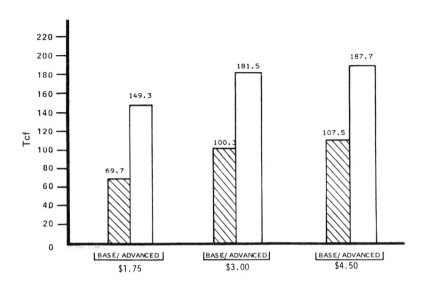

Source: DOE EF-77-C-01-2705

- For Shallow Basins—the gain due to the Advanced technology at $1.75/Mcf is only about 10%. Raising the gas price to $3.00/Mcf, however, crosses the price threshold of the tighter portions of these basins, leading to a 50% increase in anticipated ultimate recovery and yielding almost 90% of the technologically recoverable gas. The Shallow Basins require both higher prices and technological improvements to increase production.

- In the Other Tight, Lenticular Basins, the Advanced technology makes practically all the technologically recoverable gas economic at $1.75/Mcf. These basins pointedly show how improved technology can substitute for economic incentives.

- Tight, Blanket Sands respond to changes in price and technology. Advanced technology nearly doubles the projected ultimate recovery at $1.75/Mcf. At $3.00/Mcf, it adds >25 Tcf and exceeds 85% of the technologically feasible recovery.

Figure 2.33: Ultimate Recovery from the Tight Gas Basins—by Tight Gas Resource Target

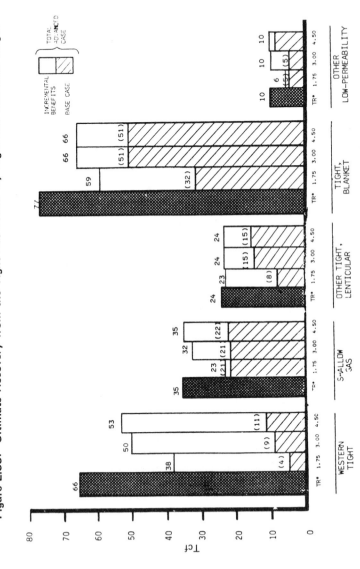

Source: DOE EF-77-C-01-2705

- The Other Low Permeability Reservoirs show a small increase for the Advanced Case at $1.75/Mcf, but a doubling of the Base Case estimate at $3.00/Mcf. Under the Advanced Case and at $3.00/Mcf, nearly all the technologically recoverable gas would be produced. In these reservoirs, the price mechanism and R&D can be mutually supportive federal strategies for increasing gas supply.

Production Rate — Figures 2.34 and 2.35 show the total annual and cumulative production from the Advanced Case as well as the increments over the Base Case, at $3.00/Mcf.

Total production rises rapidly due to acceleration of field development and improved recovery. Annual production reaches a peak in 1990 at 7.7 Tcf/yr and declines slightly after that time to 7.2 Tcf in 1995 and 6.8 in 2000. This peaking is due to the rapid development of the Western Tight and Tight, Blanket Sands. Annual production from the other targets continues to expand throughout the period to 2000.

- Under Advanced Technology at $3.00/Mcf, the tight gas basins can make a significant contribution to the nation's overall demand for gas, providing 4 Tcf in 1985 and nearly 8 Tcf in 1990.

- The cumulative recovery curve (at $3.00/Mcf) shows that 45 Tcf, or almost one-quarter of the ultimate recovery, can be produced by 1990.

Figure 2.34: Annual Production to the Year 2000 (at $3.00/Mcf)—by Tight Gas Resource Target

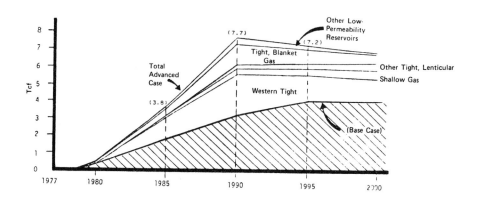

Source: DOE EF-77-C-01-2705

Figure 2.35: Cumulative Production to the Year 2000 (at $3.00/Mcf)—by
Tight Gas Resource Target

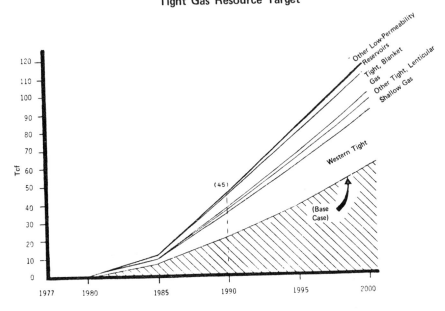

Source: DOE EF-77-C-01-2705

Well Requirements: Table 2.12 shows the number of new wells required
in each resource target under Base and Advanced Technology at $1.75 and $3.00
per Mcf.

Table 2.12: Drilling Requirements in 1990 and 1995—by Tight Gas Resource
Target

| | . .No. of New Wells—1990 . . | | | | . .No. of New Wells—1995 . . | | | |
| | . Base Case . | | Advanced Case | | . Base Case . | | Advanced Case | |
	$1.75 per Mcf	$3.00 per Mcf	$1.75 per Mcf	$3.00 per Mcf	$1.75 per Mcf	$3.00 per Mcf	$1.75 per Mcf	$3.00 per Mcf
Western tight gas sands	120	260	970	1,350	60	150	510	590
Shallow gas basins	750	750	1,290	3,170	1,210	1,210	1,300	4,740
Other tight, lenticular basins	290	450	700	700	320	510	830	830
Tight, blanket gas sands	730	1,120	1,400	1,720	630	900	880	1,020
Other low permeability reservoirs	60	60	80	120	30	30	30	40
Total	1,950	2,640	4,440	7,060	2,250	2,800	3,550	7,220

Source: DOE EF-77-C-01–2705

In 1976, approximately 7,400 successful onshore gas development wells were drilled, excluding all exploratory wells, all oil wells, and all dry holes.

As the supply of natural gas falls behind the need, it is reasonable to assume that drilling capacity will expand. It has been estimated (12) that the growth in onshore drilling capacity could be sustained at 10 to 15% between now and 1995. Since the drilling requirements in the tight gas basins would use only 10 to 20% of projected capacity, as shown below, well requirements should not pose a constraint on reaching the projected production levels.

Drilling Capacity Requirements of the Tight Gas Basins

Annual Drilling Growth Rates (%)	Requirements as Percent .... of Total	
	1990	1995
8	30	21
10	23	14
15	12	6

The Relative Roles of Price and R&D

One of the strategy decisions facing federal energy policy makers involves balancing economic incentives (price and taxes) and sponsored research and development for augmenting domestic supplies of natural gas.

The study finds that future production from the tight gas basins is highly sensitive to economic incentives and to major advances, but most sensitive to improved economics and technology used in combination.

- Gas prices up to $1.75/Mcf provide enough economic incentives to stimulate the geologically more favorable portions of all the resource targets. Under Base Case technology, a total ultimate recovery of 70 Tcf and 1990 annual production of 2.7 Tcf are projected.

- Raising the price to $3.00/Mcf permits development of additional areas, increasing ultimate recovery to 100 Tcf and 1990 production to 3.2 Tcf.

- Further increasing price—up to $4.50/Mcf—has negligible effect on the estimates of production.

- Even at the lower level of economic incentives, $1.75/Mcf, focused federal-industry R&D adds substantial potential production; under Advanced Case technology the projections are: ultimate recovery of nearly 150 Tcf; annual production in 1990 of 6.3 Tcf.

- Combining Advanced Case technology and improved economic incentives yields substantial additional recovery: at $3.00/Mcf, Advanced Case ultimate recovery is estimated at over 180 Tcf, and 1990 annual production at 7.7 Tcf; at $4.50/Mcf, however, the projections are only marginally higher than at the $3.00/Mcf levels.

Maximizing production from the tight gas basins relies on both focused R&D and increased price. While estimating the exact levels of economic incentives that would accompany the R&D program was beyond the scope of the present study, the analysis suggests that the optimal price is between $1.75 and $3.00/Mcf in 1977 dollars.

SUMMARY OF R&D STRATEGIES

Research and Development Goals

The assessments of the resource base and the current technology defined the problems and constraints that currently deter development of the tight gas basins. Overcoming these problems and constraints constitutes the goals of the R&D program. In summary, the strategic goals are:

1. Resource Characterization

   - Resource Measurement—Develop accurate, reliable methods for collecting and analyzing reservoir and geologic data.

   - Resource Evaluation—Conduct a series of reservoir measurements and production tests in the better defined (Probable) areas of the basins.

   - Resource Definition—Obtain sufficient geologic data to define the resource in the less defined (Possible) areas of the basins, thus accelerating application of advanced well stimulation technology.

2. Fracture Technology

   - Fracture Length and Height—Create efficiently propped fractures out to 1,500 ft from the wellbore, where desired; control fracture height to three times the net pay or 150 ft, whichever is greater.

   - Fracture Azimuth—Determine likely fracture azimuth to assist in spacing wells and contacting net pay.

   - Fracture Conductivity—Introduce improved proppants such that fracture conductivity is not a limiting condition.

   - Fracture Effectiveness—Improve effective fracture length and drainage area in lenticular formations by intersecting sand lenses not encountered by the wellbore. In the thin, lenticular pays of the Mesaverde, achieving this goal would increase effective fracture length from 400 to 800 ft and increase drainage area from about 50 to about 80 acres.

3. Field Development Technology

   - Effective Drainage Area—Encourage a reduction in spacing in highly lenticular sands to six wells per section. Under this spacing the wellbore would be in contact with about 50% of the pay, with no appreciable well-to-well interference. When this reduced well spacing is combined with the capacity to intersect sand lenses away from the wellbore, 80% of the gas in place would be in contact with the wellbore. (This contrasts with about 20% of the net pay in contact with the wellbore at current well spacing and fracture technology.)

   - Multiple Completions—Improve well completion technology to enable numerous formations to be fractured and produced from the same wellbore.

4. Economics and Development

   - Risk Reduction—Reduce the risk premium (the minimum

required return on investment) by 10 percentage points
to 16% from the current high risk level of 26%.

- Timing of Field Development—Accelerate the pace of
field development in the Probable (defined) area and
open for exploitation the Possible (less defined) area.

- Economic Analysis—Determine the appropriate price incentives required to develop each basin in a timely fashion.

5. Technology Transfer

- Transfer the resource definition, technology, and analysis
to the gas production industry.

Ten R&D programs were defined to achieve these goals; four for the Western Tight Gas Basins, three for the Shallow Basins, and one each for the Other
Tight, Lenticular Sands, Tight Blanket Gas Formations, and Other Low Permeability Reservoirs. These programs are summarized below.

The research and development strategy for the Western Tight Gas Basins is
organized into four segments:

- Program 1.1—Resource Evaluation and Characterization:
Obtain data on basin and reservoir properties essential for
producing tight, lenticular formations of the Greater Green
River, Piceance, and Uinta Basins.

- Program 1.2—Greater Green River Full Program: (a) Resource Evaluation and Characterization—Obtain data on
basin and reservoir properties essential for producing tight,
lenticular formations of the Greater Green River Basin; and
(b) Technology Development—Develop improved well stimulation technology for application in this basin.

- Program 1.3—Piceance Full Program: (a) Resource Evaluation and Characterization—Obtain data on basin and reservoir properties essential for producing tight, lenticular formations of the Piceance Basin; and (b) Technology Development—
Develop improved well stimulation technology for application
in this basin.

- Program 1.4—Uinta Full Program: (a) Resource Evaluation
and Characterization—Obtain data on basin and reservoir
properties essential for producing tight, lenticular formations
of the Uinta Basin; and (b) Technology Development—Develop
improved well stimulation technology for application in this basin.

The resource characterization program has certain incremental benefits on
its own, but it is also prerequisite to successful technology development strategies.
Program 1.1 can stand alone, but because technology development strategies pre-
suppose resource evaluation and characterization, Programs 1.2, 1.3, and 1.4 each
contain their essential share of Program 1.1. To the extent that Programs 1.2,
1.3, or 1.4 are selected, Program 1.1 can be reduced.

The research and development strategy for the Shallow Gas Deposits is organized into three sections:

- Program 2.1—Characterize the R&D advanced technology
in the Tight, Shallow Gas Basins.

- Program 2.2—Assess the potential of producing low permeability, shallow gas deposits with improved technology.
- Program 2.3—Optimize recovery and development in shallow, near-conventional gas sands.

The strategies in this section are organized around distinct geological targets so each contains appropriate resource characterization and technology development tasks.

The R&D program for the Other Tight Lenticular Gas Sands consists of one strategy: Develop improved well completion technology along with the means for intersecting multiple sand lenses.

This strategy contains resource characterization as well as technology development tasks.

The research and development strategy for the Tight, Blanket Gas Sands consists of one program: Develop optimum recovery strategies to fully exploit the available resource.

The above R&D strategy contains resource characterization as well as technology development tasks.

The R&D strategy for the Other Low Permeability Gas Deposits consists of one program: Assist operators to define reservoirs and use optimum fracturing technology.

The above R&D strategy contains resource characterization as well as technology development tasks.

Costs of the R&D Program

Unlocking the potential of the tight gas basins will require a concerted program of research, development, and demonstration. In addition to ongoing industry outlays, nearly $250 million is required, over the next five years, for the joint federal-industry research programs in enhanced gas recovery. DOE would provide $160 million and industry the remaining $90 million.

The yearly costs for the five-year DOE/industry joint research program are as follows (in millions of constant 1977 dollars):

	Total Costs ($)	DOE Share ($)
Total 5 yr costs (FY 79-FY 83)	249.2	158.6
Yearly costs		
FY 79	40.6	27.8
FY 80	51.0	33.7
FY 81	51.4	33.3
FY 82	61.5	33.5
FY 83	44.7	30.3

The total five-year DOE costs by resource target are as follows (in millions of constant 1977 dollars):

Resource Target	DOE Share
Western Tight Gas Basins	$102.5
Shallow Gas Basins	14.7
Other Tight Lenticular Basins	23.4
Tight, Blanket Sands	11.5
Other Low-Permeability Reservoirs	6.5
Total	158.6

Further detail is provided in Table 2.13. The types and levels of effort that these budget outlays will support are as follows:

Activity Category	Level of Effort
Resource Characterization	379 person-years
Improved Diagnostic Tools	167 person-years
Field-Based Research, Development, and Demonstration	262 projects*
Technology and Information Transfer	35 person-years

*Field-based R&D is a mix of resource characterization cores and wells, measurement calibration tests, technology improvement tests, and field demonstrations of improved technology.

Additional detail is provided in Table 2.14

Table 2.13: Level of R&D Activities—Tight Gas Basins

Target/Program	Resource Characterization .. (person-years) . . .	Improved Diagnostic Tools	Field-Based RD&D (cores/wells)	Technology/ Information Transfer (person-years)
(1) Western tight gas sands				
(1.1) Resource characterization, 3 Basins*	(145)*	(40)*	(90)*	(5)*
(1.2) Greater Green River, Full program	68	31	44	5
(1.3) Piceance, Full program	45	8	37	0
(1.4) Uinta, Full program	67	31	44	5
Subtotal	(180)	(70)	(125)	(10)
(2) Shallow gas basins				
(2.1) Tight, shallow gas sands	25	10	31	5
(2.2) Low permeability, shallow gas sands	11	7	20	3
(2.3) Shallow, near conventional gas sands	13	8	24	5
Subtotal	(49)	(25)	(75)	(13)
(3) Other tight, lenticular sands	65	37	12	12
(4) Tight, blanket gas sands	60	15	25	0
(5) Other low permeability reservoirs	25	20	25	0
Total	(379)	(167)	(262)	(35)

*Person-years, cores, and wells also counted in the three basin-specific full programs.

Source: DOE EF-77-C-01-2705

Table 2.14: Costs of R&D Strategies—Tight Gas Basins

Target/Program	5 yr Program Costs (millions)	
	Total	Federal Share
(1) Western tight gas sands		
(1.1) Resource characterization, 3 basins	(88.0)	(56.0)
(1.2) Greater Green River, full program	66.4	38.7
(1.3) Piceance, full program	44.3	25.2
(1.4) Uinta, full program	66.3	38.6
Subtotal	(177.0)	(102.5)
(2) Shallow gas basins		
(2.1) Tight, shallow gas sands	9.3	7.0
(2.2) Low permeability, shallow gas sands	5.1	3.8
(2.3) Shallow, near conventional gas sands	4.4	3.8
Subtotal	(18.8)	(14.6)
(3) Other tight, lenticular sands	35.4	23.4
(4) Tight, blanket gas sands	11.5	11.5
(5) Other low permeability reservoirs	6.5	6.5
Total	249.2	158.5

Source: DOE EF-77-C-01-2705

Production Benefits

Successful execution of the R&D program would lead to additional gas recovery and acceleration of its production. Two measurements were used to quantify these benefits:

- A long-term measure of additions (additional quantities, over the Base Case, that would accrue due to successful R&D leading to the Advanced Case) to ultimate recovery (at $3.00/Mcf) due to the DOE/industry R&D program.

- A near-term measure of additional (additional quantities, over the Base Case, that would accrue due to successful R&D leading to the Advanced Case) gas that can be produced between now and 1990 (at $3.00/Mcf) due to the DOE/industry R&D program.

The estimated additional recovery due to the joint DOE/industry R&D is shown below, by resource target:

Resource Target	Long-Term Measure Ultimate Addition to Recovery (@ $3.00/Mcf)	Near-Term Measure Cumulative Addition to Recover: 1978–1990 (@ $3.00/Mcf)*
Western tight gas basins	41	14
Shallow gas basins	11	2
Other tight, lenticular basins	9	1
Tight, blanket sands	15	7
Other low-permeability reservoirs	5	2
Total	81	25

*Totals may not add due to rounding.

Table 2.15: Incremental Benefits Due to Advanced Technology
(at $3.00/Mcf)—by Tight Gas Resource Target

Target/Program	Ultimate Recovery (Tcf)	1990 Production (Tcf/yr)	1990 Cumulative Recovery (Tcf)
(1) Western tight, gas sands			
(1.1) Resource characteri- zation, 3 basins	(2.6)*	(0.1)*	(0.7)*
(1.2) Greater Green River, full program	16.7	0.9	5.3
(1.3) Piceance, full program	8.6	0.5	2.5
(1.4) Uinta, full program	15.5	1.0	6.0
Subtotal**	(40.8)	(2.3)	(13.7)
(2) Shallow gas basins			
(2.1) Tight, shallow gas sands	7.6	0.1	0.1
(2.2) Low permeability, shallow gas sands	1.3	0.05	0.05
(2.3) Shallow, near con- ventional gas sands	2.3	0.3	1.6
Subtotal**	(11.0)	(0.4)	(1.8)
(3) Other tight, lenticular sands	8.9	0.2	1.0
(4) Tight, blanket gas sands	15.2	1.2	6.8
(5) Other low permeability reservoirs	5.1	0.3	1.9
Total**	81.2	4.4	25.2

*Program 1.1 benefits also included in the three basin-specific programs, so are not added into subtotals.
**Totals may not add due to rounding.
Source: DOE EF-77-C-01-2705

Cost-Effectiveness of the Research Program

An essential question facing officials responsible for allocating public funds is: "How cost-effective is the expenditure?" Using the two production benefit measures discussed above, the analysis indicates that the payoff from R&D in enhanced gas recovery is considerable and cost-effective:

- The long-term cost-effectiveness measure for the five resource targets combined is 510 Mcf per dollar of federal R&D.

- The near-term overall cost-effectiveness measure is 150 Mcf per dollar of federal R&D.

Individually, each of the resource targets also have favorable cost-effectiveness ratios:

Resource Gas Target	Long-Term Measure	Near-Term Measure
	. . . .(Mcf/$)	
Western tight gas basins	400	130
Shallow gas basins	150	120

(continued)

Resource Gas Target	Long-Term Measure	Near-Term Measure
	. . . .(Mcf/$)	
Other tight lenticular basins	380	40
Tight, blanket sands	1,320	590
Other low permeability reservoirs	780	290

As the five tight gas resource targets would be developed according to ten R&D programs, Table 2.16 shows the cost-effectiveness measures for the individual programs:

- Two of the programs, the Tight, Shallow Gas Sands and Tight, Blanket Sands, yield long-term cost-effectiveness ratios greater than 1,000 Mcf per dollar.

- The lowest ratio for resource characterization (without technology improvement) in the Western Tight Gas Basins returns 46 Mcf per dollar. This strategy, however, is principally designed to prepare the prerequisite methods and information for the application of the improved technology. The full benefits of this strategy, then, are seen in the other three Western Tight Gas strategies. In these, the long-term benefits range from 341 to almost 436 Mcf per federal dollar, for a weighted average of 398.

- The near-term cost-effectiveness ratios generally follow the long-term measures with the following modifications:

    Two segments of the Shallow Gas Basins show only limited near-term response.

    The Tight, Blanket Sands become relatively more cost-effective in the near-term.

In selecting the R&D strategies, the decision-maker should consider the absolute size of the incremental benefits, the time at which these benefits are incurred, and the relative cost-effectiveness of expenditures of public funds.

Table 2.16: Cost-Effectiveness of R&D—by Tight Gas Resource Target

Target/Program	Long-Term	Near-Term
	. . . .(Mcf/$)	
(1) Western tight gas sands		
(1.1) Resource characterization, 3 basins	(46)	(13)
(1.2) Greater Green River, full program	436	137
(1.3) Piceance, full program	341	99
(1.4) Uinta, full program	402	155
Average	(398)	(132)
(2) Shallow gas basins		
(2.1) Tight, shallow gas sands	1,086	14
(2.2) Low permeability, shallow gas sands	342	–
(2.3) Shallow, near conventional gas sands	605	526
Average	(753)	(123)

(continued)

Table 2.16: (continued)

Target/Program	Long-Term	Near-Term
	......(Mcf/$)......	
(3) Other tight, lenticular sands	380	43
(4) Tight, blanket gas sands	1,322	591
(5) Other low permeability reservoirs	785	292
Overall average	512	159

Source: DOE EF-77-C-01-2705

- Advanced technology offers considerable production potential at the gas prices of $1.75 and $3.00 per Mcf.

- The technological improvements for the tight basins require evolution of existing technology rather than entirely new approaches, providing greater confidence that the estimated benefits can be achieved from the five-year plan.

- Under Advanced technology, the tight gas basins offer the potential of substantial near-term additions to domestic gas supply, annually totaling nearly 4 Tcf in 1985 and over 7 Tcf in 1990. Of these totals, about 2 Tcf per year in 1985 and 4 Tcf per year in 1990 would be the increments due to Advanced Case over Base Case technologies.

- Industry should be expected to contribute a substantial portion of the total research program costs. In addition to their own, in-house R&D efforts, it is estimated that industry would provide about $90 million to the joint federal-industry research efforts in the Tight Gas Basins.

REFERENCES

(1) Booz, Allen and Hamilton, Inc., *Empirical Study of the Natural Gas Industry,* conducted for the Department of Energy, Division of Oil and Gas (1977).
(2) Elkins, L.E., "The Role of Massive Hydraulic Fracturing in Exploiting Very Tight Gas Deposits," *Gas from Unconventional Sources,* NAS, Washington, DC (1976).
(3) U.S. ERDA, *Western Gas Sands Project Plan,* (August 1, 1977).
(4) Potential Gas Committee, *Potential Supply of Natural Gas in the United States,* Colorado School of Mines, Golden, Colorado (1977).
(5) American Association of Petroleum Geologists, *Bulletin,* Vol. 60, No. 8 (August, 1976).
(6) Knutsen, C.F., "Outcrop Study of Fracture Patterns and Sandstone Geometry, Eastern Uinta Basin, Utah: Study Results and Implications to the Stimulation of Tight Gas Sands in the Area," U.S. ERDA, Nevada Operations Office (May 23, 1977).
(7) Fast, C.R., Holman, G.B., Covin, R.J., "The Application of Massive Hydraulic Fracturing to the Tight Muddy 'J' Formation," Wattenberg Field, Colorado; *Journal of Petroleum Technology* (January, 1977).
(8) U.S. Federal Power Commission, *National Gas Survey* (1973).
(9) Thomas, R.D. and Ward, D.C., "Effect of Overburden Pressure and Water Saturation on Gas Permeability of Tight Sandstone Cores," *Jour. Pet. Tech.,* Vol. 24 (1972).
(10) Holditch, S.A. and Morse, R.A., "The Effects of Non-Darcy Flow on the Behavior of Hydraulically Fractured Gas Wells," *Journal of Petroleum Technology* (October, 1976).
(11) AGA/API, *Joint Association Survey of the U.S. Oil and Gas Producing Industry,* American Petroleum Institute, Washington, DC.
(12) Mortada International, special analysis for U.S. ERDA (August, 1977).

Devonian Shale

THE RESOURCE

The information in this and the following section is based on *Status Report on the Gas Potential from Devonian Shales of the Appalachian Basin*, OTA-E-57, November 1977, prepared by the staff of the Office of Technology Assessment for the Congress of the United States. References in this chapter are at the end of each section.

Geographic Extent

Sedimentary rocks of the geological age known as Devonian are present in the Appalachian region from New York to Alabama, in an area of some 209,000 square miles. The region includes two geological provinces of unequal size. In the smaller of these, an eastern belt known as the Valley and Ridge province, all rocks have been so intensely folded and faulted that few geologists consider them important sources of oil or gas, although recent studies indicate that the southern part of the Valley and Ridge has considerable potential for gas production. In the larger area, the Appalachian Plateaus immediately to the west, the rocks are flat-lying or only gently folded.

Furthermore, the upper part of the Devonian system of rocks in this province contains dark brown or black shale, rich in organic matter, that yields some natural gas at present and reportedly has the potential of yielding a great deal more. The area of the Appalachian Plateaus is 163,000 square miles (1). These shales are covered by younger rocks but can be readily reached by the drill, and are located in southwestern New York, western Pennsylvania, eastern Ohio, most of West Virginia, and eastern Kentucky (Figure 3.1).

In addition, Devonian dark brown or black shales outcrop at the surface or beneath a few feet of glacially deposited debris in western New York, in a belt extending from Cleveland southward through central Ohio, and in a series of disconnected outcrops in north-central Kentucky.

Figure 3.1: The Appalachian Basin

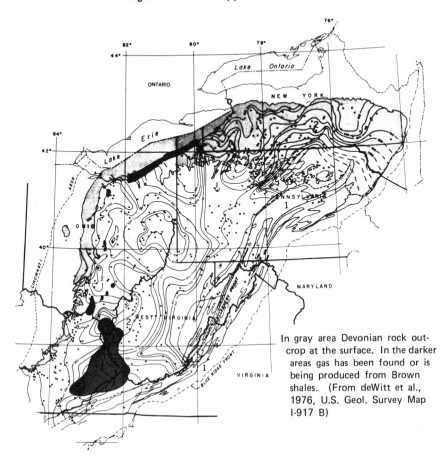

In gray area Devonian rock outcrop at the surface. In the darker areas gas has been found or is being produced from Brown shales. (From deWitt et al., 1976, U.S. Geol. Survey Map I-917 B)

Source: OTA-E-57

Terminology

The term Devonian shales refers to all the shale strata that lie beneath a widespread younger formation known as Berea sandstone and above an older limestone termed Onondaga or Corniferous. The shales are found in one-half dozen Appalachian States; similar strata are known in Indiana, Illinois, and Michigan. They occur in the subsurface, where they are encountered in wells, and at the surface, where they have been mapped and studied. Over time, they have acquired a variety of geographically based names: Chattanooga shale in the Appalachian States; Marcellus shale in New York; Ohio shale in Ohio; and New Albany shale in Kentucky and Indiana. In this report, these terms are considered to be synonymous.

The Devonian shales include strata that are gray, greenish gray, grayish brown, and deep brown to black. The deep brown to black shales contain much

organic matter, and are locally productive of gas. In most reports, they are
called Brown shale. It is important to keep in mind the distinction between the
whole thickness of Devonian shales and those parts, the Brown shale, that are
richer in organic matter and of greater interest as a source of gas.

Origin

The position of the Devonian shales in the Appalachian Basin, and their relation to other rocks, can best be understood by looking briefly at their origin.
To do this, it is necessary to erase one's mental image of present-day Appalachian
geography for a moment and substitute Late Devonian geography of some 350
million years ago. To the east of the region shown in Figure 3.1 in a position
roughly parallel to that of the present-day coastline, was a lofty range of mountains. Erosion of these mountains produced immense volumes of mud, silt, and
sand, which were carried westward by streams and deposited in a great compound delta, the Catskill Delta. The Delta was built out into a seaway that covered parts of what is now the Appalachian Basin.

In this sea, black organic-rich muds accumulated. The Devonian shoreline
was not fixed: at times, the sea level rose and marine muds were spread across
the seaward parts of the Delta; at other times, deltaic sands and silts flooded
westward into the seaway, displacing the shoreline far to the west. As a result
of these fluctuating conditions, the Brown shale interfinger to the east with much
thicker and coarser deltaic rocks (Figure 3.2). To the west, the black-mud bot-
tomed sea was at times restricted by a lowland area termed the Cincinnati arch,
and at other times, it flooded across this feature to merge with seas in the Michigan basin and the Illinois basin.

Although Brown shale deposition of Late Devonian time was not restricted
to the Appalachian Basin, this region appears to be most important from the
viewpoint of potential gas supply. In the long stretch of geologic time since the
Devonian period, both the deltaic rocks and the shales have been buried by
younger sediments. Around their edges, they have been partially uncovered by
uplift and erosion, but they remain under cover of younger rocks in much of the
Appalachian Basin.

Thickness

As indicated on Figure 3.2, Upper Devonian rocks are a great wedge-shaped
deposit, thin and shaly on the west and becoming thicker and more sandy toward
the east. In south-central Kentucky, the section consists of about 20 feet of
black shale (2). This section thickens to about 400 feet at about the Kentucky-
West Virginia line, and merges into a mass of siltstone, black silt, and sandstone
some 7,000 feet thick still farther east near the West Virginia-Virginia line. The
shale section increases in thickness from less than 500 feet at the outcrop in
southern Ohio to about 4,200 feet at the eastern edge of the state (3).

Only a fraction of these thicknesses, however, represents Brown shale of potential interest as a commercial source of gas. A generalized log of the shale section penetrated by wells in eastern Kentucky, for example, shows an overall shale
thickness of 677 feet, of which only an average thickness of 228 feet was Brown
shale (4)—one-third of the section. Of the eastern shales in general, a 1,000-foot
interval generally contains approximately 600 feet of light-colored shales and
400 feet of dark shales (5). A general idea of Brown shale thicknesses in the
Appalachian Basin is given by the contours on the map (Figure 3.1).

Figure 3.2: West-East Cross Section Across Central West Virginia

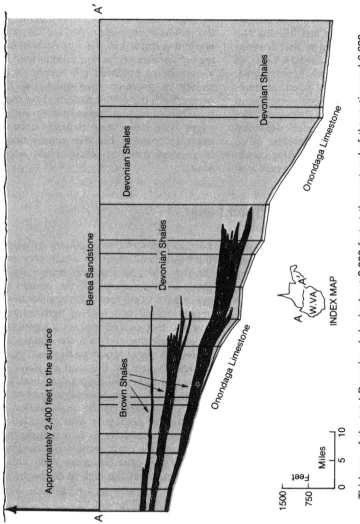

Thickness of the total Devonian shales is about 2,000 feet at the west end of the section and 6,600 feet at the east end. Brown shales disappear into thicker strata toward the east. (Modified from Martin and Nuckols, 1976, ERDA Pub. MERC/SP-76/2, Figure 4.)

Source: OTA-E-57

The thicknesses of Brown shale beds are higher in the eastern part of the Basin than is suggested on the cross section (Figure 3.2) because the geologists who compiled the map included more shale as Brown shale than those who made the cross section. Gas has been found through the entire Devonian shale, although the Brown shale has the highest concentration.

Attitude and Depth

In common with the other rocks in the Appalachian Basin, the Devonian shales have a gentle inclination, or dip, to the southeast. In eastern Kentucky, for example, they dip southeast at 30 to 50 feet per mile (7). At the surface in central Ohio, the top of the shale section has an elevation of about 800 feet, but in southeastern Ohio, this surface is some 1,400 feet below sea level (3). This is a decline of 2,200 feet in 85 miles, or a dip of 26 feet per mile. It places the top of the shale section at a depth of about 2,000 feet in the Ohio River Valley between Ohio and West Virginia. A well in Carter County, northeastern Kentucky, near the common corner of Kentucky, West Virginia, and Ohio, reached the top of the shale section at 1,173 feet; in Pike County, eastern-most Kentucky, the top of the shale lies at about 5,000 feet; the top of the shale is 12,000 feet below the surface of northeastern Pennsylvania.

At no place in the Appalachian Basin is depth to the shale too great to be reached by the drill; indeed, many wells in parts of the Basin are drilled through the shale to deeper oil- or gas-bearing strata. There are minor variations in the southeastward dip, but these do not seem to have had a significant effect on accumulation of gas.

Composition

The basic unit of the Brown shale is a pair, or couplet, of microscopically thin layers: one rich in mineral matter and the other made up chiefly of organic matter. The fineness of the resulting lamination is hard to appreciate. Samples of Ohio shale taken from the outcrop in central Ohio were found to have as many as 230 laminae (couplets) in a 5-inch thickness (6). References to hairline bedding planes and paper-thin laminations in published descriptions of the Brown shale from cores makes it clear that this characteristic persists in the subsurface as well.

Core samples of dark brown organic-rich shale from a producing gas well in the Cottageville Field, Jackson County, West Virginia, were analyzed (9). The inorganic part of the rock was found to consist chiefly of clay minerals, mainly illite, with the extremely fine grain size that is typical of clay (less than 0.004 mm). Silt-size grains of quartz and feldspar were present in amounts of 5% or more, and there were small amounts of calcite, dolomite, and pyrite. Another core from the same field analyzed at 60% clay minerals, 35% quartz and feldspar, and the remainder mostly pyrite and dolomite (10). The grains of quartz and feldspar, mostly coarser than 0.004 mm, tend to occur in very thin laminae or lenses.

The organic fraction, reddish brown to chocolate brown in color, is made up of particles of coal-like material in the micron size range (0.001 mm). There are also minute shreds of coalified woody substance, and of spores and algae. The evidence from the organic material shows that it was mostly derived from plants, and this conclusion is supported by carbon-isotope studies. In the jargon of the coal petrologist, the material consists largely of humic degradation products, which were washed into the sea from lands to the east and possibly from lowlands on the Cincinnati arch. There must have been a density stratification in the waters of the Devonian Sea, a stagnant condition that inhibited vertical circulation and prevented the organic matter from being oxidized and destroyed as it accumulated on the bottom. A reasonable assumption is that each couplet of mineral-rich and organic-rich sediment may represent an annual accumulation. Organic matter typically makes up 10 to 20% of the rock by weight, or 40 to 60% by volume (8).

It should be noted that the Brown shale is not oil shale like that of Colorado and Wyoming (11). The organic matter is not the type of kerogen that characterizes such oil shales; rather, as noted above, the Brown shale are coal-like.

At outcrops, shales almost always split into thin flakes and plates parallel to the bedding. Another interesting aspect of the Brown shale is their content of uranium. Little is known of the reasons for this, except that the association of organic matter and uranium is evidently primary; the uranium was present at the time the muddy sediment was deposited, and was not introduced in later time (12). The uranium content of the Brown shale, which ranges from 0.005% to slightly more than 0.007%, has not yet allowed them to be of use as a commercial source of uranium, although it is conceivable that they may be of such use in the future. The radioactivity of these shales, however, is a highly useful characteristic in the search for gas, as the radioactivity makes Brown shale readily recognizable on gamma-ray logs of drilled wells.

Fractures

A feature of the Devonian shales, which is of special significance in the gas-bearing Brown shale, is a system of near-vertical fractures (also known as joints). Most of these are only a fraction of a millimeter wide. In well cores, some fractures have been observed to be filled with brown crystalline dolomite, which helps make the fracture porous and permeable. Spacing of the fractures is variable, but they may occur close enough together so that two or more are often intersected in a 6-inch well bore. The fractures are not randomly oriented, but occur in sets that are alined in certain directions.

The relationship between fractures and gas production is well shown by cores taken from two wells in the Cottageville Field, Jackson County, W. VA (9)(10). The cores intersect numerous fractures and a study of the orientation of these fractures resulted in two important findings. First, the dominant direction of the fractures is North 40° to 50° East. This is the regional trend of the Appalachian Mountains (though the significance of the parallelism is not well understood); more practically, it is also the direction in which the most productive gas wells in the Cottageville Field are alined. This clearly suggests a relation between gas production and this set of fractures.

Second, in the well from which the larger flow of gas comes, there is a wide variation in fracture alinement. Only 21% of the fractures are alined North 40° to 50° East; other preferred directions are slightly west of north, slightly east of north, and nearly east-west. Parts of the core from this well are completely shattered by fractures, and the well had a natural flow of more than 1 million cubic feet (MMcf) of gas per day. The core from the second well showed few fractures, and the well had no open flow at all. The conclusion seems clear that here, as elsewhere in the Appalachian Basin, gas production from the Brown shale is controlled largely by fractures, with the production rate dependent on the number, length, openness, and direction of these fractures (13).

Although mapping of fracture patterns and intersections may well be the best guide to gas accumulation, mapping is a very difficult task. The cause of fracturing is not well understood, and at least nine theories have been suggested (14). The fracture systems may be related, for example, to the deformation that produced the Appalachian Mountains; to settling above deep-lying faults, thousands of feet below the Devonian shales (Figure 3.3); or to major zones of fracturing (lineaments), scores or even hundreds of miles long, that are known or suspected to exist in the region. Until the origin is known, a rational search for fracture-controlled gas accumulations will be difficult. A few fractures extend upward through overlying rocks and reach the surface, and their patterns can be detected by remote-sensing techniques (LANDSAT imagery). This is probably the best current approach to the problem.

Figure 3.3: Model Showing Fractures Generated by Deep Seated Basement Faulting and Propagated Upwards into the Devonian Shales

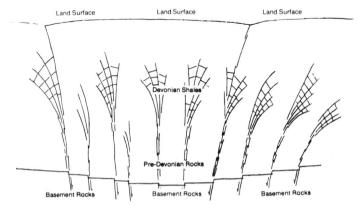

From Overbey, 1976, Energy Research and Development Administration Pub. MERC/SP-76/2, Fig. 9.

Source: OTA-E-57

Natural Gas in the Brown Shale

Although it has been known for more than 150 years that shallow wells drilled into the Devonian shales along their belt of outcrop would yield natural gas, the Brown shale was not generally considered a primary objective in exploration for natural gas until recently. In most parts of the Appalachian Basin, wells were drilled to deeper, more promising formations. If those failed to produce, the wells were plugged back to the Brown shale and attempts were made to stimulate enough gas from the formation to make a productive well (4).

Drilling for gas started in western New York as early as 1820, and moved westward along the south shore of Lake Erie across northwestern Pennsylvania and into Ohio as far as Cleveland (15). Shallow wells in the Brown shale supplied Louisville, KY with gas in the 1880s (4). Two facts about this early production stand out. First, the rate of gas production was low; only enough to supply a small local industry, or a small cluster of households for heating and cooking could be expected from a given well. Second, the wells were very long-lived. Two wells at Fredonia, NY, one drilled in 1821 and the other in 1850, had a combined annual production of only 6 MMcf (16,400 cubic feet per day), but when plugged 60 years later, they were still producing 6 MMcf per year (15). It was clear from this early experience that there was gas in the shale, but that the shale would yield it only at a low rate over a long time period. Today, it is known that gas moves readily only in fractures, and perhaps along some of the more silty bedding surfaces. The vast bulk of the gas is held in the shale mass, or matrix, from which it will move into fractures and well bores at very low rates and over long time periods.

Estimates of the total amount of gas in the Devonian shales of the Appalachian Basin range from a few Tcf to many hundreds of Tcf (16). Although the magnitude of the total resource is not known, there can be no doubt that it is large enough to be of potential importance to the eastern United States.

Present-day wells producing gas from the Brown shale recover only 2 to 10% of the original gas in place; 90 to 98% is left in the ground. The history of production in 50 Brown shale wells is indicated by the decline curves on Figure 3.4.

Figure 3.4: Averaged Production Decline Curves for 50 Devonian Shale Gas Wells

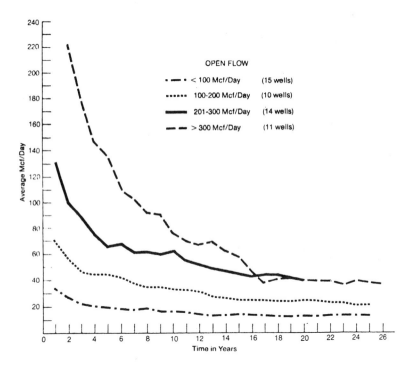

Lincoln, Mingo, and Wayne counties, West Virginia. Wells were metered on open flow after shooting or fracturing of the shale pay zone. Mcf = thousand cubic feet. (From Bagnall and Ryan, 1976, ERDA Pub. MERC/SP-76/2, Fig. 11; and W.D. Bagnall, personal communication.)

Source: OTA-E-57

Initial production is relatively high, as free gas in fractures moves to the well bore, but the flow decreases steadily to some value determined by the slow rate at which the gas in the shale matrix is released. Various techniques are being applied to the shale in an attempt to create artificial fractures extending outward from the well bore, thus potentially increasing the amount and rate of gas recovery by connecting more fracture systems to the well and exposing more surface area.

References

(1) de Witt, W., Jr., et al, "Oil and Gas Data From Devonian and Silurian Rocks in the Appalachian Basin," U.S. Geol. Survey Map I-917 B (1975).

(2) Provo, L.J., "Upper Devonian Black Shale—Worldwide Distribution and What It Means (abstract)," *Devonian Shale Production and Potential*, ERDA Pub. MERC/SP-76/2 (1976).

(3) Janssens, A., *Potential Reserves of Natural Gas in the Ohio Shale in Ohio*, Ohio Geol. Survey, prepared at the request of the Joint Select Committee on Energy of the Ohio Legislature (1975).

(4) Avila, J., "Devonian Shale as Source of Gas," *Natural Gas From Unconventional Geologic Sources*, Nat. Acad. Sci. Pub. FE-2271-1 (1976).

(5) Brown, P.J., "Energy From Shale—A Little Used Natural Resource," *Natural Gas From Unconventional Geologic Sources*, Nat. Acad. Sci. Pub. FE-2271-1 (1976).

(6) deWitt, W., Jr., Personal communication (1976).

(7) Ray, E.O., "Devonian Shale Development in Eastern Kentucky," *Natural Gas From Unconventional Geologic Sources*, Nat. Acad. Sci. Pub. FE-2271-1 (1976).

(8) Schopf, J.J., Personal communication (1976).

(9) Patchen, D.G. and Larese, R.E., "Stratigraphy and Petrology of the Devonian 'Brown' Shale in West Virginia," *Devonian Shale Production and Potential*, ERDA Pub. MERC/SP-76/2 (1976).

(10) Martin, P. and Nuckols, E.B., III, "Geology and Oil and Gas Occurrence in the Devonian Shales: Northern West Virginia," *Devonian Shale Production and Potential*, ERDA Pub. MERC/SP-76/2 (1976).

(11) Deul, M., "Discussion," *Natural Gas From Unconventional Geologic Sources*, Nat. Acad. Sci. Pub. FE-2271-1 (1976).

(12) Conant, L.C. and Swanson, V.C., "Chattanooga Shale and Related Rocks of Central Tennessee and Nearby Areas," U.S. Geological Survey (1961).

(13) Bagnall, W.D. and Ryan, W.M., "The Geology, Reserves, and Production Characteristics of the Devonian Shale in Southwestern West Virginia," *Devonian Shale Production and Potential*, ERDA Pub. MERC/SP-76/2 (1976).

(14) Hodgson, R.A., "Fracture Systems: Characteristics and Origin," *Devonian Shale Production and Potential*, ERDA Pub. MERC/SP-76/2 (1976).

(15) deWitt, W., Jr., "Current Investigation of Devonian Shale by the U.S. Geological Survey," *Natural Gas From Unconventional Geologic Sources*, Nat. Acad. Sci. Pub. FE-2271-1 (1976).

(16) Foster, J.M., "A 'New' Gas Supply," SPE-AIME #5451 (1975).

GENERAL RESERVOIR CHARACTERISTICS

The reservoir characteristics of Brown shale are vastly different from those of typical oil- and gas-producing formations. Porosity indicates how much space exists in a particular formation where oil, gas, and/or water may be trapped. A commercially oil- or gas-productive sandstone or limestone reservoir has porosities in the range of 8 to 30%. By contrast, gas-producing Brown shales have porosities of 4% or less (Table 3.1).

Much of the oil and gas in a formation may be unrecoverable because the pore structure is such that reasonable flow cannot take place. The ability of fluids and gases to flow through a particular formation, or permeate it, is called the permeability. The typical oil- and gas-producing formation has a permeability in the range of 5 to 2,000 millidarcies (md). By contrast, most of the measured permeabilities of the Brown shale in productive areas are in the range of 0.001 to 2.0 md (Table 3.1)

Table 3.1: Comparison of Core Data for Brown Shale and Reservoir Rocks from Other Gas Producing Areas

	Typical Permeability (md)	Typical Porosity	Typical Water Saturation
		. . . . (%).	
Hugoton-Anadarko Basin*	20	14	40
San Juan Basin*	1	10	30
Permian Basin*	15	12	35
Brown Shale			
Jackson County, WV**			
Whole-core analysis	2.0	3.2	65
Conventional-core analysis	0.1	3.0	70
Lincoln County, WV***			
Whole-core analysis	0.004	0.6	0.0†
Perry County, KY††			
Whole-core analysis	0.3	4.0	35

*S. Rudisell, N. Beckner and W.B. Taylor (Phillips Petroleum Co.). Personal communication, 1976.

**W.L. Pinnell (Consolidated Gas Supply Corp.), core data on well No. 11440 and No. 12041. Personal communication, 1976.

***Phase Report No. 1, Massive Hydraulic Fracturing of the Devonian Shale, Columbia/ERDA Contract E (46-1)-8014. Research Department, Columbia Gas System, October, 1976.

†Centrifuge measurement.

††Final Report—Well No. 7239, Perry County, KY, ERDA-MERC, July, 1975.

Source: OTA-E-57

Since the characteristics of Brown shale reservoirs are so different from those of the usual oil and gas reservoir, evaluations of gas-production potential of the shales by using conventional oil and gas techniques may result in erroneous conclusions. In the conventional oil and gas reservoir, it is a simple matter to measure the percentage of the total reservoir that is occupied by oil, gas, and water. However, in dealing with the Brown shale, it is very difficult to accurately determine these percentage saturations because the pores are so very small.

The manner in which natural gas is held in the Brown shale is a subject of considerable speculation. Some scientists believe that it is simply entrapped in extremely small pores. Others think the gas is adsorbed or molecularly held on the surface of the shale particles (1). Some of the natural gas may be dissolved in solid and liquid hydrocarbons in the reservoir. There is also some reason to believe that the gas may be in a liquid state in pores in the Brown shale. Available evidence (2) indicates that virtually all of the Devonian shale contains gas that is released or flows from the shale when the shale is placed in a relatively low-pressure atmosphere. However, current commercial production appears to enter the wells mainly from the Brown shale.

All subsurface reservoirs initially exist at elevated pressures, regardless of whether they contain water, oil, or natural gas (3). In conventional oil and gas reservoirs, a normal reservoir pressure (in lb/in²) is generally obtainable by multiplying the depth (in ft) below the surface of the ground by a factor of about 0.4. For example, an oil and gas reservoir at a depth of 3,300 ft in the Clinton sand in Ohio would be expected to have an initial pressure of about 1,300 lb/in² (psi). Since Brown shale formations produce gas at very low rates, it is difficult to determine an accurate initial reservoir pressure. However, shale wells that are shut in for long periods often exhibit pressures in the range of 0.125 times the depth, which is much less than would be expected in a normal oil or gas reservoir. The initial reservoir pressure is very important if the gas in the shale exists in a gaseous state, because the amount of gas in the reservoir measured at atmospheric conditions is proportional to the reservoir pressure. For example, all other things being equal, a reservoir with a pressure of 2,000 lb/in^2 absolute will contain twice as much gas in a given volume of reservoir rock when measured at atmospheric conditions as a similar reservoir at the same depth whose pressure is 1,000 psia.

Reservoir Evaluation Tools

Core Analysis: In drilling an oil or gas well with rotary tools (the drill bit rotates at the bottom of the hole as opposed to moving up and down as in cable-tool drilling), it is possible to use a special type of drill bit that works much like a doughnut cutter and permits the operator to cut plugs or cores from the formation and bring them to the surface as samples of the rock being drilled. This operation is referred to as coring. The samples so obtained can then be subjected to various types of analyses.

Geologists and engineers examine cores of Brown shale to detect fractures or joints. The visual appearance, odor, or taste of a core sample provides an indication of the presence of gas, oil, or water in the pores of the core.

After a quick gross examination, 6-inch long pieces of the core may be sealed in cans or other containers to maintain the fluid content insofar as possible. These samples are used to determine the porosity, permeability, and fluid saturation of the shale. It is important to note that the laboratory procedures used to analyze the Brown shale were designed for normal sandstone and limestone reservoirs which have much greater porosities and permeabilities (4).

Basic to an understanding of the gas production potential of the Brown shale is the need for analytical techniques capable of accurately determining critical reservoir characteristics from core samples. If it is not possible to determine accurately from the core samples (1) the physical nature of the pore structure that constitutes the reservoir (subsurface gas container); (2) the percentage of the total bulk volume of the reservoir that is made up of pore space; (3) the ability of fluids to flow through these pores; and (4) the percent of pores occupied by gas, liquid hydrocarbons, solid hydrocarbons, and water, then there is much smaller chance of determining these same parameters from less direct methods such as electrical logs. A log is a record of some physical property (e.g., electrical resistivity or radioactivity) of the rocks penetrated in a well.

The conventional type of core analysis involves cutting a ¾-inch diameter, 1-inch-long plug from the core, whereas the whole core type of analysis uses the entire sample which is 3½ to 4 inches in diameter and 6 inches long. Whole core analysis is generally thought to be more applicable than the conventional type.

Permeability, Porosity, and Saturation: The permeability, porosity, and saturation of the Brown shale are vastly different from the same parameters of most gas-producing reservoirs. A general comparison of these characteristics is given in Table 3.1. The Hugoton-Andarko, San Juan and Permian Basins represent some of the better known gas-producing areas. They tend to contain reservoirs that are on the tight (low permeability) side, as compared with offshore production, where the reservoirs may have a permeability of 1,000 md and a porosity of 35%. Nevertheless, the typical sandstone reservoir has permeabilities and porosities that are much greater than those of Brown shales. This is a strong indication that methods different from those used in conventional gas-producing reservoirs must be used to obtain commercial rates of production from the Brown shale. Development and evaluation of such methods can only come from basic research and field testing.

The characteristics of Brown shale listed in Table 3.1 vary widely, even though the data presented are all from the same geographical region in southwestern West Virginia and eastern Kentucky. This variation is probably due principally to the heterogeneity of the shale itself.

It appears that whole core analysis gives more meaningful information for the Brown shale because it includes the effects of joints and fractures. Conventional core analysis, run on a small plug, will be affected by a fracture if one exists in such a sample, but the plug may not contain one even though fractures appear to be present every few inches in the Brown shale. Fractures caused by drilling and coring operations may produce spurious data from both coring analyses.

Table 3.1 does not indicate the very high permeabilities of some of the samples. The whole core analysis of the Lincoln County well represents 19 samples distributed through 1,300 ft of shale. Three of these samples had permeabilities of 906 md, 502 md, and 93 md, whereas the other 16 samples ranged from 0.0002 to 0.023 md. Similarly, the whole core analysis of the Perry County well represented 12 samples covering 64 ft, with two permeabilities of 9 md and 23 md and the others between 0.1 and 0.9 md.

The Lincoln County, West Virginia whole core analysis shown in Table 3.1 is markedly different from the other Brown shale analyses. This is probably due to the manner in which the analysis was made. The cores from the Jackson and Perry County wells were analyzed using horizontal flow, while the analysis of the Lincoln County well was based on vertical flow. Since vertical flow is likely to encounter impermeable barriers of paper-thin laminae that would not affect horizontal flow, much lower permeabilities would be calculated. The lack of vertical communication would also result in reduced measured porosities.

This Lincoln County core analysis also indicated a water saturation of 0.0% (5)(6), whereas the other shale analyses showed substantial water content. The Lincoln County analysis was based on centrifuge measurements. The centrifugal force created apparently did not exceed the capillary or other forces holding the water in the very small pores; hence, it appeared that the water saturation of the shale was 0.0%. These examples clearly emphasize the need for research in the area of core analysis of the Brown shale.

The Brown shale is characterized by a porosity of about 3%. However, a 3% porosity estimate may be too low. The operator who drilled the Lincoln County well canned whole core samples throughout the entire 1,300 ft. All of these samples liberated sufficient natural gas to cause the pressure in the can to increase considerably. Although it took about 3 weeks for most of the cans to reach a static gas pressure, some of the cans containing the tighter sections of the shale were still increasing in pressure after a 2-month period (7). The gas liberated in the cans had a volume greater than could be accounted for by the measured porosity and the assumed initial reservoir pressure. In other words, the gas-occupied porosity may be greater than the 3% currently indicated by the core analysis.

Because it takes as much as 2 months for the gas to escape or flow from a core sample 3.5 or 4 inches in diameter and 6 inches long, it may be that the amount of gas in the Brown shale can be most accurately determined by measuring the gas that escapes from core samples. In a normal oil and gas formation, this would be impossible because most of the gas would escape from the core during normal canning or handling operations. However, in dealing with a material with such a low permeability as the Brown shale, it is obvious that very little gas is lost during the period of time necessary to remove the core from the bottom of the hole and place it in a container. The amount of gas lost from the cores during the canning operation would apparently be limited to the gas in the permeable fractures and would be negligible compared with the gas in the matrix of the Brown shale. Use of this method of determining the gas in the shale might eliminate the necessity of measuring the porosity, saturations, and reservoir pressure. A technique similar to this is used by the U.S. Bureau of Mines to determine the amount of natural gas in coal (8).

Core Data Distribution: The gas-producing potential of the Brown shale cannot be realistically evaluated until its physical and chemical characteristics throughout the area have been determined. Even though there are about 10,000 wells currently producing gas from the Brown shale, coring to date has been limited almost entirely to the better-producing areas shown in Figure 3.5. The data of Table 3.1 relate only to wells in the producing area of Kentucky and West Virginia. Recent research has involved the coring of 12 experimental wells, but only 4 of these are very far outside currently producing areas.

Figure 3.5: Major Devonian Shale Gas Production Areas

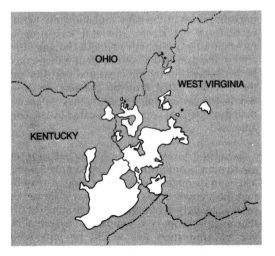

From Energy Research and Development Administration.

Source: OTA-E-57

An expanded shale inventory by the Energy Research and Development Administration (ERDA) will provide core samples from wells distributed across a wide expanse in the Appalachian Basin and areas to the west and northwest (Figure 3.6). Such data are needed to evaluate the extent of the natural gas resource in the Devonian shales.

Flow Tests: The actual significance of core analysis data and visual observation of core quality can only be obtained through flow tests of the wells, which determine how fast the gas can move through the shale. Due to the extremely low permeability of the shale, it may take several years to detect drainage of the potential drainage area of a well.

Figure 3.6: Location of Core Wells in a Proposed
Inventory of Brown Shales by ERDA

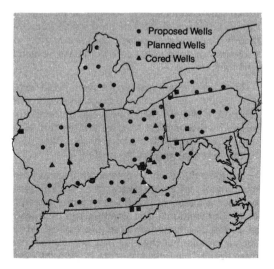

From Energy Research and Development Administration

Source: OTA-E-57

To reduce the time required to determine flow rates in Brown shale, a special type of test is required. The so-called isochronal flow test involves determining flow rates under conditions where the entire drainage area of a well has not yet been affected and extrapolating the resulting data in order to estimate what the well behavior will be after the well has affected the entire drainage area.

Pressure buildup and drawdown tests are conducted to determine the significance or accuracy of the core-analysis or log-measured permeability, thickness, and saturation data (6)(9). A pressure drawdown analysis is a mathematical analysis of the pressure that results in the well due to continued production at a constant rate, whereas a pressure buildup analysis is a mathematical analysis of the increase in well pressure that results when the well is shut-in after being produced at a constant rate. The increase in wellhead pressure is determined at regular intervals for a specific number of days, weeks, or months.

Determining the initial pressure in the Brown shale is difficult and time consuming because of its low permeability. Reservoir pressures are normally determined by temporarily shutting in a well and then measuring the pressure in the well bore at the depth being investigated. Using this procedure after shutting in a well in the Brown shale will provide an accurate measure of the reservoir pressure only after weeks or months because of the time required for equilibrium pressure to be reached between the well bore and the adjoining shale pore space (9). Much of the variation in formation pressure gradients (i.e., pressure per foot of depth) that has been observed and recorded might be caused by measurements taken before reservoir well bore pressures are equalized (10)(11)(12).

Logging: The term logging is applied to a variety of measurements made in a well by lowering a measuring device on an electric cable and recording variations of the particular physical property being measured. The plot of the data versus depth is known as a log. After permeabilities, porosities, and gas saturations have been determined from core analysis, logging techniques are used to measure various physical properties of the subsurface formations in place. Interpretation of well logs permits the determination of porosities, saturations, and permeabilities of the formation.

A wide variety of physical properties are traditionally measured in oil and gas wells in this manner. Some of these (4) are electrical resistivity, difference in electrical potential between mud in the well and the fluid in the rock (self-potential log), natural radioactivity (gamma-ray log), induced radioactivity (neutron log), speed of sound in the formation (sonic log), formation density, hole size (caliper log), temperature, sound intensity (13) (sibilation log), earth gravity (2) and formation dip.

Most of these logs may be made either in empty holes or holes containing drilling fluid or water. Only a few types of logging can be done after casing has been set and cemented in the hole.

Whether or not water-based liquids damage the Brown shale by reducing its permeability is a subject of controversy (13)(14). This potential water damage is not only a problem in logging but also causes difficulty in drilling the well and in stimulating production by fracturing. Various combinations of logs must be run to obtain the porosity, water saturation, oil saturation, gas saturation, and organic content of formations. It may be possible to obtain logs in an empty hole, but it appears to be somewhat easier and simpler to use a series of wet-hole logs to determine these parameters (15).

The sibilation, temperature, and Seisviewer logging techniques have special applications in the Brown shale (13). The sibilation log is a high sensitivity, high frequency noise detector that can be used to determine where gas is entering the bore hole. The temperature log measures changes in temperature to detect where gas is entering the well bore. Both of these logging techniques are useful to determine which part of the well in a massive shale section should be treated. The Seisviewer log produces an acoustic picture of the bore hole. Such pictures often detect formation fractures and this is, of course, useful in the completion of the well.

Stimulation Techniques

Knowing that there is a great amount of gas in the Brown shale, where it is geographically, and which vertical portion of the formation is capable of producing it, is of no commercial use unless some method can be devised which will permit production of the gas at an acceptable rate. In other words, it makes little difference how much gas is in the shale unless some method can be developed to permit its production at an economic rate.

Evaluation of any drilling, stimulation, or production method is very difficult, because no two wells are the same. This problem is magnified considerably in dealing with the Brown shale, since its characteristics vary so widely from well to well even in the same area. Various techniques have been used to stimulate or increase the flow of gas from the shale. Early gas wells were stimulated by explosions (shooting) (16). More recently, hydraulic fracturing has become a useful technique. There is no clear-cut experimental evidence concerning the relative merits of shooting and fracturing, although hydraulic fracturing generally produces slightly higher flow rates. Some companies reportedly continue to shoot their Brown shale wells while others claim fracturing gives superior results (16). Other techniques are now being tested. Descriptions of several stimulation methods follow.

Explosive Stimulation: Explosions tend to develop fractures and shatter a formation, due to the rapidity with which the force is applied. Explosive stimulation does not affect a formation to as great a depth as does hydraulic fracturing.

Conventional Shooting — Prior to about 1965, stimulation of oil and gas production from Brown shale was mostly limited to shooting (16). This entails setting and cementing casing in a drilled hole with its bottom above the prospective producing formation, then detonating explosives in the open (uncased) hole opposite the prospective producing formation. The explosion cracks and/or shatters the formation, thereby increasing the size of the well bore and the permeability of the formation around the enlarged well bore due to the cracks therein. Improving the permeability of even a few feet of the formation around the well bore normally greatly improves the capacity of that well to produce (9). Explosive stimulation is the method that has been used in the completion of most existing Brown shale wells (16).

An explosion in the well tends to fill the uncased well bore with shattered rock. The general consensus seems to be that rubble in the well reduces the productivity of the well (15). Therefore, most operators attempt to remove the loose material from the well before trying to produce gas from it.

Most prospective Brown shale wells produce little or no gas before treatment. Consequently, a typical percentage increase in production cannot be predicted from stimulation efforts. Some wells have a dramatic increase in gas production after shooting, whereas others are not benefited.

Explosive Fracturing — This technique combines some of the features of hydraulic fracturing and shooting (17). The well is first fractured hydraulically and into those fractures explosives are injected and detonated. The explosion creates additional small fractures away from the large hydraulically induced fracture and may also shatter some of the material near the hydraulic fracture. It is theorized that the shattered material will hold open the fractures and make a system with a much higher productivity than a simple hydraulic fracture would create. The outward explosive force of the artificial hydraulic fracture also tends to open up natural fractures that were encountered by the artificial hydraulic fracture. There has been very little experience with this technique in Devonian shales and it is therefore necessary to classify it as experimental. One of three tests involving ERDA and the Petroleum Technology Corporation has been performed.

Dynafrac — Dynafrac is an experimental process in which several radiating fractures from the well bore are created and extended by using a slow-burning solid propellant above a column of fluid (18). Mechanically, the shooting takes place as follows:

> A small diameter solid propellant is centralized in the hole opposite the producing formation;

> This solid propellant is covered with a liquid that extends upwards into the casing;

> A slow burning solid propellant is placed in a trapped airspace above the fluid level in the casing;

> Both the small diameter charge and slow burning solid propellant are fired at the same time;

> The small diameter charge communicates its force quickly to the surrounding formation and causes several radiating fractures to form;

The slow burning solid propellant develops pressure more slowly and applies this pressure to the fluid beneath it; and

The fluid is forced out through the fracture formed by the explosion of the small diameter charge and the fractures are extended out into the formation.

The result of the Dynafrac treatment is several radiating fractures through the formation with a minimum of rubble in the well bore. Developing several radiating fractures from the well bore will give a better opportunity to encounter additional vertical fracture systems in the Devonian shale.

Nuclear Explosives — The use of nuclear explosives in the Brown shale is a possible stimulation technique. However, the minimal success achieved in stimulating gas production in formations in the West is not encouraging (19). The lack of successful nuclear shots and the sociopolitical difficulties of conducting nuclear explosions largely negate the possibility of using this technique to stimulate Devonian reservoirs.

Hydraulic Fracturing: Hydraulic fracturing (hydrofracturing) became available in the Appalachian Basin in the late 1950s. This technique involves injecting fluid into the formation at a rate and pressure sufficient to shatter and fracture the formation. The plane of the resulting fractures is generally vertical, except at very shallow depths (Figure 3.7). This fracture greatly increases the capacity of a well to produce (20).

Figure 3.7: Diagram Showing Relationship of Maximum Principal Stress and Least Principal Stress to the Plane of an Induced Hydraulic Fracture

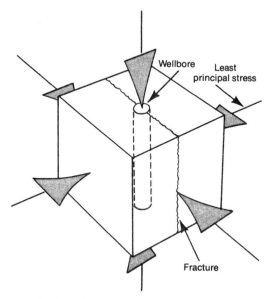

From Overbey, 1976, Energy Research and Development Administration Pub. MERC/SP-76/2, Fig. 3.

Source: OTA-E-57

Hydraulic fracturing of a formation can often be made more effective by using a fluid that has a high viscosity. In order to keep a fracture open, sand normally is added to fracture fluids, as it can prop open the fracture and give it high permeability. Because the Brown shale has extremely small-sized pores, it has been assumed that any contact of the formation by liquids, particularly water, would result in a great reduction in the permeability of the formation to gas. It is theorized that the liquid would be held by capillary attraction in the extremely small pores and the threshold pressure of this adsorbed liquid would be so high that much of the liquid would block the gas from flowing into the well bore. Also, water-based fluids might swell the clay particles in the shale and thus further reduce the permeability (21).

Consequently, until recent years, Devonian shale wells were not hydraulically fractured but stimulated entirely by shooting. Recently, however, some hydraulically fractured wells have performed better than adjacent wells shot with explosives (16).

One of the disadvantages of fracturing a gas well with a liquid is the length of time required for the fracture liquid to flow back into the well bore. In low capacity gas wells, fracture fluids may interfere with the gas production for long periods of time.

Normal Hydraulic Fracturing — Normal hydraulic fractures are defined and differentiated from massive hydraulic fractures by the amount of fluid in the treatment. Any fracture requiring less than 100,000 gallons is defined as a normal fracture. On the other hand, the use of foam or gas as described later in this section is differentiated from a normal fracture treatment by reason of the unusual fluids being used for fracturing.

Most fracture treatments of the Brown shale are now made using water-based fluids with chemicals added to minimize the effect of water on the clays or minimize reductions in permeability.

It is very difficult to quantify the effect of fracturing on gas production, because most Brown shale wells produce little or no gas before treatment. Generally, increased gas production results from fracturing Brown shale.

Massive Hydraulic Fracturing (21)-(24) — A massive hydraulic fracture is defined as one in which more than 100,000 gallons of fluid are used in the fracture treatment. Some massive hydraulic fractures have used over 1 million gallons of fluid.

Questions continue to exist concerning the lateral extent of fractures resulting from massive hydraulic treatment (15). In many cases, subsequent flow tests have not corroborated the formation of a large fracture. Conflicting opinions exist concerning the advisability of massive fracturing. A major difficulty has been the tendency of the fracture to leave the target area of a formation and migrate into portions of a formation that do not contain oil or gas (21). Fluids moving into nonproductive parts of the shale sequence will not increase gas production. This problem may be minimal in Brown shale, since shale fractures more readily than most formations above and below it.

Another difficulty with massive hydraulic fractures is the long cleanup time required. As much as 6 months may be required to get all of the mobile fracture fluid out of a well (15)(21). An additional problem is that more than an acre of surface space is needed to accommodate the equipment required for a massive treatment. In hilly Appalachia, flat sites of more than an acre are not easily found or constructed, particularly if the well is located on a steep mountain side or in a narrow gorge.

In spite of all the problems inherent in massive hydraulic fracturing, this stimulation technique may still have potential in the Brown shale (21).

Fracturing With Foam (25)(26)(27) — There is considerable question about the extent of the damage done to Devonian shale formations when liquids, especially water, come in contact with the shale (16)(21). Mixing of appropriate chemicals with the treating water minimizes the damage to the shale (21). Foam, a mixture of nitrogen, water, and a foaming agent, tends to minimize the leak-off of the fracture fluids into associated permeability zones (6).

A properly compounded foam can shorten the time needed to recover fracture fluid after a treatment (21). When injected, the foam is compressed; after fracturing, it expands towards the lower pressure at the well bore and helps expel the fracture fluid from the rock into the well. A time- and/or temperature-effective emulsion breaker can be added to the foam so that by the time the well is ready to produce, the foam has broken into a mixture of gas and liquid, which facilitates cleaning the well bore (6).

Fracturing With Gas (28) — Using a liquefied gas as a fracturing agent overcomes cleanup difficulties and potential damage to the formation by liquids; no water is used and the liquefied gas vaporizes as the pressure in the well bore is dropped. However, this technique is quite expensive.

Dendritic Fracturing (29) — Instead of obtaining one long fracture, the Dendritic fracture method attempts to form a fracture that branches in many directions (29). After one small fracture has been created, the well system is placed on production for a very short time to reverse the stress in the formation. Additional small fractures along the main fracture are thought to form due to this reversal of stress. When a new fracture force is applied, one or more small fractures branching from the large fracture are extended. This procedure of fracture-relaxation is continued to develop a Dendritic-shaped fracture.

Assertions that such a Dendritic fracture can actually be formed by this technique still require confirmation (29). If the technique does cause fractures to develop in a variety of directions and thus intercept a large number of the natural parallel fractures in the Brown shale, the technique might have potential for increasing gas production from them.

Directional Drilling (12)(30) — Directional drilling is another production stimulation technique that may have potential in the Brown shale. Because most natural fractures in the Brown shale appear to be parallel vertical fractures (2), it is theorized that a well drilled diagonally across this vertical system of fractures would encounter more of the fractures and thus result in substantially greater production. Very little directional drilling has been done in the potential producing area of the Brown shale (12).

Considerable difficulty was encountered in an experiment with directional wells in the Brown shale (12). Although the mechanics of the drilling operation were successful (Figure 3.8), gas production did not meet expectations and, therefore, only one of three planned wells was drilled.

Other Stimulation Methods: Many other techniques have been proposed for recovering gas from the Brown shale, although most of these are techniques that have been used to recover oil rather than gas.

Microbial — It has been proposed that bacteria could be introduced into oil reservoirs to form gases and/or change the interfacial tension and viscosities to make the trapped oil more mobile. Microbial techniques do not appear to have great potential for gas recovery where the gas mobility is limited by the tight matrix of the Brown shale. Although there are bacteria able to withstand temperatures and pressures found at a depth of 3,000 to 4,000 ft, none are known that will both successfully generate useful modifying products in sufficient amounts and also tolerate the chemical and thermal environments at those depths.

Figure 3.8: Deviated Wells and Earth Fracture Systems Process

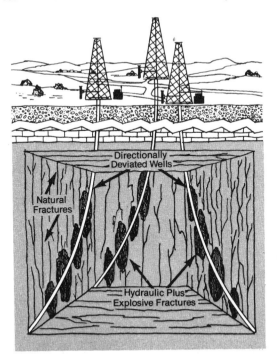

From Energy Research and Development Administration.

Source: OTA-E-57

The job of inoculating a large area of very low-permeability shale would be very difficult, if not impossible, unless a microbial hydrofracture technique could be perfected. Further, any strain of bacteria developed would need to be carefully screened for potential environmental impacts. Even should the conceptual process be feasible, it is unlikely that the necessary strains could be developed, field tested, and put into commercial operation within time to influence shale gas recovery by the year 2000.

Thermal — A variety of thermal methods have been successfully used to increase recovery of oil from various formations. The value of these methods for reducing the viscosity of gas would appear to be minimal, although laboratory results indicate that gas is released from Brown shale faster when the shale is heated. This appears to be due to the expansion of gas in the shale and the resulting increase in pressure which forces the gas from the shale at a higher rate. It seems possible that such an effect might be useful in the Devonian shale reservoir. Burning of gas in the Devonian shale (or applying heat by other means) could increase gas pressures locally and cause the gas to move more rapidly toward the well. The cost of supplying oxygen to the formation to maintain a fire, and the poor heat conductivity of shales in general, make it unlikely that thermal processes would be economical.

Mining — Brown shale outcrops cover an extremely wide area in the Appalachian Basin. It is technologically possible to mine the Brown shale, then recover the gas from the shale by means of various thermal-chemical methods. Such methods might also recover any liquid hydrocarbons contained in the Brown shale. Because of the low volume of gas in the Brown shale, costs of mining and retorting probably would be great. Likewise, environmental problems associated with processing the shale and disposing of the spent shale could be obstacles to any large-scale mining venture. It appears the most proposed approaches to recovering gas from strip mined Brown shale will not result in net energy gains. Producing shale gas by subjecting mined shale to various thermal-chemical processes will probably result in costs of $5.00 to $6.00 per Mcf, comparable to, or higher than the cost of producing high Btu gas from coal.

None of the thermal, microbial, or thermal-chemical methods proposed for recovering gas from the Brown shale appear to have a high potential for recovering a significant amount of gas within the next 20 years. It has been shown that thermal, microbial, and thermal-chemical techniques are capable of recovering gas from the Brown shale under very limited and controlled conditions, but the physical and economic feasibility of commercial operation has not been demonstrated to date.

Brown Shale Production in Three Localities

Production data have been obtained from three gas-productive locations in the Appalachian Basin. These localities, in descending order of general investment attractiveness, are Cottageville, WV (high quality); Clendenin, WV (medium quality); and Perry County, KY (lower quality). The quality designations reflect geologic and economic characteristics of each region and are not intended to reflect any differences in the actual Btu content of the natural gas in the fields. The 15-year production profiles are given in Table 3.2. In the high-quality area, production data were available only from shot wells. These figures are the averages of actual production data for 13 wells in this field for the 15-year period.

Table 3.2: Production Statistics of Natural Gas from Brown Shale in Three Localities

	High Quality	Medium Quality		Lower Quality			
				Good		Bad	
Year	Shot	Frac*	Shot	Frac*	Shot	Frac*	Shot
				(Mcf/yr)			
1	36,318	17,989	17,858	21,250	18,750	11,400	6,800
2	29,490	20,227	16,053	20,850	15,880	10,900	5,000
3	23,883	17,978	12,342	20,600	13,600	9,200	5,600
4	20,071	18,570	11,001	17,700	11,480	9,600	5,100
5	17,439	17,000	10,000	18,350	11,170	8,300	5,400
6	15,980	16,000	9,000	17,290	11,080	7,500	5,300
7	14,879	15,000	8,200	17,000	10,000	6,900	5,000
8	13,464	14,500	7,500	16,300	9,300	6,300	4,950
9	12,772	13,800	7,000	15,600	8,700	5,800	4,900
10	12,498	13,500	6,500	15,000	8,200	5,050	4,800
11	11,661	12,700	6,100	14,500	7,800	4,750	4,700
12	11,304	12,200	5,800	13,800	7,500	4,500	4,600
13	11,131	11,700	5,500	13,300	7,200	4,250	4,550
14	10,842	11,300	5,200	12,800	7,000	4,050	4,450
15	9,766	10,800	5,000	12,300	6,800	3,850	4,400

Note: Production on Brown shale operations are averages from over 200 wells in Kentucky and West Virginia (Columbia Gas Co., Ray Resources Corp., and Consolidated Gas).

*Hydrofracturing.

Source: OTA-E-57

In the medium-quality shale, data were available for both shot and hydro-fractured wells, but only for 5 years. The rest of the profiles were extra-polated using production decline curves for Brown shale wells developed for the region (31).

Because of great variability in the production from the wells in the lower-quality location, the wells were separated into two groups, good and bad, based solely on their production rates. Shot and hydrofractured wells fell into both groups. Fifty-nine percent of the wells in this locality fell into the good group, while the remaining forty-one percent were in the bad group. One might be misled by looking only at the good groups in this locality for comparison with the high- and medium-quality locality, because one assumes a risk of having a bad well in this locality forty-one percent of the time. So, while one can get a good well from the lower-quality locality, the investment potential on aver-age is less attractive than in the other localities.

Extent of the Economically Producible Area

Estimates of the natural gas in the Brown shale are subject to great vari-ability. The question involves not only the total resource present but also the portion that can be economically produced. Until the Brown shale resource of the Appalachian Basin is more fully characterized, there will continue to be great uncertainty in any attempt to estimate the extent of the Appalachian Basin which might sustain commercial development of shale gas production.

Production data from 490 shale wells in the gas-productive area of the Appalachian Basin were used to estimate the potential production from other areas of the Appalachian Basin where shale gas production might be economi-cally feasible.

ATNPV (after-tax net-present value) analyses indicate that under many price and tax scenarios, drilling for and producing shale gas from localities in the known shale gas productive area is economically feasible. However, it is unrealistic to assume that the current gas-productive area is representative of the whole Appalachian Basin. A number of general observations about re-source deposits are relevant. First, the distribution of resource deposits in na-ture tend to be highly skewed, i.e., there are fewer very high-quality resource deposits than medium-quality deposits, and fewer medium-quality deposits than low-quality deposits (32)(33). Second, the better-quality resources tend to be developed first (34). There being no strong evidence to the contrary, OTA assumes that these principles apply to gas-bearing shales of the Appalachian Basin.

In a marginal resource base, such as the Brown shale, the definition of better-quality resource includes, as a determinant, location relative to existing production and pipelines. Until recently, the Brown shale have not been a pri-mary target of drilling except in the Big Sandy area. The current areas of shale development were initially by-products of other activity.

While it might appear that this fact would blunt the operation of the prin-ciple that the better prospects are drilled first, this is not the case. Even if the initial knowledge of Brown shale prospects was developed as a by-product of other activity, the better Brown shale prospects (by-products or not) are de-veloped first. "Better" here, however, involves a strong element of location rel-ative to existing pipelines. This is particularly true for historical wellhead price levels. Evidence of this is that much of the Brown shale production in West Vir-ginia is served by existing interstate pipelines.

All of this suggests that there may be other areas which are geologically as promising as the three localities examined here. These other areas, although more remote relative to existing pipelines, may become economically feasible at $2.00 to $3.00 per Mcf price levels.

There might be a temptation to extrapolate the production results from the three sample locations in the currently productive area directly to the entire Appalachian Basin. Results of such an extrapolation are not likely to be valid primarily because:

- The existing wells are not located randomly in the Appalachian Basin, but rather are clustered in a known producing area;

- The gas-productive area sampled (98.6 square miles) is less than 0.06% of the 163,000-square-mile Appalachian Basin;

- The 490 sample wells are but a very small (5%) portion of the 10,000 producing wells in the Appalachian Basin, and do not represent a random sample; and

- Average production data from producing wells are biased because dry holes and plugged and abandoned wells are not included in the average production.

OTA assumed that the production potential in the currently producing area is much higher than is characteristic of the Appalachian Basin as a whole.

Based on the following information, OTA estimates that about 10% of the 163,000-square-mile extent of the Appalachian Basin might be of high enough quality to produce shale gas economically at a price of $2.00 to $3.00 per Mcf.

Production History: The wells which have a potential of producing more than 240 to 300 Mcf of shale gas over a 15- to 20-year period tend to be clustered in a few locations in the Appalachian Basin. This type of distribution of commercially productive wells indicates that not all of the Appalachian Basin is composed of the same resource quality. No doubt additional locations exist which have commercial potential, but it is unlikely that these areas will comprise a significant portion of the 163,000-square-mile extent of the Appalachian Basin.

Shale Depth: The Brown shale outcrops at the surface in central Ohio and is 12,000 ft below the surface in northeastern Pennsylvania. Because drilling and stimulation costs increase with depth, commercial production of shale gas in the volumes encountered in the best wells is generally limited to depths less than 5,000 ft. A considerable extent of the Brown shale of the Appalachian Basin is deeper than 5,000 ft and is, therefore, less likely to sustain commercial shale gas production.

Shale Thickness: The total thickness of the gas-productive Brown shale sequence in the Devonian rocks varies from less than 100 ft to more than 1,000 ft across the Appalachian Basin. It is not generally economical to stimulate Brown shale layers which are less than 100 ft in thickness unless multiple layers in one well can be treated. The Brown shale resource in a significant portion of the Appalachian Basin consists of thin layers of Brown shale which may not be amenable to modern hydrofracture techniques.

Fractures: The fracture system (number, length, openness, and direction of fractures or joints) in the Brown shale is not uniform across the Appalachian Basin. The much-fractured areas of the Brown shale tend to be more gas-productive than the less-fractured areas. Extensive areas of the Appalachian Basin have limited fracture systems and, therefore, are potentially poorer areas for shale gas production even with modern stimulation techniques than the much-fractured areas.

Drilling Experience: Drilling and production records of independent operators in the Appalachian Basin have reflected vast areas where shale gas production is uneconomic unless new stimulation techniques can more than double shale gas production rates without significant increases in cost. Poor shale gas production experience over extensive areas probably is a result of a combination of the circumstances outlined above.

An Estimate of Readily Recoverable Reserves

The Appalachian Basin has an areal extent of about 163,000 square miles. If 10% of this area is of high enough quality to be economically attractive for shale gas production at prices of $2.00 to $3.00 per Mcf, it provides a potential production area of 16,300 square miles. (The present-gas-productive area is less than 5% of the 163,000-square-mile area.) With a spacing of 150 acres per well, this area would support approximately 69,000 wells. Production data presented in Table 3.2 show that wells economically feasible at $2.00 per Mcf will produce approximately 240 MMcf of shale gas per well over a 15-year period, and about 290 MMcf per well over 20 years. In a similar analysis for a smaller producing area, ultimate recoverable reserves were used at levels of 300, 350, and 400 MMcf per well (1). Readily recoverable reserves were determined by multiplying the number of potential wells by the average production per well as follows:

15-year readily recoverable reserve

69,000 wells x 240 MMcf/well = 16.6 Tcf

20-year readily recoverable reserve

69,000 wells x 290 MMcf/well = 20.0 Tcf

If the entire undeveloped gas-productive area were a medium-quality resource and all wells were shot treated, the 15-year readily recoverable reserve would be 9 Tcf; use of hydrofracturing rather than shot treatment would increase this figure to 15 Tcf. The 20-year readily recoverable reserves would approximate 11 and 19 Tcf, respectively. If 10% of the 163,000-square-mile (16,300 square miles) gas-productive area were all high-quality resource and all wells were shot treated, the readily recoverable reserve would be about 17 Tcf over a 15-year period and about 20 Tcf over a 20-year period. Assuming that hydrofracturing results in a 50% increase in shale gas production (as is suggested by production data in Table 3.2), 69,000 hydrofractured wells on high-quality Brown shale sites might produce 26 Tcf of gas over a 15-year period, and approximately 30 Tcf over a 20-year period.

It is highly unlikely that all of the undeveloped gas-productive Brown shale resource will be high quality, and also unlikely that all of it will be medium or low quality. For this reason, a 15 to 25 Tcf estimate of readily recoverable shale gas reserves appears justified until the Brown shale resource base is more thoroughly characterized. This range clearly indicates that the Brown shales do in fact have a potential for making a significant contribution to the U.S. natural gas supply.

Some estimates of the total amount of gas-in-place in the Brown shale range in the hundreds of Tcf (35). However, such estimates of the resource base should be distinguished from estimates of readily recoverable reserves, which represent the fraction of the total resource whose recovery is feasible under reasonable assumptions about costs, taxes, and geologic formations. OTA's 15 to 25 Tcf estimate of readily recoverable reserves is consistent with a total resource estimate of hundreds of Tcf because of the fact that under present technology the average shale well recovers only 3 to 8% of the calculated gas in place (35).

Obstacles to Development Using Available Technology

Development of the Brown shale using available technology is hindered by problems associated with high local drilling costs, difficulties in lease acquisition, and title clearance. Drilling costs are substantially higher in southwestern West Virginia and eastern Kentucky because of the rugged terrain and poor roads, which make equipment movement difficult and expensive and increase the costs of installing gas gathering and distribution systems.

Areas with multiple coal seams require additional casing for each coal seam which, in turn, increases drilling rates per foot and tangible expenses for casing for wells that penetrate coal seams. The problems associated with drilling through minable coal seams will increase in the future due to the increased value of coal, and more intensive exploration and development efforts by coal operators. Additionally, leasing and purchasing of coal mining rights by investors far removed from the site will result in delays in acquiring approval to drill through coal seams. To gain approval to drill through a coal seam, a plat of the drill-site location must be submitted to the operator holding the mining rights on the potential drill-site property. If a drill site is approved by the coal operator, that operator must agree to leave a pillar of coal around the drill hole to provide an unbroken well bore through the seam.

This procedure results in a loss of recoverable coal. If, in areas of low- to medium-quality Brown shale, minable coal seams are numerous and thick, the amount of coal left as pillars around the well bore may have a greater value than the potential gas from the proposed well and, therefore, the coal operator will refuse to permit a gas well to be drilled through the coal seams.

Environmental constraints do not pose serious deterrents to Brown shale development. Fluids produced from wells must be contained by on-site tanks to prevent stream pollution, all pits are required to be closed, disturbed land must be reseeded, and surface erosion from access roads and the drilling site must be controlled by drainage ditches. Recent legislation imposing stringent controls on potential stream pollution and land degradation has increased drilling costs by $2,000 to $5,000 per well. This increase in well cost is minimal, representing between 1 and 4% of the cost of drilling and completing a typical gas well in the Brown shale.

Shortages of drilling and well-completion rigs could pose a temporary constraint on development of gas production from the Brown shale of the Appalachian Basin. If all of the rigs currently in the Appalachian Basin were used exclusively to drill and complete new shale wells, it would not be possible to develop enough wells (69,000) to produce 1.0 Tcf per year of shale gas over the next 20 years. Favorable economics could possibly overcome the drilling and completion rig constraint over a 3 to 5 year period.

Obstacles to Advances in Shale Gas Technology

An important barrier to advances in Brown shale gas production technology is the lack of resource characterization. Even though approximately 10,000 wells produce gas from the Brown shale, very few quantitative data are available to adequately characterize the resource. Only a few of the 10,000 wells in the Brown shale have core samples available for examination, and those that do come from a relatively small portion of the 163,000-square-mile extent of the Appalachian Basin. Until the Brown shale resource is adequately characterized, focusing on specific targets for technology development is very difficult. Lack of specific research targets could result in haphazard hit-and-miss and trial-and-error experimentation with only limited chance of significant success in the near future. Detailed chemical-petrophysical data are needed for the Brown shale before significant progress can be expected in technology capable of releasing gas from those shales. Additionally, basic research is required to determine the manner in which gas is held by the Brown shale, i.e., is it only in the fractures, in the pores, adsorbed on the shale surface, or contained within the matrix porosity?

Characterization of the Brown shale involving shale petrography, core analysis work, geochemical research, and other pertinent data collection by different people in separate localities and agencies must be carefully coordinated to be effective.

Resource characterization is the initial and most pressing step for advancing technology for the purpose of increasing gas production from the Brown shale. Without an intimate knowledge of what the resource is, it is almost impossible to program research efforts in stimulation technology, logging methods, or any of the various satellite research needs dependent on reservoir characterization.

References

(1) Brown, P.J., "Energy From Shale—A Little Used Natural Resource," *Natural Gas From Unconventional Geologic Sources*, National Academy of Science Publ. FE-2271-1 (1976).

(2) Phase Report No. 1, *Massive Hydraulic Fracturing of the Devonian Shale*, Columbia/ERDA Contract E (46-1)-8014, Research Department, Columbia Gas System (October 1976).

(3) Slider, H.C., *Practical Petroleum Reservoir Engineering Methods*, Petroleum Publishing Co., Tulsa, OK (1976).

(4) Galtin, C., *Petroleum Engineering: Drilling and Well Completion*, Prentice Hall, Englewood Cliffs, NJ (1960).

(5) Pinnell, W.L., Consolidated Gas Supply Corp., Core Data on Well #11440 and #12041, Personal communication (1976).

(6) *Final Report—Well #7239, Perry County, Ky.*, ERDA-MERC (July 1975).

(7) Phase Report No. 1, *Massive Hydraulic Fracturing of the Devonian Shale*, Columbia/ERDA Contract E (46-1)-8014, Research Department, Columbia Gas System (October 1976).

(8) *Hydrocarbon Evaluation Study of Shales From Wells #R-109, #12401, and #11940*, Geochem Laboratories, Inc., prepared for ERDA-MERC (October 1976).

(9) Slider, H.C., *Practical Petroleum Reservoir Engineering Methods*, Petroleum Publishing Co., Tulsa, OK (1976).

(10) Overbey, W.K., "Effect of In Situ Stress on Induced Fractures," *Proc. Appalachian Petroleum Geology Symp., Morgantown, W. Va.* (1976).

(11) Martin, P. and Nuckols, E.B., III, "Geology and Oil and Gas Occurrence in the Devonian Shales: Northern West Virginia," *Proc. Appalachian Petroleum Geology Symp., Morgantown, W. Va.* (1976).

(12) Overbey, W.K. and Ryan, W.M., *Drilling a Directionally Deviated Well to Simulate Gas Production from a Marginal Reservoir in Southern West Virginia*, ERDA-MERC/TPR-76/3 (1976).

(13) Myung, J.I., "Fracture Investigations of the Devonian Shale Using Geophysical Well Logging Techniques," *Proc. Appalachian Petroleum Geology Symp., Morgantown, W. Va.* (1976).

(14) Douglas, J., Schlumberger Manger, Personal communication (1976).

(15) Ranostaj, E.J., Columbia Gas System Service Corp., Personal communication (1976).

(16) Ray, E.O., "Devonian Shale Development in Eastern Kentucky," *Natural Gas From Unconventional Geologic Sources*, National Academy of Sciences Publ. FE-2271-1, pp. 100–112 (1976).

(17) Komar, C., ERDA Morgantown Energy Research Center, Personal communication (1976).

(18) "Blasting: State-of-the-art," U.S. Bureau of Mines Internal Document, San Francisco Research Office (1969).

(19) Lemon, R.F. and Patel, H.F., "Formation Permeability and Gas Recovery at Project Gasbuggy," *Journal of Petroleum Technology*, Vol. 24 (October 1972).

(20) Prats, M. and Levine, J.S., "Effect of Vertical Fractures on Reservoir Behavior Incompressible Fluid Case," *Journal of Petroleum*, Vol. 15, No. 10 (October 1963).

(21) Ranostaj, E.J., *Massive Hydraulic Fracturing the Eastern Devonian Shales*, ERDA Contract No. E (46-1)-8014 (1976).

(22) Holditch, S.A. and Morse, R.A., "Large Fracture Treatments May Unlock Tight Reservoirs," *Oil and Gas Journal*, Vol. 69, No. 13 (Mar. 29, 1971), Vol. 69, No. 14 (Apr. 5, 1971).

(23) Carney, M.J. and Wieland, D.R., *Stimulation of Low Permeability Gas Wells in the Rocky Mountain Area*, SPE Paper No. 4396 (1973).

(24) Fast, C.R., Holman, G.B., and Covlin, R.J., *A Study of the Application of MHF to the Tight Muddy J. Formation, Wattenburg Field, Adams and Weld Counties, Colo.*, SPE Paper No. 5624 (1975).

(25) Blauer, R.E. and Mitchell, B.J., *New Technology in Formation Fracturing With Low Density Sand-Ladened Foam*, SPE Paper No. 5003 (1974).

(26) Blauer, R.E. and Holcomb, N.L., "Foam Fracturing Shows Success in Gas Oil Formations," *Oil and Gas Journal*, Vol. 73, No. 31 (1976).

(27) Frohne, K.H., *Comparison of Conventional Hydraulic and Water/Nitrogen Foam Fracturing in Two Ohio Devonian Shale Gas Wells*, MERC/TPR 76/1 (February 1976).

(28) Anonymous, "Tailored Treatments Promise Better Frac Economics," *Oil and Gas Journal*, Vol. 71, No. 12, p. 82 (1973).

(29) *Hydraulic Fracturing Process Using Reverse Flow*, U.S. Patent No. 3,933,205 (Jan. 20, 1976).

(30) Strubhar, M.K., Fitch, J.L., and Glenn, E.E., *Multiple, Vertical Fracture From an Inclined Wellforce—A Field Experience*, SPE Paper No. 5155 (1976).

(31) Bagnall and Ryan, "The Geology, Reserves and Production Characteristics of the Devonian Shale in Southwestern West Virginia," Figure 11, *Devonian Shale-Production and Potential*, ERDA (1976).

(32) McKie, J.W., "Market Structure and Uncertainty in Oil and Gas Exploration," *Quarterly Journal of Economics*, Vol. 74, pp. 543-571 (November 1960).

(33) Kaufman, G., "Statistical Decision and Related Techniques in Oil and Gas Exploration," Prentice Hall, Englewood Cliffs, NJ (1973), and Aitchison, J. and Brown, J.A.C., "The Lognormal Distribution," Cambridge University Press, New York, NY (1957).

(34) Kaufman, G., Bulcer, Y., and Kryt, D., "A Probabilistic Model of Oil and Gas Discovery," *Studies in Geology*, Vol. 1, American Association of Petroleum Geologists, Tulsa, OK, pp. 113-142 (1975).

(35) DeWitt, W., "Current Investigation of Devonian Shale by the U.S. Geological Survey," *Natural Gas From Unconventional Geologic Sources*, National Academy of Sciences (1976).

ALTERNATE ANALYSIS OF ECONOMIC POTENTIAL

The information in the remainder of this chapter is based on Enhanced Recovery of Unconventional Gas: The Program—Volume II (of 3 volumes), October 1978, prepared by V.O. Kuuskraa and J.P. Brashear of Lewin and Associates, Inc., T.M. Doscher of University of Southern California and L.E. Elkins for the U.S. Department of Energy under DOE Contract No. EF-77-C-01-2705.

Areal Extent

Overall Area: The Appalachian Basin covers approximately 210,000 square miles and includes all or parts of ten states (1). Approximately 100,000 square miles of the area has a low probability of being productive and has been excluded from the study described in the rest of this chapter (Figure 3.9) for the reasons shown on the following page.

- Area A—The eastern portions of the Basin between the Allegheny and Blue Ridge Fronts (45,000 square miles). The brown/black Upper Devonian shales outcrop and the deeper Middle Devonian shales have been metamorphosed to such an extent that existing gas deposits are improbable.

- Area B—The northern area of the Basin from Upper Erie across the State of New York and northeastern Pennsylvania (19,000 square miles). Like the eastern portion of the Basin, the shales have likely been metamorphosed to the extent of losing volatile hydrocarbons.

- Area C—The western flank of the Basin between the outcrops of Devonian brown/black shale and the Cincinnati Arch (15,000 square miles). The Devonian shales have played out and cannot be found from cross sections.

- Area D—The southern portion of the Basin between the Allegheny Front and Cincinnati Arch from eastern Kentucky to Alabama (20,500 square miles). The shales are thin, averaging 25 to 50 ft in thickness, and represent a submarginal resource.

Figure 3.9: Geological Distribution of the Devonian Shales of the Appalachian Basin

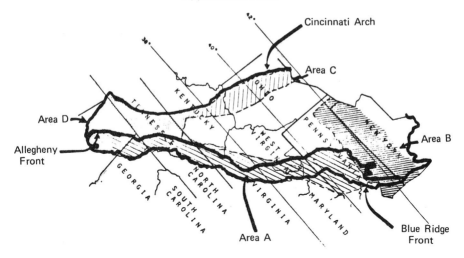

From deWitt, W., Perry, W.J., and Wallace, L.G., "Oil and Gas Data from Devonian and Silurian Rocks in the Appalachian Basin," U.S.G.S., Map I-197B (1975).

Source: DOE EF-77-C-01-2705

The use of the term brown/black shales probably originated with drillers to identify dark colored shale formations. In current literature, the terms are used interchangeably, or to distinguish the Upper Devonian shales (brown) from the deeper Middle Devonian shales (black). The USGS uses the term black shale to designate all Devonian shales of rich organic content. Typically, brown shale may be defined as younger shale, which has more hydrocarbons in the organic material than black shale where the organic material is closer to elemental carbon. Brown and black shales may be interbedded. In this study, brown/black shale refers to organic shales in general. Where the discussion involves "brown shale" as used in driller's nomenclature, it is in quotation marks.

It is possible that some of the excluded areas have small gas traps as are occasionally found along the Allegheny Front near the Pine Mountain Overthrust. However, these small discoveries have been episodic and provide no systematic evidence of continuous shale sequences.

An additional 48,000 square miles on the periphery of the examined acreage were classified as speculative, as they had no known productive wells, and were excluded from the analysis.

The remaining 62,000 square miles of the potential shales gas region of the Appalachian Basin were studied in detail. Of this, 5,000 square miles have already been drilled or found dry, leaving 57,000 square miles as the future target. This assessment of the total 210,000 square mile area is summarized below:

Definition of the Area	Areal Extent	Treatment by Study
Areas A, B, C and D	100,000	Excluded because geology indicated the shale is thin or absent or the likelihood of gas is low
Speculative	48,000	Insufficient data is available to define the economic potential of the area*, excluded from this study
Probable/possible	57,000	Included in the study as the source of additional gas or potential gas
Proved/developed or found dry	5,000	Gas potential already included in proved reserves or past production

*By definition, speculative resources were excluded from analysis.

Analytic Areas: The 62,000 square miles of drilled and future potential area was divided into twelve analytic areas (Figure 3.10). Their principal characteristics are as follows:

- Area I—Eastern Kentucky, within the heart of the Big Sandy Field; 2,164 square miles, of which 1,050 square miles are undrilled.

  "Brown Shales" dip as they reach the Allegheny Front and become thin.

- Area II—Eastern Kentucky, southwestern periphery of the Big Sandy Field; 2,089 square miles, of which 1,609 square miles are undrilled.

  "Brown Shales" in northern area merge into the oil producing zones without significant gas shows.

  Southern "Brown Shales" approach the outcrop area and are beginning to thin into less than 100-ft intervals.

  The Pine Mountain Overthrust in the southeastern region serves as a boundary for development drilling.

- Area III—Southwest West Virginia, northeastern periphery of the Big Sandy Field; 1,771 square miles, of which 1,098 square miles are undrilled.

Figure 3.10: Areal Extent of Study

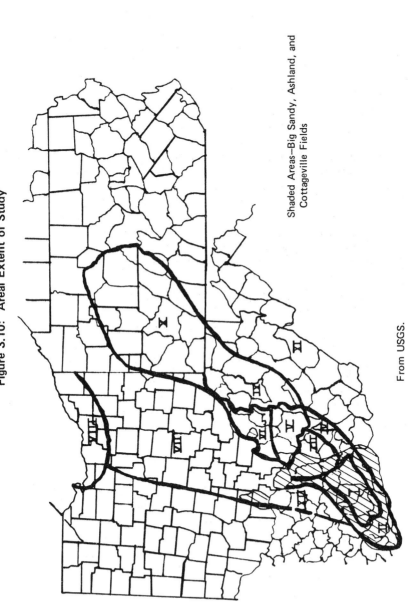

Shaded Areas—Big Sandy, Ashland, and
Cottageville Fields

From USGS.

Source: DOE EF-77-C-01-2705

The three extensive "Brown Shales" of Ohio and eastern Kentucky (Cleveland, Upper, and Lower Huron) appear in the western portion but are lost before reaching the eastern section, reducing the sequences to only one producing horizon.

Southeastern boundary is the limit of known drilling.

- Area IV—Southwestern West Virginia, east of Area III, northeast of the Big Sandy Field, known to have brown shale deposition; 871 square miles, of which 848 square miles are undrilled.

  "Brown Shales" thin as they outcrop along the Allegheny Front.

- Area V—Central West Virginia, north of the Big Sandy Field; 1,811 square miles, of which 1,690 are undrilled.

  Eastern boundary is found where the brown shale deposits thin and a facies change occurs with the shale going into coarser sands.

  Southern boundary approximates the division between known production in Area III and very scattered sporadic production of Area V.

  Northern boundary is outside the Cottageville Field, where production is scattered and mostly undrilled.

- Area VI—Northwest West Virginia; 2,067 square miles, of which 2,032 are undrilled.

  Eastern boundary lies along the facies change of "Brown Shales" into sands.

- Area VII—Central and eastern Ohio; 15,575 square miles, of which 15,261 square miles are undrilled.

  The area contains continuous deposits of Upper and Lower Ohio shales, the "Big and Little Cinnamon."

  Little or no observable major faulting, anticlines, or synclines.

- Area VIII—Lake Erie and northern Ohio; 3,090 square miles, of which 2,955 are undrilled.

  Shallow shale deposition.

  Isolated small gas wells.

- Areas IX and X—Northcentral West Virginia and southwestern Pennsylvania; 17,776 square miles are undrilled.

  Thinning out and disappearance of "Brown Shales."

  Continuous deposition of deeper Marcellus/Harrel shales.

- Area XI—Eastern West Virginia, 12,332 square miles are undrilled.

  Deep Marcellus/Harrell shales.

  No Upper Devonian "Brown Shales."

- Area XII—Northeastern Kentucky, 2,634 square miles, fully drilled.

  Shales are oil bearing without associated gas.

  Gas deposits depleted in northeastern segment.

Each of the analytic areas was assigned a unique set of production curves, developed from empirical data on shot well performance, which were increased to represent incremental additional production from fracturing. The 30-year recovery curve was applied to the central part of each area and reduced in the periphery areas to represent field playout. Table 3.3 summarizes the location, the total, and the undrilled acreage in each analytic area.

Table 3.3: Areal Extent of Analytic Units

Analytic Area	Location	Acres (square miles) Total	Undrilled
I	Eastern Kentucky	2,164	1,050
II	Eastern Kentucky Extension	2,089	1,609
III	SW West Virginia	1,771	1,098
IV	SE Extension	871	848
V	Central West Virginia	1,811	1,690
VI	NW West Virginia	2,067	2,032
VII	Central and Eastern Ohio	15,575	15,261
VIII	NE Ohio	3,093	2,955
IX and X	NW West Virginia and Pennsylvania	17,776	17,776
XI	Eastern West Virginia	12,332	12,332
XII	NE Kentucky	2,634	—
	Total	62,183	56,651

Source: DOE EF-77-C-01-2705

Estimates of the Resource and Its Productive Capacity

Traditional Methods of Resource Analysis: The traditional means for estimating the original size and productive capacity of a gas field is to first collect the key volumetric data (e.g., porosity, net pay, areal extent, etc.) that provide an estimate of the size of the resource in place; second, to analyze core and well test data to develop the properties that govern rates of gas flow (e.g., permeability, pressure, etc.); and third, to apply reservoir engineering analysis to estimate production and ultimate recovery.

Given the limited data on the resource and particularly where the producible resource is the free gas in the fracture porosity, such a traditional approach is not yet possible for the Devonian shales. Past estimates using volumetrics or cannister off-gas analysis fail to define the dominant feature that governs production in the Devonian shales, the natural fracture system. However, some data has been gathered on certain of the key parameters and provide a departure point for further inquiries through reservoir simulation and history matching.

Porosity and Permeability Data from Core and Pressure Build-Up Analysis: The Morgantown Energy Research Center measured the permeability and porosity on a Lincoln County well core under confining pressure approximating that of the reservoir. The cores had an effective permeability of less than 0.1 md and a porosity of less than 1%.

A pressure build-up analysis, the first known to be run in the Devonian shale gas reservoir, indicated an effective permeability between 0.009 and 0.18 md and a gas filled porosity between 0.8 and 1.9%. Although the permeability value derived from this analysis is that of the particular cores for the subject well, these observations are of considerable value in indicating the nature of the reservoir parameters for the particular area being studied.

Estimating Reservoir Data from History Matching: A standard procedure in reservoir engineering is to estimate reservoir parameters that govern production by matching the historical performance of the wells with the theoretical response of a hypothetical reservoir with assumed reservoir parameters—history matching.

The Essential Parameters — In order to predict the performance of such a gas reservoir using history matching, the following parameters must be known: drainage area, net thickness, initial reservoir pressure and temperature, and well pressure. For purpose of the analysis, it is assumed that the drainage area is 150 acres, the net thickness 500 ft, and the initial pressure and temperature 500 psi and 100°F, respectively. The wellbore pressure over most of the life of the well is assumed to be 100 psi. It is then possible to vary porosity (ϕ), permeability (k), and wellbore radius (r_w) to achieve a match with actual data.

The Analytic Approach — The preliminary analysis (performed for Lewin and Associates, Inc. by Holditch and Moore at Texas A & M) of the shales was done using an isotropic, linear flow, reservoir simulation model modified to accept homogeneous and nonhomogeneous flow. There is some thought that a fractured reservoir might be anisotropic, but no hard empirical evidence for this was found. Therefore, the reservoir analysis was not unduly complicated by including anisotropy.

Homogeneous Model: The assumption was made that the Devonian shale reservoir comprises a network of fine, isotropic fractures through which the gas flows to the producing well. The flow through such a network is conceived to be approximating that for the radial flow of fluids through a uniformly distributed interstitial porosity such as that of a sandstone. It is then possible to analyze the performance of a Devonian shale gas reservoir much as one would analyze the performance of a gas filled sandstone reservoir.

Because the reservoir is tight, it can be inferred that the wells are producing in an unsteady state flow regime, with the boundary pressure being virtually unaffected despite the length of time during which the well has been producing. Therefore, the analysis can be made using the analytical expression for unsteady state flow that can be derived from the continuity equation, Darcy's law, and the solutions for the radial diffusivity equation developed by Hurst and vanEverdingen. The analytic equations used in the reservoir simulator differ from the Hurst and vanEverdingen solution by making an adjustment to account for the compressibility drive effects prevalent in low pressure formations. The compressibility of the gas under such flow conditions is that of the gas at original reservoir conditions, and the average pressure used in the solutions is the arithmetic average of the initial pressure and the wellbore pressure. A reservoir with an enlarged wellbore radius, due to shooting, was used for the homogeneous case.

Nonhomogeneous Model: As a test of the analytic approach, a second set of reservoir geometrics was introduced—a nonhomogeneous reservoir consisting of a system of natural, horizontal fractures having radius of 15 to 50 ft spaced throughout the pay interval.

The History Matching — Three typical Devonian wells (a high, medium, and low), each having 30 years of production (shown on Figure 3.11) were history matched using the two reservoir geometrics.

The analytic approach provided an extremely close history match to long term production and potentially a unique solution. According to the analyst, it is possible that slightly different values of gas in place and recovery efficiency could be obtained, if the parameters such as original pressure, reservoir temperature, flowing bottom hole pressure, etc. were altered. However, assuming the input data are valid, the established withdrawal rates and reservoir decline rates are of sufficient duration to provide unique solutions for ϕh and kh.

A history match of long term production data can be realistically divided into two parts. The early time data (0 to 5 years) is dominated by the permeability-thickness product (kh) of the reservoir and by the wellbore geometry. Therefore, using different values for wellbore radius, fracture length, etc. will result in different values of formation permeability. However, all combinations of kh and r_w that can be used to match actual production data will result in a common value of productivity index. (It would be possible to fix the value of r_w at any reasonable value and find a value of kh that would match the early time production data.) To obtain a unique match of early time data requires a series of well tests and/or a long term pressure buildup test.

**Figure 3.11: History Match of Simulation and Field Data—
Three Typical Devonian Wells**

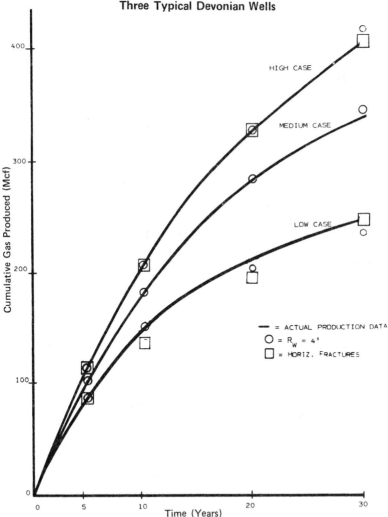

Source: DOE EF-77-C-01-2705

The reservoir decline rate, however, is dominated by the formation porosity-thickness product (ϕh), which determines the original gas in place. Normally, if 30 years of production data can be history matched, the value obtained for gas in place is considered to be a unique solution.

The values used for porosity were insensitive to the wellbore geometry. Permeability, however, did slightly vary according to the wellbore geometry, but in general, the range of variation was not significant.

The Results of the Analysis — Several major conclusions emerge from this history matching:

- The Devonian shale gas reservoirs have porosities of less than 1%; higher production is a direct function of higher porosity as would be accounted for by more intense fractures:

Sample Wells	ϕ
High production	0.8%
Medium production	0.6%
Low production	0.3%

- The permeabilities average about 0.02 md ranging from 0.018 to 0.027 md.

- Combining the three parameters that most directly affect production, porosity (ϕ), permeability (k), and net pay (h), the following parameters best describe the Devonian shale:

Sample Wells	ϕh	kh
High production	4	12
Medium production	3	11
Low production	1–2	10

- Considering the high values of kh compared to the low values of ϕh, the analysis shows that a typical Devonian shale gas reservoir consists of a vast network of interconnected, natural fractures. Moreover, since all of the production could be accounted for by the free gas in the fractures, the contribution of gas in matrix porosity, to the extent such matrix porosity exists, would be small.

- The wellbore appears to be in contact with the full horizontal extent of the fractures in the drainage area; however, it may be possible for additional fracture systems to exist in a vertical plane and not be connected to the wellbore.

- The reservoir data from history matching of well performance are in accord with the initial reservoir values obtained from pressure build-up in the field and in the laboratory using restored state pressure conditions:

Reservoir Properties	Pressure Build-Up Data	Reservoir Simulation
Porosity, %	0.8–1.9	0.3–0.8
Permeability, md	0.01–0.18	0.02–0.03

Summary of the Productive Capacity of the Devonian Shales: The analysis strongly indicates that the Devonian shales are essentially a traditional, low pressure and low permeability gas reservoir where the natural fracture system provides the permeability and gas storage porosity. It may be that the matrix contains additional porosity and gas. However, this matrix porosity does not appear to be in contact with the wellbore. Adding one more fracture plane to a well drainage area containing several hundred such fractures could not be expected to appreciably interconnect the now occluded areas. Further, under current production practices, the shales are already being efficiently drained of the gas in place. (Further improvements are possible, however, at higher gas prices, as discussed in the next section.)

The major unknown is whether the wellbore is in contact with full vertical extent of the natural fracture system in a drainage area.

Estimation of Production: Additional reservoir data and improved capacity to model this resource may allow a more traditional approach for estimating the productive capacity of the Devonian shale to be pursued in the future. Until that time, however, the analysis must rely on an empirical approach using a representative sample of production histories from the nearly 10,000 wells that have produced from these shales.

The basis of the production estimates was empirical well data acquired from various gas companies on 250 individual wells. These data include annual production, rock pressures, and line pressures which enabled development of specific production decline curves for the areas defined as having potential shale gas. Each well chosen had to meet five criteria:

- It was individually metered.
- The shale gas production was distinguishable from other producing horizons.
- The well needed to have a minimum historical production of 25 years for older fields and 15 years for newer fields.
- The sample needed to include high, average, and low producers.
- The sample had to contain at least 4 to 6 wells for each county defined within an areal unit.

The wells were classified by the eleven areally defined analytic units (Area XII was excluded because it appears thoroughly depleted). For each area, production from the sample wells was averaged and fit to a cumulative 30-year production curve using Marquardt's Algorithm, providing the base production profiles required for economic analysis.

These base profiles were then modified to reflect two factors: that fields tend to produce less as drilling moves into extension and stepout areas; and that production from wells stimulated by hydraulic fracturing is generally higher than production from wells stimulated by shooting. Production multiplier values (PMVs) were created to account for these two factors.

Development of the field play-out value was based on experience in the eastern Kentucky Big Sandy Field. Historically, Kentucky wells averaged initial open flows near 330 Mcf/d. Recent wells are clustered in the range of 100 to 275 Mcf/d, suggesting less production as the better portions of the reservoirs are depleted. The decreasing production as reservoir boundaries are approached is known as field play-out. Sample wells from the four primary counties comprising the Big Sandy Field were found to have a 30-year cumulative recovery of

Table 3.4: Basis of Devonian Production and Recovery Estimates

Area	Undrilled Acreage	No. of Areal Unit	30 Year Recovery Shot Well Data (MMcf)	Field Playout (PMV)	Production Increase— Fracturing	Final (PMV)	Estimated 30 Year Recovery/Fractured (MMcf)
I	1,050	50	411	1.0	15	1.15	472
		25		0.8	25	1.00	411
		25		0.5	40	0.70	288
II	1,609	33	349	1.0	15	1.15	411
		33		0.7	30	0.91	318
		33		0.5	40	0.70	244
III	1,098	50	348	1.0	15	1.15	400
		25		0.8	25	1.00	348
		25		0.5	40	0.70	244
IV	848	33	376	1.0	15	1.15	434
		33		0.7	30	0.91	343
		33		0.5	40	0.70	264
V	1,690	33	338	1.0	15	1.15	389
		33		0.7	30	0.91	308
		33		0.5	40	0.70	237
VI	2,032	33	264	1.0	15	1.15	337
		33		0.7	30	0.91	267
		33		0.5	40	0.70	205
VII	15,261	33	96	1.0	40	1.40	134
		33		0.7	50	1.05	100
		33		0.5	50	0.75	71
VIII	2,955	33	362	0.9	15	1.04	377
		33		0.7	25	0.87	315
		33		0.5	40	0.70	253
IX and X	17,776	33	338	1.0	—	1.00	338
		33		0.7	—	0.70	237
		33		0.5	—	0.50	169
XI	12,332	33	267	1.0	—	1.00	267
		33		0.7	—	0.70	211
		33		0.5	—	0.50	102

Source: DOE EF-77-C-01-2705

512 MMcf, while those in outer counties were found to have 30-year cumulative recoveries of 255 MMcf, about one-half of the best area.

Using these values as indicators, each of the eleven areal units was divided into three portions and assigned a field play-out factor (a percentage of the base production curve) that reflected the shale thickness and extensiveness of fracturing, as estimated from the stratigraphy and tectonic activity.

Next, the base production curves (derived from shot well data) were increased to reflect anticipated improved performance from using hydraulic fracturing rather than shooting. Work done by Ray (2), described above, suggests that wells having initial open flows between 100 to 200 Mcf would increase cumulative production by 46% after 5 years with hydraulic fracturing; while wells with an initial open flow of 200 to 300 Mcf would increase cumulative production by 17% after 5 years. Thus, production was increased by 15 to 50%, depending on the initial open flow.

Table 3.4 shows the area, the original shot well production, and the field play-out adjustment and fracturing adjustment that provide the final 30-year cumulative recovery estimates. (Because of extremely limited production history, Areas IX, X and XI were estimated from other areas without adjustment for stimulation technology.) The annual production profiles developed through this procedure were used in the economic analysis.

Improving Gas Recovery from Devonian Shales—The Federal Role

As the Devonian shale resource base becomes further understood, the major challenge will be to increase the amount and advance the timing of gas production from these shales. For this, one needs to pose the basic question:

What appears to constrain rapid development and economic production from the Devonian shales:

- Are the natural gas reservoirs within the Devonian brown-black shale sequence being effectively drained?

- Are there additional reservoirs of natural gas in the brown-black shales from which recovery of the resource is constrained by the lack of suitable technology and/or economic incentives?

An analysis of the resource base through reservoir engineering indicates that three basic strategies should be followed for increasing gas recovery from the Devonian shales:

- Identifying additional Devonian shale areas in the Appalachian Basin having economic potential for gas recovery.

- Testing dual-completion practices for recovering gas from low production, economically marginal formations in central Ohio (and if possible in other parts of the Basin).

- Improving recovery efficiency in the identified and productive parts of the Basin.

Additional Potential: The research effort involved in the first strategy, identifying additional areas of Devonian shale potential, is straightforward. One needs to drill and test a number of resource characterization wells to establish that a sufficiently intense natural fracture system is present and that gas fills this fracture system. This is what the gas production industry has done, in the Basin, for over 50 years. The special aspect is that the area under consideration is deep, expensive to drill, and costly to produce. Economic recoverability in these areas, even if the shales are found to be fractured and gas containing, will likely require gas prices considerably in excess of today's levels. Thus, the gas production industry has relegated resource definition in these areas to a lower priority. It appears that public R&D is required to accelerate a definition of the productive potential of these deeper, higher cost areas. Traditional exploration and production analysis was used to assess the potential of this first R&D strategy.

Dual-Completion Practices: A second target for research and development involves areas where noncommercial, but physically productive, shale sequences overlie deeper, traditional producing horizons. Here, it may be possible to use dual-completion (with improved stimulation) and produce the now uncommercial shale by having the primary formation bear the major drilling and operating costs.

Such an opportunity exists in central and eastern Ohio. Given the need to tailor special completion practices as well as work over existing wells, a demonstration of the technical and economic feasibility of such an approach appears required before industry could be expected to vigorously pursue this second target. A marginal cost analysis was used to assess the potential of this second R&D target.

Improved Recovery Efficiency: A third and traditional target for R&D is improving recovery efficiency to a level above that being attained by traditional production practices. Three approaches appear worthy of consideration:

- Closer well spacing in the more productive areas.
- Improved stimulation practices, such as using more efficient and cost-effective stimulation and fracture technology.
- Placing the wellbore in contact with the full vertical fracture system, should current completion practices be falling short of this goal.

History matching with a reservoir simulator using a two dimensional, finite difference model, described previously, was used to assess the potential of this third R&D target.

Three actual sets of field data from typical Devonian gas wells having the properties previously shown on Figure 3.11 were used in the analysis. The findings from this analysis are discussed below.

The Efficiency of Small Fractures — At 150-acre spacing and with small fractures, in general, the field is already being drained efficiently with recovery efficiencies of 50 to 60%.

Using an economic limit of 8 Mcfd, the following recovery efficiencies are found for shot versus fractured wells at 150-acre spacing:

- Shot wells (4' radius or 6 horizontal fracs):

	Gas in Place (MMcf)	30 Year Production (MMcf)	Recovery Efficiency (%)
High production	880	410	47
Medium production	666	330	50
Low production	352	220	61

- Small vertical fractures (100 ft):

	Gas in Place (MMcf)	30 Year Production (MMcf)	Recovery Efficiency (%)
High production	880	500	57
Medium production	666	430	65
Low production	352	220	62

Using small vertical fractures improves 30-year recovery by about 20% over the shot case for high and moderate production wells, but only accelerates recovery in the low production wells.

Additional analysis was conducted on the wells in Ohio, however, since recovery efficiencies in these areas appear to be about 60%, significant improvements in recovery efficiency did not appear feasible.

The Potential of Infill Drilling — A first step toward assessing the potential for further improving 30-year recovery in low permeability formations would be to examine closer well spacing. For this, analytic areas were developed on 75 rather than 150 acres, using shooting as well as small fractures. The simulation showed that 30-year recovery in the high production case could be further improved by 20%, from a base of 56% to a range of 65 to 70%. Smaller improvements were realized in the medium production case, while the low production case had (because of the economic limit) essentially no change. This is shown below:

- Reducing spacing (75 acres) with small fracs:

	Gas in Place (MMcf)	30 Year Production (MMcf)	Recovery Efficiency (%)
High production	440	300	68
Medium production	333	230	70
Low production	176	120	66

Under reduced spacing, recovery efficiencies are in 60 to 70% range and production reaches economic limits within the 30-year period. At this spacing, further recovery efficiency improvements do not appear feasible, although acceleration of production in the earlier years would be possible through improved stimulation practices and enlarged wellbores. The simulation model was run to test the potential of using larger, 500 to 1,000 ft fractures. The results of the analysis showed no significant difference in 30-year recovery efficiency between using 100 to 200 ft fractures and using larger 500 to 1,000 ft fractures. This result is due to the fact that recovery efficiency using smaller fractures is already quite high and additional production can only be achieved at what appears to be less than economic flow rates.

Other Potential Improvements — The major unknown is: what portion of the total natural fracture system in a drainage area is in contact with the wellbore? Since no unusual boundary effects are evident from matching field data, the model gives strong evidence that the wellbore is in contact with the full horizontal plane of the drainage area. However, the Devonian shale comprises several depositional cycles, and vertical communication between intervals may not be naturally present or included by conventional completion procedures. The relatively narrow range of response of wells in any particular area argues against this. Nevertheless, a high priority must be assigned to determining whether it is possible to enhance Devonian gas production by selective and multiple completions within the total Devonian shale section.

Potential of the Devonian Shale of Appalachia

Approach: Estimating the economic potential of the Devonian shales requires two sets of assumptions, namely:

- Specifying the level of technology:

  Base Case: The level of the resource development expected to be attained by industry during the next 5 years without active federal involvement.

  Advanced Case: The level of the resource development expected to be attained by virtue of active federal and industry collaboration.

- Defining field development costs, return on capital requirements, prices, and timing.

These two key steps are further described below.

The Technological Assumptions — Two levels of resource development and technology were assumed, the Base Case and the Advanced Case. The Base Case conditions were derived from discussions with producers in the Basin concerning their development plans and the status of the technology over the next 5 years. The Advanced Case was specified as the result of successful execution of a Federal R&D program summarized later in this chapter.

Table 3.5 shows the principal differences between the Base and Advanced Cases. These differences can be grouped under four broad headings:

- Resource Characterization
- Technology
- Economics
- Development

Resource Characterization: Industry's efforts are centered currently on the probable areas, essentially extension drilling and step-outs.

A focused program of resource characterization could obtain sufficient data on the poorly defined, possible areas of the Devonian shale to stimulate further drilling and development, where economic. Such a program would be directed at a 30,000 square mile area (Areas IX, X and XI) of Devonian shale, and should sufficient fracture intensity and gas be discovered, this could appreciably increase the potential in the Appalachian Basin.

A second purpose of the resource characterization effort would be to apply improved geologic and reservoir engineering measures to the probable areas to reduce the number of marginal and dry wells from the current rate of 20%.

A third purpose would be to quantify the distribution of free gas porosity between matrix and fractures and to obtain a range of gas permeability contrast between matrix and fracture porosity systems.

The remainder of the resource characterization effort would be directed at obtaining sufficient reservoir data to support improved production technology. This becomes a prerequisite to the technology improvements discussed below.

Technology: The Base Case assumes that producers will use small fracturing technology. Further, technological improvements due to R&D are twofold:

- Achieving improved well completion practices to enable the dual completion of economically marginal, low producing pays with underlying economically producible gas horizons (particularly in Ohio).

● Increasing recovery efficiency in the high production areas (the heart of the Big Sandy) by 20% by using infill drilling and improved stimulation technology.

Economics: Producers currently view gas production in the Devonian shales as a moderate risk venture. The intended outcome of the R&D program is to reduce the financial risk premium to conventional levels.

Acceleration: The final purpose of the R&D program is to accelerate the time by which the Devonian resource could be drilled and produced.

Table 3.5: Summary of Major Differences Between Base and Advanced Cases in Devonian Shale Analysis

Strategy/Item	Base Case	Advanced Case
Resource characterization		
Eligible areas	Probable	Probable and possible
Dry hole rates	20%	10%
Technology		
Completions	Single	Dual where low producer underlain by other productive pay
Stimulation	Standard 1,000 bbl fractures	Optimized fractures
Recovery efficiency per unit area	Current levels	Improved by 20% in higher producing areas
Economics		
Risk, reflected in discount rates*	21%	16%
Development		
Start year for drilling		
Probable areas	1978	1981 (R&D effect begins)
Possible areas	1987	1987
Development pace		
Probable areas	17 years**	13 years**
Possible areas	17 years**	15 years**

*Discount rates include a constant ROR base of 10 to 15% and an inflation adjustment of 6%.
**To completion.

Source: DOE EF-77-C-01-2705

Economics and Timing — A net present value (discounted cash flow) model was used to simulate the economics of production. The unit of analysis was the individual well and its drainage area representing specific areal units. The estimated investment and operating costs of the well were offset by the revenue stream generated by gas production times its price. This net cash flow was then discounted by the specified return on investment. The areal units for which this discounted (present) value exceeded zero were developed, according to the timing model.

State-level drilling and completion costs were drawn from the API Joint Association Survey of Costs. Well stimulation costs were provided by major service companies. Surface equipment and operating costs were developed from studies by Gruy Federal, Inc. (3). Dry hole and exploration costs were functions of drilling and completion costs and the areas involved. All costs were varied as a function of depth and geographic region. Taxes, royalties, burden rates, and accounting procedures were drawn from actual applications in each region. Constant 1977 dollars were used throughout the analysis.

These costs were then compared with company records and modified to incorporate any unusual features of Appalachian shale wells.

The analysis was conducted for three gas prices—$1.75, $3.00, and $4.50 per Mcf. Where the well, representing a specific areal unit, was found to be economic, it was extrapolated to the full area. Each area was assumed to be developed according to a fixed schedule. The better defined (probable) areas were assumed to be developed first, followed by the less well defined (possible) areas.

Base Case Estimates: The amount of gas production and its rate in the Base Case is highly sensitive to gas price. As shown in Figures 3.12 and 3.13, additional recovery could range from less than 2 Tcf at $1.75 per Mcf to over 10 Tcf at $4.50 per Mcf; the production rate in 1990 would range from about 0.1 Tcf per year (at $1.75/Mcf) to 0.3 Tcf per year (at $4.50/Mcf). This is discussed further below.

Ultimate Recovery — In terms of ultimate recovery (total production or reserve additions in 30 years of well life):

- About 2 trillion cubic feet will be economic at $1.75/Mcf.

- Increasing the price of gas to $3.00/Mcf could raise the estimate to about 8 Tcf. This significant increase is contingent on the conclusion that apparent field play-outs have been primarily dictated by sheer economic considerations.

- At $4.50/Mcf, estimated ultimate recovery could only rise another 2.5 trillion, to 10.5 trillion cubic feet.

Production Rates —

- At $1.75/Mcf, the Base Case production rates would peak at 0.1 Tcf in 1990 and decline thereafter.

- Higher prices could sustain production over a longer period, providing 0.3 Tcf per year in 1990.

The Advanced Case Estimates: Considerable amounts of additional gas could accrue from a successful R&D program in the Devonian shales—the Advanced Case as shown on Figures 3.14 and 3.15. (The difference between the Advanced Case and the Base Case estimates are the production benefits attributable to the R&D program.)

Ultimate Recovery — Under the Advanced Case assumptions, ultimate recovery at $1.75 rises from the Base Case estimate of 2 to 4 Tcf. At higher prices, considerably more recovery could be forthcoming.

- At $3.00/Mcf, ultimate recovery could rise to 16 Tcf (versus about 8 Tcf in the Base Case).

- At $4.50/Mcf, ultimate recovery could range from 18 to 25 Tcf (the range reflects geological uncertainties in the possible areas where little is known about the intensity of the natural fracture system).

Figure 3.12:   Devonian Shale Ultimate Recovery at Three Prices,
Base Case

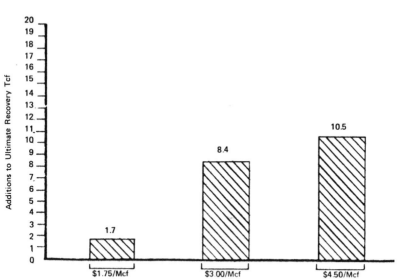

Source: DOE EF-77-C-01-2705

Figure 3.13:   Annual Production from the Devonian Shale to the Year 2000
at Three Prices, Base Case

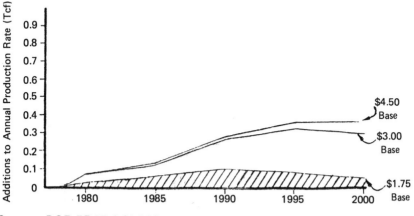

Source: DOE EF-77-C-01-2705

Production Rate —

- In 1990, annual production under the Advanced Case and at
  $1.75/Mcf is projected at about 0.2 Tcf.

- At $3.00, 1990 annual production is estimated at 0.6 Tcf.
- At $4.50/Mcf, annual production rate continues to climb reaching a range of 0.7 to 0.9 Tcf in 1995.

Figure 3.14: Devonian Shale Ultimate Recovery (at Three Prices)

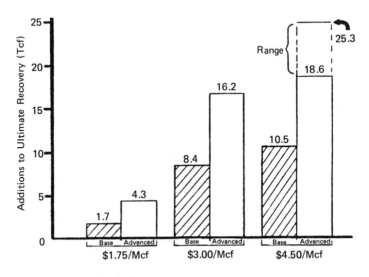

Source: DOE EF-77-C-01-2705

Figure 3.15: Annual Production from the Devonian Shale to the Year 2000
(at $1.75 and $3.00/Mcf)

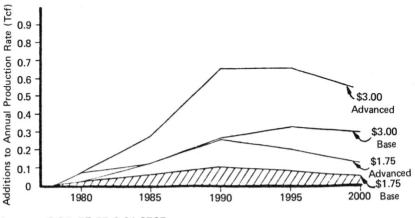

Source: DOE EF-77-C-01-2705

The Essential R&D Programs — The improvements in production in the Advanced Case come from successful execution of three highly different programs of technological advance.

- The first program in eastern West Virginia and Pennsylvania is directed at extension drilling in the deep shale that appears economic only at prices higher than $3.00/Mcf. (The benefits of this program are expressed as a range to reflect present uncertainty about the extensiveness of the natural fracture network and the presence of producible gas.)

  Ultimate recovery due to this R&D program would range from 0 to 7 Tcf at a gas price of $4.50/Mcf.

  The 1990 production rate would be below 0.1 Tcf.

- The second program (in Ohio) depends on dual completions of wells in the shales, underlying sandstone and limestone gas reservoirs, thus permitting the shales to be produced at the marginal costs of stimulation.

  Ultimate recovery from this second program would range from 3 to 6 Tcf, depending on gas price.

  The 1990 production rate from this program would be 0.2 Tcf.

- The third program (in eastern Kentucky and western West Virginia) is to improve recovery efficiency in the heart of the currently being developed area through optimizing stimulation, well spacing, and development practices.

  Increasing recovery efficiency would add about 2 Tcf at gas prices of $3.00/Mcf.

  The 1990 production rate would be about 0.1 Tcf.

Figures 3.16 and 3.17 show the Base Case and the Advanced Case for the three R&D programs at three gas prices.

Figure 3.16: Base Case—Ultimate Recovery from the Devonian Shales at Three Prices by Program

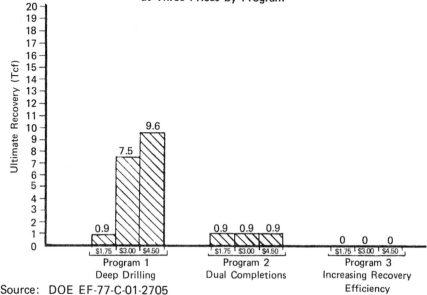

Source: DOE EF-77-C-01-2705

Figure 3.17: Advanced Case—Ultimate Recovery from the Devonian Shales at Three Prices by Program

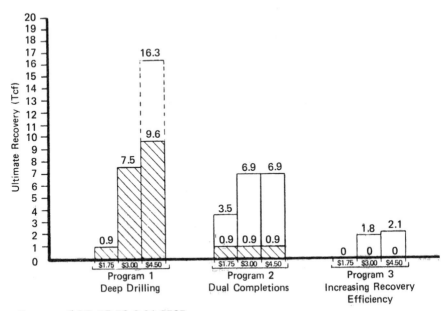

Source: DOE EF-77-C-01-2705

Summary: The Devonian shales could add from 2 to nearly 25 Tcf of gas to domestic production, depending on the technology and price assumptions. From a public policy perspective, a combination of technological advances and higher prices appears to provide the most cost-effective strategy for ensuring additional quantities of gas for portions of the country that have been seriously imperilled by curtailments in interstate gas supplies.

Summary of R&D Program and its Potential

Research and Development Goals: The three technological challenges—new areas, dual-completion, and improved recovery efficiency—form the R&D goals of the Devonian shale programs in the Appalachian Basin listed below:

- Define the Potential of Deep Devonian Shales—Drill exploratory wells and employ improved stimulation technologies to the deep (over 5,000 ft), currently nonproducing shales in northern West Virginia and southern Pennsylvania.

- Produce Marginal Devonian Shales Through Dual Completion—Use dual completion and improved stimulation technologies in areas where the currently nonproducing shale sequences overlie deeper producing horizons (the Clinton formation) in central and eastern Ohio.

- Improve Recovery Efficiency—Infill drill and apply improved stimulation techniques in areas of major historical production, particularly the Big Sandy Field and its extension.

Activities and Costs: The activities required to achieve these goals can be grouped in four broad categories:

- Resource Characterization
- Improved Diagnostic Tools
- Field-Based Research, Development, and Demonstration
- Technology and Information Transfer

The amount of professional time and the number of tests required to carry out these activities are summarized in Table 3.6. A total of 65 person-years are devoted to resource characterization, 25 to improved diagnostic tools, and 20 to technology transfer. Field-based R&D is a mix of resource characterization of cores and wells, measurement calibration tests, technology improvement tests, and field demonstrations of improved technology. A total of 106 such projects are required.

Table 3.6: Summary of Program Activities for Devonian Shales of Appalachia (Five-Year Strategy Totals)

Target/Program	Resource Characterization	Improved Diagnostic Tools	Field-Based R&D	Technology/ Information Transfer
	. . . . . (person-years)		(cores/wells)	(person-years)
Program 1				
Define potential of deep Devonian shales	15	10	19	5
Program 2				
Produce marginal Devonian shales through dual completion	30	15	30	10
Program 3				
Improve recovery efficiency	20	—	57	5
Total	65	25	106	20

Source: DOE EF-77-C-01-2705

Table 3.7 summarizes the costs of these strategies for the five-year period FY 79 to FY 83. The total five-year cost, in constant 1977 dollars, is $38.1 million. The program is designed to take advantage of industry's interests in production from the Devonian shales, thus presenting cost-sharing opportunities. Of the $38.1 million total, industry is expected to contribute $8 million, leaving a federal cost of about $30.1 million.

Table 3.7: Costs of the R&D Program for the Devonian Shales

| | 5 Year Program Costs (millions) | |
Target/Program	Total ($)	Federal Share ($)
Program 1		
Define potential of deep Devonian shales	10.6	8.6
Program 2		
Produce marginal Devonian shales through dual completion	13.0	11.5
Program 3		
Improve recovery efficiency	14.5	10.0
Total	38.1	30.1

Source: DOE EF-77-C-01-2705

Incremental Production Benefits: Successful execution of these R&D strategies—improvements in resource characterization and advances in technology, stimulated by federal/industry research—yield substantial benefits (Table 3.8):

- At $3.00/Mcf, about 8 Tcf of additional gas could be ultimately recovered through the research program, providing 0.4 Tcf of additional annual production in 1990.

- At a higher, $4.50/Mcf, gas price, the gas potential from Program 1 could also become economic, raising the ultimate benefits of the three programs to a range of 8 to 15 Tcf.

Table 3.8: Production Benefits Due to Successful R&D (at $3.00/Mcf)

Target/Program	Ultimate Recovery	1990 Annual Production	1990 Cumulative Production
	.... (Tcf)		
Program 1			
Define potential of deep Devonian shales	*	*	*
Program 2			
Produce marginal Devonian shales through dual completion	6.0	0.3	1.7
Program 3			
Improve recovery efficiency	1.8	0.1	0.3
Total	7.8	0.4	2.0

*Gas production from Program 1 is economic only at $4.50/Mcf; the production benefits at $4.50/Mcf are: ultimate recovery, 0 to 6.7 Tcf; 1990 production, 0 to 0.3 Tcf; and cumulative production, 0 to 0.1 Tcf.

Source: DOE EF-77-C-01-2705

Cost-Effectiveness Measures: Two measures were used to assess the relative cost-effectiveness of the R&D strategies, stated in terms of Mcf of incremental recovery per dollar of federal expenditure (in constant 1977 dollars).

- Long-term measure—incremental ultimate recovery (at $3.00/Mcf) per dollar of federal cost.

- Near-term measure—incremental cumulative recovery to 1990 (at $3.00/Mcf) per dollar of federal cost.

Table 3.9 shows the cost-effectiveness measures for the three Devonian shale strategies:

- The long-term benefit of the program would be about 270 Mcf per dollar of federal R&D.

- In the near-term, between now and 1990, about 110 Mcf could be added to production per dollar of R&D costs.

Table 3.9:   Cost-Effectiveness Devonian Shale Strategies

Target/Program	Long-Term	Near-Term
	. . . . (Mcf/$).	
Program 1		
Define potential of deep Devonian shales	*	*
Program 2		
Produce marginal Devonian shales through dual-completion	520	150
Program 3		
Improve recovery efficiency	180	30
Weighted average	270	110

*For Program 1, the benefits can only be calculated at a price of $4.50/Mcf: long-term is 0 to 780 and near-term is 0 to 1.

Source: DOE EF-77-C-01-2705

References

(1) American Petroleum Institute, *Reserves of Crude Oil, Natural Gas Liquids and Natural Gas in the United States and Canada as of December 31, 1976*, API, Vol. 31 (May 1977).
(2) Ray, E.O., "Devonian Shale Production, Eastern Kentucky Field," The Future Supply of Nature-Made Petroleum and Gas, UINTAR Conference, Pergamon Press, New York (1976).
(3) Gruy Federal, Inc., study for the Department of Energy (1978).

Methane from Coal Seams

HISTORICAL RESEARCH EFFORTS

The information in this and the following five sections is based on *National Gas Survey Report to the Federal Energy Regulatory Commission by the Supply-Technical Advisory Task Force on Nonconventional Natural Gas Resources*, DOE/FERC-0010, June 1978, "Methane in Coal" prepared by Sub-Task Force II, A. Warner, Sr. of the U.S. Bureau of Mines, T. Jennings of National Gas Survey, G. Denton of The Pittston Company Coal Group, M. Deul of the U.S. Bureau of Mines, S.S. Galpin of Consolidated Gas Supply Corp., W. Laird of Gates Engineering Co., W. La Londe III of Elizabethtown Gas Co., W.E. Matthews IV of Southern Natural Gas Co., D. Patchen of West Virginia Geological Survey and M.L. Skow of the U.S. Bureau of Mines for the U.S. Department of Energy. References for these sections are on p 161.

Coal Mine Safety

The occurrence of methane in coal was the subject of investigations by R.T. Chamberlin (1) and N.H. Darton (2) conducted for the U.S. Department of the Interior from 1907 to 1915. The frequency and severity of coal mine disasters caused by the accumulation, ignition, and explosion of indigenous methane gas established the need to investigate the occurrence and disposal of methane in coal mines. For many years, improving ventilation or reducing the rate of coal extraction were the only methods employed for preventing dangerous concentrations of methane in underground mines. Later the European practice of piping methane-rich air mixtures from mine workings to the surface atmosphere was developed where longwall mining methods are employed for extracting coal (3).

In 1964, a comprehensive applied research program was initiated by the U.S. Bureau of Mines (4) to better understand and control methane emission in coal mines. The methane drainage research program is described below.

Methane Program of the Bureau of Mines: The Bureau of Mines methane program currently includes two separate but related objectives. The original and continuing objective is to develop the technology necessary for safe and economic mining of methane-laden coalbeds so as to prevent the mine disasters caused by accidental ignition of methane-air mixtures. Recently the objective of demonstrating the feasibility of recovering commercial quantities of methane from virgin coal beds and from underground gob areas to increase the production potential and productivity from United States coal deposits has been added. Selected major accomplishments are as follows.

Developed technology for drilling long horizontal holes (more than 1,000 ft) into a coalbed for degasification.

Demonstrated degasification of Pittsburgh coal through horizontal holes into the coal bed from the bottom of a shaft.

Demonstrated degasification of permeable coal beds through vertical boreholes with hydraulic stimulation in advance of mining.

Developed simple, inexpensive direct method for estimating the methane content of a coal bed from a core sample.

Demonstrated control of methane at the face with water infusion.

Established the dependency of methane emission rate during mining on numerous factors including nature of coal bed and surrounding strata, depth, fracture permeability, reservoir pressure, and coal production rate.

Demonstrated control of methane in gob areas by vertical holes from the surface ahead of pillaring and longwall operations.

Initially, the Bureau of Mines research effort was concerned mainly with identifying fundamental principles regarding methane associated with coal. The basic questions were:

Why are some coal beds "more gassy" than others?

What are the physical factors that control the migration and retention of methane in coalbeds?

To what extent does the geology of the coal bed influence these factors?

The research projects include laboratory studies of sorption and diffusion properties of coal samples from various coalbeds (5)-(10). Such projects helped to explain the methane emissions observed in deep mines (11)-(17). The first comprehensive survey of the amount of gas emitted from United States coal mines was completed in 1972 (18). An updated report followed in 1974 (19).

Measuring Methane Content of Coal Beds

The study of such physical properties of coal as porosity, permeability, diffusion constants relative to methane and water, fine structure, and composition of sorbed gases (20)(21) made it possible to determine directly the gas content of a coal bed (22). Additional refinements have made it possible to quantify the in-place gas content of coal beds (23)(24). The direct determination technique consists of putting a cored coal sample in a sealed canister, and periodically removing the desorbed gas and measuring the quantity by water displacement; the process is continued until the rate of methane desorption declines to 0.5 cm^3 of gas per gram (0.87 in³/oz) of coal in 24 hr. The gas remaining in the coal is estimated either by using an empirical curve for blocky or friable coals or by measuring directly the gas desorbed from the finely pulverized coal core.

Based on considerable data using the direct determination technique, it was concluded that coalbeds may constitute an important potential methane resource (25), and further development of this resource has been urged (26)-(29). It is estimated that the cumulative total of coal bed gas vented from deep coal mines ranges from 200 to 250 MMcf per day or between 73 and 91 Bcf per year (19). This is equivalent, in calorific value (based on the calorific value of methane and coal of 1,000 Btu/ft³ and 25 million Btu/ton, respectively), to more than 3 million tons of coal, or about 1% of the annual production from U.S. underground coal mines. The methane normally is diluted by air as it is vented from coal beds; 200 parts of air per 1 part methane. The total quantity of gas vented to the atmosphere will increase as deeper coal beds are developed (30), unless measures are taken to capture the coal bed methane.

Chemical Composition

The chemical composition of the gas drained directly from coal beds is primarily methane. The quantity of other hydrocarbons found in coal tested to date does not exceed 2% by volume. The few impurities, found in significant quantities, consist of carbon dioxide and nitrogen (15)(20)(21). No hydrogen sulfide or sulfur dioxide has been detected in coal bed gas. Thus, the composition of most coalbed gas is somewhat similar to and compatible with natural gas. Table 4.1 shows the analysis of gas from several selected coal beds.

In addition, the thermal quality of methane associated with coal is practically identical to a high methane-content natural gas. Pure methane has a heat content of 1,012 Btu/ft³ at 60°F and atmospheric pressure. Pipeline grade natural gas normally ranges from 950 to 1,035 Btu/ft³.

Table 4.1: Composition of Coal Bed Gas* Compared with Natural Gas**

| | .Coal Bed Gas, %. . . . . | | | | | Natural |
	Pocahontas No. 3	Pittsburgh	Kittanning	Lower Hartshorne	Mary Lee	Gas, %
CH_4	96.87	90.75	97.32	99.22	96.05	94.40
C_2H_6	1.39	0.29	0.01	0.01	0.01	3.80
C_3H_8	0.0147	–	–	–	–	0.6
C_4H_{10}	0.0008	–	–	–	–	0.3
C_5H_{12}	–	–	–	–	–	0.2
O_2	0.17	0.20	0.24	0.10	0.15	–
N_2	1.7	0.59	2.3	0.6	3.5	0.4
CO_2	0.36	8.25	0.14	0.06	0.10	–
H_2	0.01	–	–	–	–	–
He	0.03	–	–	–	0.27	–
BTU/scf	1,059	973	1,039	1,058	1,024	1,068

*Deul, M. and Kim, A.G., "Methane in Coal: From Liability to Asset," *Mining Congress Journal.*

**Moore, B.J., et al., "Analyses of Natural Gas of the United States," USBM IC 8302, 1966.

Source: DOE/FERC-0010

Quantifying Methane in Coal Beds

The quantity of methane in a specific coal bed, estimated in the laboratory by the direct determination method, indicates that the methane contents of bituminous coal beds range from nearly zero to 600 ft³/ton (23)(29). However, data for all types of coals are limited. Most of the data that are available have been derived only from bituminous coal beds; very few data are available for anthracite and subbituminous coals.

The total volume of methane in a coal bed may be determined by multiplying the estimated quantity of coal in-place by the gas content measured in a core of the specific coal. If a core is not available or cannot be obtained from a coal bed to make a laboratory analysis, the gas content may be estimated by using the relationship established by Kim (31). That relationship, illustrated by Figure 4.1, is based on the fixed carbon and volatile matter content of coal, the depth of the coal, and constants determined experimentally. Specifically, the chart shows the relationship between the depths of coal, the rank of coal, and the methane content for depths to 2,300 ft (700 m).

Table 4.2 provides the conversion factors for calculating the volume of gas contained in coal beds of different thicknesses and gas content. By using Kim's chart and Figure 4.1 with Table 4.2, it is possible to estimate the amount of methane in a coal bed of known rank, thickness, areal extent, and depth.

Based on Averitt's coal resource estimates and the calculated or postulated gas content of coal per ton, it is estimated that there are 300 Tcf of methane gas associated with coals classified as remaining identified sources, a total of 1.5 T tons of minable coal with less than 3,000 ft of overburden, excluding strippable coal beds, and with an average gas content of 200 ft³/ton. In addition, it is estimated that the gas content of coal classified as estimated hypothetical resources in unmapped and unexplored areas, amounting to 1.8 T tons, and 0.4 T tons of coal classified as estimated additional hypothetical resources in deeper structural basins, situated between 3,000 to 6,000 ft below the surface, may be as great as 550 Tcf. This volume is derived by assuming 200 cf of gas per ton of coal for the 1.8 T tons and 480 cf/ton for the 0.4 T tons of coal. In total, therefore, it is estimated that 850 Tcf of methane may be present in the shallow and deep coalbeds of the United States.

Table 4.2: Volume of Gas in Coal Beds*

Gas Content (ft³/ton)	. . . . Coal Volume				
	Acre-foot	1 ft/mi²	3 ft/mi²	6ft/mi²	50 ft/mi²
	. . . . Coal, tons				
	1,800	1.15×10^6	3.45×10^6	6.9×10^6	57.5×10^6
50	90×10^3	58×10^6	173×10^6	345×10^6	2.9×10^9
75	135×10^3	87×10^6	255×10^6	515×10^6	4.3×10^9
100	180×10^3	115×10^6	345×10^6	690×10^6	5.8×10^9
200	360×10^3	230×10^6	690×10^6	1.4×10^9	11.6×10^9
300	540×10^3	345×10^6	1×10^9	2×10^9	—
400	720×10^3	460×10^6	1.4×10^9	2.7×10^9	—
500	900×10^3	575×10^6	1.7×10^9	3.4×10^9	—
600	1.1×10^6	690×10^6	2×10^9	4.1×10^9	—
650	1.2×10^6	748×10^6	2.2×10^9	4.5×10^9	—

*In cubic feet, assuming various gas contents (ft³/ton) and thicknesses of coal (ft/acre and ft/mi²).

Source: DOE/FERC-0010

Figure 4.1: Estimated Methane Content with Depth and Rank

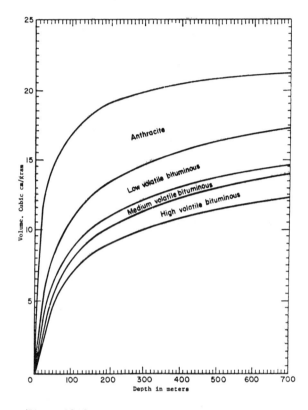

Source: DOE/FERC-0010

The estimated quantity of methane associated with coal is roughly equivalent to the total reserves and undiscovered recoverable natural gas resources in
the United States as estimated by the U.S. Geological Survey (USGS). The
USGS estimates of natural gas reserves and undiscovered recoverable resources
range from 761 to 1,094 Tcf (33).
    In specific regions of the U.S., reports have been published that show the
gas resources in the Pittsburgh coal bed in two counties in southwestern Pennsylvania alone contain 500 Bcf of methane (34)(35). The Mary Lee group of
coal beds in a part of the Warrior Basin in Alabama contain more than 1 Tcf of
methane (36) and the gas content of the Beckley coalbed in six mine properties
in Raleigh County, West Virginia, contain more than 100 Bcf (37).

INDUSTRY CRITERIA FOR COMMERCIALIZATION OF COAL BED GAS

    None of the methane being vented from mines is utilized by coal mining
companies for their own use. For example, the mines owned by Island Creek
Coal Company in Buchanan County, Virginia, and Bethlehem Mines Corporation

in Cambria County, Pennsylvania, vent in excess of 10 MMcf per day. Only one mining company, in cooperation with a Bureau of Mines' mine health and safety demonstration program, is producing and selling commercial quantities of methane (16)(30).

Commercial Criteria

Coal companies and public utilities both have limited experience in the commercialization and the economic evaluation of coal bed gas. Also the criteria for such important considerations as the deliverability rates, the quality of the supply, and the terms of purchase (sale) of such gas have not been established. Any utility decision relative to the commercialization of coal bed gas at this time is dependent upon fragmentary data developed primarily for mine health and safety purposes. The data are not sufficiently detailed or precise to be useful.

It is the view of the utility industry that the limited statistical base, with regard to such data as the composition and heating value of coal bed gas for example, preclude planning of commercial development. In preparing a gas purchase agreement, a utility must consider gas quality in terms of market acceptability, compatibility with conventional supplies, and Federal and State specifications.

Tables 4.1 and 4.3 illustrate the kind of information which is essential when preparing purchase contracts. The above data also show that coal beds locally may contain a large percentage of carbon dioxide, which reduces the Btu value of the gas, as indicated. If a substantial volume of a gas with such impurities is commingled with conventional natural gas of a higher Btu content, the heating value of the resultant product would be reduced and thus would adversely affect customer utilization. A high content of impurities in coalbed gas especially carbon dioxide, also increases the probability of corrosion to gathering and transmission lines, and to distribution equipment. A utility will incur additional costs to remove such impurities.

Table 4.3: Composition of Pittsburgh Coal Bed Gas*

Methane	Ethane	Oxygen	Nitrogen	Carbon Dioxide	Btu/scf
. . . . (%)					(Dry Basis)
87.01	0.14	0.01	0.29	12.55	885
88.49	0.07	0.02	0.37	11.04	899
88.36	0.07	0.02	0.37	11.17	897
86.88	0.08	0.34	1.35	11.35	882
87.78	0.05	0.02	0.21	11.94	891
Average 87.70	0.08	0.08	0.51	11.61	891

*Letter, Consolidated Gas Supply Corporation to FPC Supply Technical Advisory Task Force, January 12, 1976.

Source: DOE/FERC-0010

The moisture content of coal bed gas also is of concern to utilities. The moisture content must not exceed 7 lb of water per 1 MMcf of gas. Coalbed gas analyses indicate that as much as 400 lb of water per 1MMcf may be typical. The moisture content, therefore, must be reduced, thus further adding to the cost of producing methane from coal.

Deliverability: The deliverability criteria for coalbed gas are important, not only to the utility and to the ultimate purchaser, but to the coal company as well. A coal company's principal interest is to vent the methane as quickly as possible to maintain mine safety. Consequently, a coal company cannot be restricted by the deliverability criteria of the utility company. A coal company operating in a gassy mine, therefore, will be compelled to reserve the right to vent coal bed gas to the atmosphere, at a rate it deems essential regardless of the utility company's ability to accept such rates into its pipeline system. Utilities on the other hand also must consider the long term availability of such supplies.

Agreements between coal operators and gas purchasers will have to be specifically written to take into account unusual environmental conditions, unusual operating situations, and unknowns relative to extraction technology.

Research efforts are being made by various Federal and State agencies to utilize coal bed gas for power generation purposes, as an internal source of electrical power for a mine complex. With proper design of gas-burning equipment, coalbed gas could be used for these purposes.

Utilization Experience

Natural gas companies operating in the Appalachian area have long noted the presence of gas in coal-bearing rock formations. A case in point is the Big Run gas field, an anticlinal structure, located in Wetzel County, West Virginia. The discovery well of the Big Run field was drilled in 1905. In 1931, an application was made to the West Virginia Department of Mines to abandon the well. When the casing was cut at the depths of 1,078 to 1,080 ft and the well was plugged back to a depth of 1,118 ft, methane began issuing from the Pittsburgh coal bed. The well was tested and found to have an open flow of 28 Mcf per day with a bottom hole pressure of 92 lb/in^2 in 16 hr. Consequently, the well was reinstated as a gas well in January 1932, and remained productive until 1968. During this period the well produced in excess of 212 MMcf of coal bed gas.

Prior to the abandonment of the well, a total of 23 wells were drilled and completed as gas producing wells from the Pittsburgh coal bed. The cumulative production through 1974 was 1.7 Bcf. The wells are productive because of the natural permeability caused by the tension cracks on the crest and lee flank of the anticline. At least 783 wells have been drilled through the Pittsburgh coal bed. Drilling densities into other coalbeds have been greater, but fewer productive wells have been completed in the other coal beds. This may be due to the lack of permeability in such coal beds.

The first major effort to produce methane from coal beds as part of a degasification program was at the Federal No. 2 Mine in Monongalia County, West Virginia, where a 839 ft deep shaft was enlarged to permit seven holes to be drilled horizontally into the coal bed. Since September 1972 after 1,438 days of operation, 834 MMcf of gas has been produced, of which 427 MMcf has been commingled in a natural gas distribution pipeline and marketed. Another shaft, at the same mine, was made available for additional experimental drilling tests. Five holes were drilled and after 1,073 days, some 784 MMcf of methane was produced of which 121 MMcf was delivered to a distribution pipeline.

In less than 4 years, more than 1.6 Bcf of gas was produced from the West Virginia mine with no appreciable decline in the flow rate.

TECHNOLOGY FOR RECOVERING METHANE FROM COAL BEDS

Methane can be recovered from coalbeds by draining either in advance of mining or during mining. Methods for removing the methane ahead of mining employ vertical boreholes drilled from the surface, horizontal boreholes drilled from shaft bottoms, and directional slant holes drilled from the surface. In active mines methane can be removed from the coalbed through horizontal holes drilled in appropriate areas of the mine.

Transportation of methane produced from coal beds to United States markets should be a minimal concern. Most major coal basins are either near major population centers or are presently crossed by or near existing natural gas pipelines. The existing gas transportation systems for the most part could be utilized for marketing methane from coal.

Vertical Surface Boreholes: One technique of recovering methane from coal several years ahead of mining actively consists of drilling small diameter (less than 9 inches) vertical boreholes from the surface to the coal bed. The hole is cased and cemented, and existing fractures in the coal are widened and extended by the application of hydraulic pressure and controlled injection of gelled water. Sand grains in the treatment fluid serve as a propping agent to keep the fractures open when pressure applied from the surface is released. The hydraulic stimulation increases the permeability of the coal, and enlarges the gas drainage area. When stimulation is applied to minable coal beds, the effect of this treatment on the future minability of the coal must be considered.

The production rate from the vertical holes depends on the effectiveness of the stimulation treatment. A production rate of 50 to 100 Mcf per day per well appears feasible.

Horizontal Boreholes from Shaft Bottoms: Another method of producing gas from coal beds is by drilling horizontal holes from the bottom of shafts sunk to the coal beds. The rate of gas flow per linear foot of 3 inch diameter horizontal hole drilled in permeable coal beds ranges from 100 to 450 ft^3 per day. In one demonstration seven horizontal holes were drilled from the bottom of a small-diameter shaft. The aggregated length of the horizontal holes was 4,325 ft. After the coal bed was sufficiently dewatered, gas flowed from the horizontal holes at an initially high rate and continued with only a modest decline over a period of almost four years. In another demonstration of this technique, five horizontal holes, with an aggregate length of 5,830 ft, were drilled from the bottom of an 18 ft ventilation shaft.

The initially high average gas flow declined only slightly over a three year period. The production history of these two field demonstrations in Monongalia, West Virginia, was discussed above under Utilization Experience.

The gas produced in this way is manifolded to a pipe that carries it to the surface. The possible rate of production from each such shaft is estimated conservatively at 1 MMcf of gas per day.

Directional Slant Holes from the Surface: The directional drilling technique for removing methane from coal combines features of horizontal and vertical degasification boreholes. A small diameter borehole is drilled from the surface and intentionally deflected to penetrate the coal bed and continue into the coal parallel to the bedding plane, thereby maximizing the penetration of horizontal fractures. The Bureau of Mines has conducted two preliminary tests with sufficient success to warrant further testing of this method of recovering methane from coal.

Horizontal Boreholes in Active Mines: Methane can be produced from the active working sections of a developing mine that has not been advanced to the property limits. The process consists of drilling small diameter horizontal boreholes into the virgin coal as the working face is advanced. Coal bed gas flowing from these holes can be conducted to the surface through a suitable piping system in the mine.

IDENTIFICATION OF PROBLEMS

Methane Capture: A major problem in commercializing methane produced from active mines is transporting the gas through pipes in underground coal mines. A recent study (38) contracted by the Bureau of Mines proposed a design for a safe piping system for transporting drained gas through a mine.

Methane produced through wells completed in coal beds may be handled conventionally. However, the two-phase flow of methane and the water, associated with the coal (39), must be considered in production planning.

Depending on the ultimate market supplied, the compression of methane for transmission may be essential. The gas pressures in coalbeds seldom exceed hydrostatic pressures. Pressures in relatively shallow beds where there is some producing experience have never exceeded 670 psig at atmospheric pressure. Such pressures, however, drop rapidly as production begins; the gas is held by sorption rather than by compression (16).

Sampling of Coal Beds: Historically, the importance of coal as a fuel resource necessitated detailed quantification and delineation of its geographic location and depth. Little interest existed for quantifying the amount of methane associated with such coal resources. Subsequent efforts to quantify the amount of methane in coal have been related to coal mine health, safety, and productivity concerns. However, with the declining availability and the increasing cost of domestic conventional natural gas, the identification and quantification of methane in coal may take on added importance.

Methods for measuring methane associated with relatively shallow coal beds and the technology for extracting such gas for commercial use generally are understood, but little is known of the methane in the thick coal beds at the depths of more than 3,000 ft. Specifically, there is no experience, and there are no active programs directed towards developing technology for methane production from deep coal beds.

Before it is possible to assess reasonably the resource potential of methane in deep coals, core samples must be obtained and the areal limits must be determined. The most likely source for recovering such cores is from wells drilled for oil and natural gas. However, the coring of coal sections, particularly deep coal, is not the common practice when drilling for oil and gas.

For the most part, oil and gas drilling programs have not provided detailed data regarding the deep coal beds that may have been penetrated, although coal beds are used as marker-beds. It is likely that the depths and thicknesses of important deposits are identified on the mechanical and electric logs maintained by petroleum companies. It is possible that, as the value of methane from coal increases, further drilling programs by the petroleum industry will consider the coring and testing of coal beds an essential part of their exploration program.

Required Technology Improvements: Commercial production of methane from coal has not been realized in part, because the techniques and the technology demonstrated in research efforts have not been applied on a commercial scale.

Continuous In-Hole Surveying — No satisfactory method for continuous surveying of nonvertical boreholes during drilling has been developed. There is a need for a system that would provide near continuous drill orientation while drilling is in progress. This would eliminate the need to interrupt drilling to make measurements for determining borehole azimuth and inclination.

Drill Guidance — A companion technology to the continuous surveying system would be an improved drill bit guidance system to operate effectively in the hard sediments associated with coal. The special tools developed for off-shore directional drilling have not been totally compatible with the objectives of coalbed gas production.

Increasing Coal Bed Permeability — The low permeability of coal beds is the main impediment to producing methane at commercial rates-of-flow. Coal beds are anisotropic, that is they have directional permeability. Coal beds can be drained most efficiently if horizontal boreholes intersect the face cleat perpendicularly. Cleat is the natural occurring vertical fracture in coal. The face cleat cuts across bedding surfaces and extends for many feet. The butt cleat is a short poorly developed fracture commonly truncated by the face cleat. Blocky coals, generally found at depths of 800 to 1,200 ft, have prominent cleats. The friable, medium- to low-volatile bituminous coals generally have complex cleat systems, thus directional permeability is difficult to measure, and these coal beds require unique stimulation methods to increase their permeability.

Economic Uncertainty: There is no typical situation upon which to base the economics of methane production from coal beds. It is apparent, however, that certain unique factors can contribute to making the utilization of methane from coal feasible, such as proximity to existing pipelines, shallow depth, effective coal bed permeability, favorable geologic structural conditions which minimize water production and disposal, and a mutual commercial interest between the producers of coal and methane.

Another consideration which bears indirectly on the economics of methane extraction from coal beds is the benefit derived by reducing coal mine hazards caused by methane-air mixtures.

In view of the limited experience in utilizing methane and the experimental nature of methane production, the cost of developing such supplies and the ultimate cost to the consumer cannot be established with certainty at this time.

Institutional Constraints: The institutional constraints to developing coal bed methane may be identified as legal and proprietary. The legal considerations are discussed below. The proprietary issue, as used herein, refers to the conflict that may arise between the objectives of a coal operator to mine coal and retain, as confidential, information concerning his operation and coal ownership, with the objectives of a gas developer. The gas developer's interests very likely depend on information concerning coal reserves, as this permits quantification of methane resources. In addition, both the coal operator and the gas developer have an interest in the application of technology to extract methane; the technology must be effective, but should not diminish or impede the opportunity to mine coal.

LEGAL CONCERNS

Status of Developer

If coal bed gas utilization is both technically feasible and desirable from an energy conservation and the all important mine safety point-of-view, both the State and Federal Governments must become active in quantifying and defining the potentially large methane resource. Cooperation will be required between State and Federal agencies in order to recommend appropriate new legislation which prescribes the rights of all interests.

Eventually the legal question as to methane ownership must be resolved by State courts. Until then methane utilization probably will be practical only in instances where one company owns both the coal and the mineral rights or where a coal company and a gas company can cooperate in a methane production program.

At such time as it becomes economic to produce methane from active mines, such Federal agencies as MESA (Mining Enforcement and Safety Administration of the U.S. Department of the Interior), and the Department of Transportation, and State agencies governing mine safety and pipeline construction may require developers to apply for permits and meet construction standards.

In the event methane is produced through vertical boreholes drilled several years in advance of mining, the operating company may be classified and subject to a separate set of regulations applicable to gas producers. For example, the State of West Virginia does not require permits for air shafts and gob ventilation holes (vertical holes drilled from the surface into caved and abandoned mined areas). However, permits are required for any shaft or hole dug or drilled to produce gas. Thus, a coal company wishing to degasify its mine and sell the methane would be required to meet all of the obligations currently imposed on gas producers concerning drilling, completing, producing, and plugging of wells, obtaining permits for other operations, and posting performance bonds.

A developer also will be affected by State and Federal laws which require a support pillar to be left around a well that penetrates a minable coal bed. In Pennsylvania the pillar requirement is 200 ft in diameter. In West Virginia permission must be obtained before mining closer than 200 ft. Thus, clarification and/or modification of existing laws governing natural gas drilling and production may be necessary to accommodate methane extraction and not unduly limit the extraction of coals.

In essence, total cooperation is essential between natural gas and coal companies involved in this common interest. They, with State and Federal regulatory agencies, must provide appropriate rules and regulations that will enhance methane utilization while preserving mine safety and coal productivity.

It may be necessary for the Federal Power Commission, insofar as its jurisdiction is concerned, to regulate coal bed gas production operations, particularly in granting pipeline permits, determining cost of service, and the abandonment requirements for pipelines and wells. A determination must be made whether separate rules other than those currently in effect for conventional natural gas are appropriate.

Status of Resource

The legal issue of ownership of methane in coal beds could develop into a major impediment to its commercial development. Generally, mineral rights, including the right to explore for, develop and produce natural gas, may be contracted for separately from surface rights. A land owner may by deed or lease transfer acreage, and the intent of the parties governs what is conveyed. Thus, each deed and lease having a bearing on the right to the minerals must be examined separately before answering the question: Who has the right to the methane underlying a given acreage?

Usually, coal leases are held separate from oil and natural gas leases, when the land is owned in fee simple, or when the land is owned outright by a governmental body. Over the years, the courts have been called upon to construe deeds and leases relating to minerals. In some places it will be found that certain general principles of interpretation have evolved. Under Pennsylvania law, according to the "Dunham" rule, a deed conveying all coal and other minerals of every kind and character under a described tract of land grant does not include natural gas subsequently discovered on the land [New York State Natural Gas Corp. vs Swan-Finch Gas Development Corp., 278 F. 2d 577 (3d Cir. 1960)]. But whether the same courts would treat methane associated with coal beds in the same way is open to speculation. In short, one cannot expect that every court in each state where coal deposits are to be found would resolve in the same way a dispute as to whether a particular deed or lease conveyed the right to develop and produce methane.

There are other considerations which have contributed to the present lack of clarity regarding rights to methane. The presence of methane in coal beds creates a hazard for coal mining. Methane is something to be evacuated from the mine to avoid explosions, damages, injuries, and/or death. Methane is hardly considered an asset in these circumstances. Given the fact that coal has been mined in the United States for over one hundred and fifty years, one can expect to find many deeds and leases which refer to rights to coal without making any particular reference to methane found in conjunction with it.

This ambiguity about rights to methane was raised before the legislature of West Virginia in a recent session, but was not resolved. In Pennsylvania a bill was proposed to have gas in coal beds not specifically covered by existing natural gas leases revert to the Commonwealth. This proposal has not yet been introduced to the legislature. The general consensus is that only after the methane in coal becomes more valuable can any reasonable settlement on the issue be expected.

Some coal companies claim ownership of coal bed gas if they own the coal. The opinion is at odds with the petroleum industry's position and several legal opinions which state that the coal bed gas belongs to the holder of the natural gas rights. An oil and natural gas lease gives one the right to drill and produce oil and natural gas from any formation, while the ownership of the reservoir remains with the owner of the land surface. Ownership of the natural gas is not perfected until the gas is under control, i.e., in the casing at the wellhead.

Mine operators must remove the methane from the mine in order to meet stringent safety laws, and consistent with that purpose, must, therefore, control the coal bed gas to that extent. It has been ruled by the Pennsylvania Supreme Court in Chartiers Block Coal Company vs Mellon, 152 Pa. 286, 296 (1893) that this control does not give title to the coal grantee. The court stated that the

> . . . grantee of coal owns the coal but nothing else, save the right of access to it and the right to take it away. This is not to say, however, that the coal mine operator may not expel methane gas into the atmosphere. To deprive him of this right would, in effect, be depriving him of his access to the coal, since coal cannot be mined without expelling the methane from the mine shaft. Thus, the right to mine for coal necessarily includes the right to perform those actions necessary to insure the safety of such mining. Since the coal owner or grantee only retains the right to extract coal, however, the right to access to, and economic control of, the methane gas belongs to the owner or grantee of the gas rights.

The above was quoted from Official Opinion No. 53 of the Office of the Attorney General, Commonwealth of Pennsylvania, October 31, 1974, by Israel Packel, Attorney General, and Theodore A. Adler, Deputy Attorney General, who concluded that "only those persons who own or have obtained the right to extract gas have the right to assert title thereto."

Beyond the question of ownership, another legal question involves the jurisdictional responsibility of the Federal Power Commission. Conventional wisdom would suggest that the FPC has jurisdiction over methane when it is transported and sold for resale in interstate commerce because it is likely to be considered a natural gas within the meaning of the Natural Gas Act. See Deep South Oil Co. of Texas vs FPC, 247 F. 2d 882, 888 (5th Cir. 1957).

ENVIRONMENTAL CONSIDERATIONS

There is little recorded experience with commercial gas production from coal beds, except that discussed by Tilton (40), to provide a meaningful understanding of the environmental aspect of such production. Some of the problems encountered in drilling to coal beds from the surface to produce gas are common to all drilling operations.

Water in Coal Beds: The composition of coal bed waters varies widely, from slightly acid to slightly alkaline, and from potable to saline. The composition of water from three separate coal beds is shown in Table 4.4. The planning of pollution control and water disposal systems must be determined on the basis of the water at each site. Water from some coal beds, especially in the Western part of the United States, may be of a higher quality than the alkaline surface waters.

Drilling Effects on Surface: The environmental impact of drilling to recover gas from coal beds should be minimal. The sludge and mud pits are usually small in size (about 25 yd^3), and the quantity of drillpipe casing and tubing is relatively minor; thus problems, not already resolved by the petroleum industry are not expected.

Except for the laying of gathering lines and site preparation, surface damage should be minimal. Abandonment can be accomplished by plugging the holes according to State or Federal regulations. In some instances, boreholes drilled from the surface may be used by the coal mining companies as vent holes, for powerline entry, or rock dust supply holes, etc.

Favorable Impact on Mining: Improved mine safety and increased productivity make drainage of gas from coal beds worthwhile and essential. One of the long range benefits of methane drainage is that deep coal beds, below 3,000 ft, which might otherwise not be mined for many years, may well prove economic for mining if the hazards of excessive methane emissions are eliminated.

Table 4.4: Composition of Water from Three Coal Beds

| | .Coal Bed. . . . . | | | | |
	. . . . .Pittsburgh			Mary Lee	Pocahontas No. 3
pH	7.45	7.65	8.15	8.05	8.35	6.75
Acidity, ppm	0	0	0	0	0	110
Alkalinity, ppm	1,825	790	2,043	876	355	0
Dissolved solids, ppm	4,478	9,774	17,246	3,108	1,428	156,440
SO$_4$, ppm	63	133	ND	ND	ND	2
Ca, ppm	159	477	127	162	12.5	2.95%
Mg, ppm	132	193	482	29	8	0.67%
Fe, ppm	0.5	ND	ND	0.13	ND	1
Chlorides, ppm	2,356	7,700	13,600	2,200	700	13.97%

Note: ND is not detected.

Source: DOE/FERC-0010

REFERENCES

(1) Chamberlin, R.T., *Explosive Mine Gases and Dusts*, Bureau of Mines Bulletin 26 (1909).

(2) Darton, N.H., *Occurrence of Explosive Gases in Coal Mines*, Bureau of Mines Bulletin 72 (1915).

(3) Venter, J., and Stassen, P., *Drainage and Utilization of Firedamp*, Bureau of Mines Information Circular 7670 (1953).

(4) Deul, M., "Methane Drainage from Coalbeds: A Program of Applied Research," *Proc. 60th Meeting, Rocky Mountain Coal Mining Institute, Boulder, CO, June 30–July 1, 1964.*

(5) Perkins, J.H. and Cervik, J., *Sorption Investigations of Methane on Coal*, Bureau of Mines Technical Progress Report 14 (1969).

(6) Kissell, F.N., *Methane Migration Characteristics of the Pocahontas No. 3 Coalbed*, Bureau of Mines Report of Investigation 7649 (1972).

(7) Kissell, F.N., *The Methane Migration and Storage Characteristics of the Pittsburgh, Pocahontas No. 3, and Oklahoma Hartshorne Coalbeds*, Bureau of Mines Report of Investigation 7667 (1972).

(8) Kissell, F.N. and Bielicki, R.J., *An In-Situ Diffusion Parameter for the Pittsburgh and Pocahontas No. 3 Coalbeds*, Bureau of Mines Report of Investigation 7668 (1972).

(9) Bielicki, R.J., Perkins, J.H. and Kissell, F.N., *Methane Diffusion Parameters for Sized Coal Particles. A Measuring Apparatus and Some Preliminary Results*, Bureau of Mines Report of Investigation 7697 (1972).

(10) Thimons, E.D. and Kissell, F.N., "Diffusion of Methane Through Coal," *Fuel*, v 52, (October 1973).

(11) Cervik, J, *Behavior of Coal-Gas Reservoirs*, Bureau of Mines Technical Progress Report 10 (1969).

(12) Hadden, J.D. and Sainato, A, *Gas Migration Characteristics of Coalbeds*, Bureau of Mines Technical Progress Report 12 (1969).

(13) Krickovic, S. and Findlay, C., *Methane Emission Rate Studies in a Central Pennsylvania Mine*, Bureau of Mines Report of Investigation 7591 (1971).

(14) Findlay, C.S., Krickovic, S. and Carpetta, J.E., *Methane Control by Isolation of a Major Coal Panel–Pittsburgh Coalbed*, Bureau of Mines Report of Investigation 7790 (1973).

(15) Fields, H.H., Krickovic, S., Sainato, A. and Zabetakis, M.G., *Degasification of Virgin Pittsburgh Coalbed Through a Large Borehole,* Bureau of Mines Report of Investigation 7800 (1973)

(16) Fields, H.H., Perry, J.H., and Deul, M., *Commercial-Quality Gas from a Multipurpose Borehole Located in the Pittsburgh Coalbed*, Bureau of Mines Report of Investigation 8025 (1975).

(17) Jeran, P.W., Lawhead, D.H. and Irani, M.C., *Methane Emissions from an Advancing Coal Mine Section in the Pittsburgh Coalbed*, Bureau of Mines Report of Investigation 8132 (1976).

(18) Irani, M.C., Thimons, E.D., Bobick, T.G., Deul, M., and Zabetakis, M.G., *Methane Emission from U.S. Coal Mines, A Survey*, Bureau of Mines Information Circular 8558 (1972).

(19) Irani, M.C., Jeran, P.W. and Deul, M., *Methane Emission from U.S. Coal Mines in 1973, A Survey. A Supplement to Information Circular 8558*, Bureau of Mines Information Circular 8659 (1974).

(20) Kim, A.G. and Douglas, L.J., *Gases Desorbed from Five Coals of Low Gas Content*, Bureau of Mines Report of Investigation 7768 (1973).

(21) Kim, A.G., *The Composition of Coalbed Gas,* Bureau of Mines Report of Investigation 7762 (1973).

(22) Kissell, F.N., McCulloch, C.M. and Elder, C.H., *The Direct Method of Determining Methane Content of Coalbeds for Ventilation Design*, Bureau of Mines Report of Investigation 7767 (1973).

(23) McCulloch, C.M., Levine, J.R., Kissell, F.N. and Deul, M., *Measuring the Methane Content of Bituminous Coalbeds*, Bureau of Mines Report of Investigation 8043 (1975).

(24) McCulloch, C.M. and Diamond, W.P., "Inexpensive Method Helps Predict Methane Content of Coal Beds," *Coal Age*, v 81, No. 6 (June 1976).

(25) Deul, M., Fields, H.H. and Elder, C.H. "Degasification of Coalbeds: A Commercial Source of Pipeline Gas," presented at Illinois Institute of Technology Symposium, Clean Fuels from Coal, Institute of Gas Technology Chicago, IL, Sept. 10-14, 1973, *American Gas Association (AGA) Monthly*, v 56, No. 1 (January 1974).

(26) Deul, M. and Kim, A.G., "Coal Beds: A Source of Natural Gas," *Oil and Gas Journal*, v 73, No. 24 (June 16, 1975).

(27) Deul, M., "Recover Coalbed Gas," *Hydrocarbon Processing* (July 1975).

(28) Deul, M., "Gas Production from Coalbeds—Accomplishments and Prospects," *Proceedings of Transmission Conference, Operating Section, AGA, Bal Harbour, FL May 19-21, 1975.*

(29) Deul, M. and Kim, A.G., "Degasification of Coalbeds—A Commercial Source of Pipeline Gas," *Proc. Symp. on Clean Fuels from Coal, II, Institute of Gas Technology, Chicago, IL, June 22-27,* (1975); *AGA Monthly*, v 58 No. 5 (May 1976).

(30) Deul, M., Cervik, J., Fields, H.H. and Elder, C.H. "Methane Control in Mines by Coalbed Degasification," *Preprints, International Conference, Coal Mine Safety Research, Washington, D.C., Sept. 22-26, 1975.*

(31) Kim, A.G., *Extrapolating Laboratory Data to Estimate the Methane Control of Coalbeds,* Bureau of Mines Report of Investigation (1976).

(32) Averitt, P., *Coal Resources of the United States, January 1, 1974,* U.S. Geological Survey Bulletin 1412 (1975).

(33) Miller, B., Thomsen, H.L., et al, *Geological Estimates of Undiscovered Recoverable Oil and Gas Resources in the United States,* U.S. Geological Survey Circular 725 (1975).

(34) Kim, A.G., *Methane in the Pittsburgh Coalbed, Washington County, PA,* Bureau of Mines Report of Investigation 7969 (1974).

(35) Kim, A.G., *Methane in the Pittsburgh Coalbed, Greene County, PA,* Bureau of Mines Report of Investigation 8026 (1975).

(36) Diamond, W.P., Murrie, G.W. and McCulloch, C.M., *Methane Gas Content of the Mary Lee Group of Coalbeds, Jefferson, Tuscaloosa, and Walker Counties, AL,* Bureau of Mines Report of Investigation 8117 (1976).

(37) Popp, J.T. and McCulloch, C.M., *Geological Factors Affecting Methane in the Beckley Coalbed,* Bureau of Mines Report of Investigation 8137 (1976).

(38) Tongue, D.W., Schuster, D.D., Niedbala, R. and Bondurant, D.M., *Design and Recommended Specifications for a Safe Methane Gas Piping System,* Bureau of Mines Open File Report, 109-76, (1976), 90 pp; available for consultation at the Bureau of Mines libraries in Pittsburgh, PA, Denver, CO, Twin Cities, MN, Spokane, WA; Library of Natural Resources, U.S. Department of the Interior, Washington, DC; and from National Technical Information Service, Springfield, VA.

(39) Kissell, F.N. and Edwards, J.C., *Two-Phase Flow in Coalbeds,* Bureau of Mines Report of Investigation 8066, (1975).

(40) Tilton, J.G., "Gas from Coal Deposits," *Natural Gas from Unconventional Geologic Sources,* National Academy of Science-National Research Council (1976).

AN EVALUATION OF POTENTIAL METHANE RECOVERY FROM MINABLE AND UNMINABLE COAL

The information in the rest of this chapter is based on *Enhanced Recovery of Unconventional Gas: The Program—Volume II (of 3 Volumes),* October 1978, prepared by V.O. Kuuskraa and J.P. Brashear of Lewin and Associates, Inc., T.M. Doscher of University of Southern California and L.E. Elkins for the U.S. Department of Energy under DOE Contract No. EF-77-C-01-2705.

Background

Since the inception of underground coal mining, the release of methane from coal beds (coal gas) has posed a hazard to mining safety. When methane combines with air, it forms a flammable mixture and is the cause of countless mine explosions.

In response to this hazard, the Federal government has undertaken numerous efforts to install improved safety measures and regulations. While considerable information has now been gathered on methane emissions, most of this has been gathered from a perspective of safety, disposing of the unwanted methane in mines, rather than from a perspective of supply, capturing the methane for increasing domestic gas supplies.

Recovering this now vented methane could provide an important augmentation to local supplies of natural gas. Moreover, these supplies could be made immediately available since the technology is relatively simple and is commonly used in several European countries, notably Great Britain and Belgium.

In addition to methane recovery in association with mining, additional potential sources of methane are in the deep, currently unminable coal seams of the West. Such coal seams are currently considered too thin or too deep for mining, and may contain considerable methane resources that could be economically produced.

Recent efforts by the Bureau of Mines, Pittsburgh Mining and Safety Research Center, under the direction of Maurice Deul, have initiated research and demonstration efforts toward the dual objectives of increasing mining safety and adding to domestic gas supply through methane recovery from coal seams. These sections draw on the considerable knowledge base that has resulted from these past R&D efforts.

Nature of the Problem

The methane released from coal mining stems from three sources: (1) from the coal seam itself where the methane is held by adsorption in the structure of the coal and is released when the coal is mined; (2) from the thin sand lenses adjacent to the coal seams that also serve as a reservoir for desorbed gas; and (3) from fractures where methane has accumulated by desorption.

Since coal is impermeable, the gas must flow either through the natural fracture system in the coal (the butt and face cleats) or must flow through the microporous structure of the coal. Unless an area is naturally highly fractured, such as in the Big Run Field of West Virginia, unstimulated vertical holes drilled into the coal will not release appreciable amounts of methane.

Although several approaches have been tried and have produced gas, none as of yet, have demonstrated economic feasibility as purely a gas recovery project. Each approach must rely upon the safety benefits and mining production efficiencies that accrue from lowered emissions (particularly instantaneous gas bursts). The production rates and duration of production have, to date, been too low or too uncertain to offset the considerable costs of well drilling, water removal, compression, piping, stimulation, gas purification, and gathering costs associated with commercial recovery of methane from coal.

In addition to methane recovery in currently mined coal seams, a considerable resource could exist in formations too thin or too deep for economic mining, at least at present. Should these unminable coal beds have geological features that are favorable to efficient recovery, e.g., high intensity of natural fractures, relatively thick pays, high gas content, and uniform beds and seams, they could represent a commercial source of natural gas supply.

At this time, however, rapid exploitation is hampered by an inadequate definition of the resource, uncertainties in recovery technologies for the deep coal, and the marginal economics of collecting and marketing the resource. These limitations are further described in the following sections. Thus, evaluation of potential commercial production of methane from coal should consider both:

- Methane recovery from minable coal where the major drilling costs are allocated to improved safety and productivity so that the proceeds from the sale of the methane may be used to pay for the incremental costs of purifying, compressing, and gathering the gas from the wellbores.

- Methane production from deep, currently unminable coal seams having geological highly favorable characteristics.

METHANE FROM MINABLE COAL SEAMS

Resource Base

Basic studies of methane emission from U.S. coal mines (1) prepared by the Bureau of Mines, Department of Interior, supplemented by contacts with major coal mining companies, served as the data source for methane resources associated with mining. The recovery estimates for this now vented methane were based on a series of investigative reports by the Bureau of Mines (Reports of Investigation 8195, 7703, 8047, 8174, 7968 and 8195, among others) and from personal correspondence and discussion with major coal production companies. Companies contacted as part of this study included: Bethlehem Mining Corp., Island Creek Coal Co., Lydes Resources, and United Gas Pipeline. This information showed that in 1975:

- About 200 bituminous coal mines in the U.S. emitted at least 100 Mcfd of methane each.

- Total emissions from these mines in 1975 were 216 MMcfd, or about 80 Bcf per year, and have remained stable at this level since 1971 (Figure 4.2).

- Methane emissions are concentrated in a limited number of mines; the 60 mines with emissions of 1 MMcfd or more accounted for 79% of the total methane emissions; the 20 mines with the largest daily emission rates accounted for 48%, as shown in Table 4.5.

- The Appalachian region accounted for 86% of the total emissions. Figure 4.3 provides the location of the high methane emission mines.

- Mines in the Pittsburgh coal bed in southwestern Pennsylvania, eastern Ohio, and northern West Virginia accounted for almost one-half of total emissions; the second highest source, mines in the Pocahontas No. 3 coal bed in southern West Virginia and northwestern Virginia accounted for about 13% of the total.

These estimates indicate that research into methane recovery in association with mining should focus, at least initially, on the Appalachian area.

Figure 4.2: Methane Emissions from U.S. Bituminous Coal Mines (1)

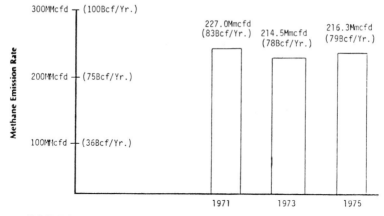

Source: DOE EF-77-C-01-2705

Table 4.5: Methane Emissions from U.S. Coal Mines—20 Largest
Methane Emission Coal Mines (1)

Mine Name	Location	Coalbed	Methane Emission (MMcfd)
Loveridge	Marion Co., WV	Pittsburgh	11.6
Humphrey No. 7	Monongalia Co., WV	Pittsburgh	9.3
Federal No. 2	Monongalia Co., WV	Pittsburgh	8.1
Blacksville No. 2	Monongalia Co., WV	Pittsburgh	6.3
Osage No. 3	Monongalia Co., WV	Pittsburgh	5.9
Beatrice	Buchanan Co., VA	Pocahontas No. 3	5.6
Concord No. 1	Jefferson Co., AL	Pratt	5.1
Olga	McDowell Co., WV	Pocahontas No. 4	5.0
Robena	Green Co., PA	Pittsburgh	4.9
Blacksville No. 1	Monongalia Co., WV	Pittsburgh	4.5
Bethlehem No. 32	Cambria Co., PA	Kittanning	4.5
Robinson Run No. 95	Harrison Co., WV	Pittsburgh	4.3
Arkwright	Monongalia Co., WV	Pittsburgh	4.0
Federal No. 1	Marion Co., WV	Pittsburgh	4.0
Virginia Pocahontas No. 1	Buchanan Co., VA	Pocahontas No. 3	3.9
Cambria Slope No. 33	Cambria Co., PA	Kittanning	3.9
Virginia Pocahontas No. 2	Buchanan Co., VA	Pocahontas No. 3	3.4
Virginia Pocahontas No. 3	Buchanan Co., VA	Pocahontas No. 3	3.3
L.S. Wood	Pitkin Co., CO	Basin B	3.3
Gateway	Green Co., PA	Pittsburgh	2.7
		Total Top 20	103.5
	Percentage of total coal mine methane emissions in 1975 (U.S.)		47.8

Source: DOE EF-77-C-01-2705

Figure 4.3: Location Map of Counties with Bituminous Coal Mines Emitting at Least 100,000 cfd of Methane in 1975 (1)

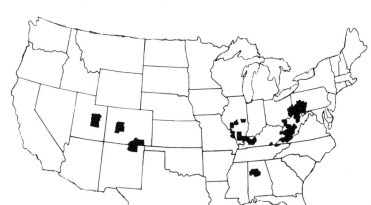

Source: DOE EF-77-C-01-2705

Applicable Recovery and Conversion Technology

Capturing the methane now produced in association with mining requires an efficient recovery and conversion technology. In turn, efficient recovery requires a highly permeable interconnection through the plane of the major fracture system, the face cleats. Several approaches have been undertaken to obtain this interconnection, including:

- Drilling wells into the beds above the coal seams, mining the coal with a longwall machine, and having the structure collapse to create a connected fracture system linked to the wellbore.
- Drilling deviated wells from the surface into the coal seam to intersect the planes of the face cleats.
- Using hydraulic fracturing to intersect the natural, face and butt cleat, fracture system.
- Drilling horizontal holes at 90 degree angles to the face cleats (horizontal holes drilled parallel to the face cleats generally have much lower production).

The first three of these entail drilling from the surface of the earth, whereas horizontal drilling takes place from the mine shaft. All four methods have produced gas, but none, as yet, have demonstrated economic feasibility as a stand alone project. Economically, all depend on the benefits of more rapid and efficient mining and mine safety for their justification. The technological problems and limitations of each of these recovery approaches are described below.

Degasification by Vertical Drilling: *Using Vertical Wells With Longwall Mining* — One means for degasifying mines involves combining longwall mining with vertical wells, drilled above coal seams.

The technique is essentially designed to draw off the methane accumulating in mined-out areas (gob gas). Data based on 10 years of experience in the Lower Kittanning coalbed in central Pennsylvania indicate that:

- This approach is particularly effective when a considerable portion of the methane is trapped in the rock strata above the coal seam (as well as within the micropores and fractures of the coalbed).

- Little, if any, flow can be expected until mining passes beneath the well, allowing settling of the mine ceiling with corresponding fracture of the gas-bearing overburden.

- Exhaust fans or vacuum pumps are required to extract the methane.

In terms of methane flow, the initial rate was above 1,000 Mcfd and decreased steadily after its initial peak. The results of field investigations of this technique by the Bureau of Mines (2)(3) showed that appreciable quantities of methane and methane/air mixtures can be produced. The data from the field test showed:

- Methane flow from one borehole started at 1,400 Mcfd, declined to 300 Mcfd in 5 months.

- Production from a second borehole started at 1,000 Mcfd, decreased to about 200 Mcfd in 5 months, and slowly declined to about 100 Mcfd in 2 years.

- The two boreholes together vented about 150 MMcf of methane while in service.

However, for the methane to be economically convertible for traditional use, it needs to have a concentration of at least 50% of the total gas produced. The produced methane/air mixtures may have a sufficient concentration during the first 3 to 6 months of life of the well to be economic.

The past field tests show that about one-third to two-thirds of the gas produced will have methane concentrations of 50% or greater. In the Bureau of Mines test, reported above:

- The first well started with a 90% methane concentration and declined to 50% in three months.

- The second well started with a 60% methane concentration that declined below 50% in 2 months but experienced an increase in concentration to over 50% that was maintained during its full second year of operation.

Ten year experience in the Lower Kittanning coalbed shows that a vertical borehole with longwall mining can produce from 60 to 100 MMcf of total methane at a concentration of 50% or more of the potentially explosive gas. The two wells in the Bureau of Mines study produced an immediate 75% decrease in the methane emission identified within the mine.

In terms of economically recoverable methane (having methane concentration of 50% or greater), a borehole may recover from 40 to 60 MMcf based on Bureau of Mines and other company data.

As such an approach begins to be further used in the gassier coal beds (such as the Pocahontas No. 3), a single borehole may be able to produce over 200 MMcf of methane, with 120 MMcf having a concentration of 50% or greater.

Using Deviated Wells and Stimulated Wells — Two additional surface based techniques have potential for recovering methane in association with and in advance of coal mining, namely:

- Using wells drilled vertically from the surface and then deviated to horizontally intersect the coal face.
- Using vertical wells from which hydraulic fractures are induced in the coal seam.

Although each approach has been tried in the Appalachian Basin, the results to date have been disappointing. Deviated wells are expensive and technically difficult to control in the often thin and discontinuous coal seams of the Appalachian Basin. Vertical wells hydraulically stimulated and drilled in advance of mining may be useful for draining water and reducing the hazards of instantaneous bursts of gas and thus useful in terms of mine safety and productivity. However, this approach provides only limited recovery efficiency of the methane contained in the coal seam and thus offers a low potential alternative for methane recovery and utilization.

As discussed further in the following section, the fracture will tend to parallel the face cleats and thus will provide low recovery efficiencies, on the order of 10 to 20% of the amount that would result from intersecting the face cleat system. (Should the gas be trapped in the strata above the coal bed, it may be possible to recover this using stimulated vertical wells; however, since this involves fracturing the overburden, it may prove a significant safety hazard to subsequent mining operations.)

Degasification by Horizontal Boreholes: A second means for recovering methane in conjunction with mining is by using a horizontal borehole drilled into the coal face. This approach appears more efficient when the larger portion of the methane is contained within the microporous structure of the coal rather than in the rock strata adjacent to the coal bed.

Horizontal boreholes have been used widely in the major, gassy coal mines, including the Pittsburgh and Pocahontas No. 3 coal beds in Appalachia and the Sunnyside coal bed in Utah.

In tests conducted by Bureau of Mines (4)(5) (using a total of 15 horizontal boreholes from a multipurpose borehole and an air shaft) in the Pittsburgh coal bed, two projects recovered 1.8 Bcf of methane in 4 years, for an average of over 100 Mcf per day, or 100 MMcf per life of each horizontal borehole.

A more common approach is to use horizontal boreholes of about 500 to 1,000 ft drilled about 2 to 4 weeks ahead of mining. In this case, the production rate may range from a negligible amount (less than 1 Mcfd) to over 200 Mcfd, depending on the orientation of the borehole with the face cleat system, the continuity of the pay, and the gas content of the coal. Based on the past data, an average production of 100 Mcfd for 30 days, for a total of 3 MMcf per service life, may be a reasonable target for a 1,000 ft horizontal borehole. Under a higher priority methane recovery program, it may be possible to drill the horizontal borehole 6 months or more ahead of mining, thus possibly raising the target recoveries to 20 MMcf or more per horizontal borehole.

A final option for using horizontal boreholes would be to drill the borehole so that it intersects other coal seams overlying or underlying the mined coal bed. While considerable geological study would be required to define the potential, this could provide an important means for increasing the potential of methane recovery from coal seams.

Additional Knowledge Required to Optimize the Recovery Technology:
The recent work by Bureau of Mines has greatly advanced the knowledge base
in methane recovery from coal seams, particularly by defining the methane content of the coal and orientation of the cleat system. Designing the appropriate
recovery technology now requires that the following research tasks be undertaken:

- Determining the precise diffusion constants applicable to the
  different coalbeds, particularly in relation to the water and
  existing pressure in the coalbed (to assist in predicting recovery).

- Identifying the location of the bulk of the methane to be
  drained, as to whether it is in the coalbed or in the overlying
  structure (to assist in choosing between surface versus underground boreholes).

- Mapping the uniformity of the coal beds (to assist in designing borehole length and direction).

- Determining the intensity and dominant direction of the
  cleat systems (to assist in predicting recovery and designing
  the drilling program).

- Designing the appropriate auxiliary equipment (e.g., pumps,
  gathering lines, etc.) essential for recovering a larger percentage of the methane now being vented.

Converting the Methane for Commercial Use

Once the coal gas has been recovered from the coal bed (or overlying rock
strata), it can be diverted toward three uses:

- Directly into a natural gas pipeline, when the quality and
  methane content of the recovered gas is sufficiently high,
  the coal gas can be gathered, purified, pressurized, and injected into a pipeline. Pipeline specifications require methane contents of at least 95%, with carbon dioxide content of 3% or less. Much of the coal gas will need to be
  first purified and upgraded before it can be injected into
  a commercial natural gas pipeline.

- Conversion into liquified natural gas (LNG), particularly
  for the lower methane concentration gob gas collected as
  part of longwall mining. Here the gas is first upgraded
  (e.g., dewatered and stripped of the CO_2) and then pro-
  cessed through a series of heat exchangers and a rectifica-
  tion column to produce pure LNG. Under current technology about 30 to 80% of the energy value of the coal
  gas, depending on its purity, would be used as fuel for
  the process. The energy consumption for producing LNG
  is about 300 to 400 ft^3/1,000 ft^3 of 90% plus concentration methane feedstock and establishes the minimum required concentrations for technical and economic feasibility.

- Other end uses; finally, it may be possible to use the coal
  gas, particularly coal gas with methane concentration be-

tween 50 and 80%, for local power generation, coal drying, or after upgrading to about 80%, for local industrial uses such as in the production of ammonia.

Other methane recovery and conversion approaches, beyond the above three, have been proposed, such as the use of membrane separation, centrifugation, or solvent extraction. While each of these requires further basic study, they do not appear economic nor highly efficient. For example, using membrane separation, even in five stages, would provide a recovery efficiency of about 2% and the energy input requirements of centrifugation are far above the energy output.

Economic Issues in Recovery and Conversion

The economics of recovering methane in association with mining rely greatly on the residual safety and efficiency benefits that accompany mine degasification. The costs are as follows.

Drilling and completing a vertical hole from the surface costs from $20 to $30 per foot; adding other components, which will vary from one installation to another (exhaust fan, pump, pipe, access roads, etc.), makes the cost of a typical 1,000 ft vertical hole about $50,000 to $60,000. These and all costs in the remainder of the chapter are in constant 1977 dollars. Drilling and completing a horizontal hole from the mine face costs $6.00/ft. Adding a collection system, pump, exhaust fan, etc. makes the cost of a typical 1,000 ft horizontal hole about $10,000 to $15,000.

However, even though the initial drilling and completion equipment and well operating costs are relatively low and can be readily justified on the basis of safety and mine efficiency (for example, a 1% increase in mine efficiency will readily pay back the capital costs of degasification), the surface collection, upgrading, and transportation costs can be considerable. Preliminary estimates show that without a charge for drilling or well equipment or well operations, it may be economic to convert the methane to LNG, to use it directly for generating local power or direct heat, or to collect and upgrade it (where necessary) for delivery directly into pipelines at $1.75 to $3.00 per Mcf. Thus, the major barriers to making recovery economic (in addition to the technological and geological uncertainties described above), would include:

- Obtaining sufficient quantities of gas in each location to justify the building of gathering and transportation facilities.
- Designing the most appropriate recovery system for a given type of mining operation and coal seam, including (1) using horizontal boreholes drilled into the coal seam from the bottom of a shaft or an air vent; (2) using vertical wells drilled above coal seams, generally with longwall mining, to capture gob gas.
- Developing economical small-scale means for purifying and upgrading the produced air/methane mixture using cryogenic liquefaction to generate LNG, particularly for individual rural and industrial consumers not now served by a utility.

Potential Production from Appalachian Coal Seams

The recovery of methane from the Appalachian Basin will need to follow the pace of current mining and the opening of new mines. Currently, the mine

shaft and its progress through the coal bed provide the most ideal wellbore im-
aginable to a reservoir engineer and serves as the point of release for the total
of the methane emitted in association with mining. Since this is the maximum
that would be emitted through a predrainage program, all estimates of methane
recovery and production are scaled from this base figure. Drawing on this base,
the following assumptions were used to establish the recovery target.

- Current emissions: 80 Bcf/year (217 MMcf/d)

- Proportion of emissions estimated from Appalachian
  Basin: 90%

- Rate of growth in mining: (a) doubling of coal production by
  1990 and (b) an equal (absolute) amount of increase by 2000.

- Ultimate recovery: equivalent to recovery over the next 30
  years.

- Proportion of mines with methane recovery facilities:

1980	10%
1985	30%
1990	50%
2000	50%

- Methane recovery efficiency for mines with methane capture facilities:

1980	30%
1985	50%
1990	50-75%
2000	50-75%

Using these assumptions, the yearly and cumulative rates are as follows:

	. . . . .Rate.		Cumulative
	(MMcfd)	(Bcf/year)	(Bcf)
1980	8	3	3
1985	50	18	50
1990	100-150	36-54	190-230
1995	123-186	45-68	390-530
2000	142-214	52-78	640-910
2008	148-222	54-81	1,080-1,560

Thus, the thirty year ultimate recovery target for the Appalachian Basin
is 1.1 to 1.6 Tcf.

In that there is currently little private sector activity in recovering methane
from minable coal seams, it is likely that publicly supported research and development programs will be required to overcome the present problems and limitations. A substantial research, demonstration and implementation program of
methane recovery from coal seams could add important quantities of natural
gas supplies in the Appalachian Basin area and accelerate the implementation
of the technology, clear the legal and gas ownership issues, and gain mine operators acceptance and support.

Sensitivity of the Recovery Estimates to Key Variables: The Appalachian
Basin coal seams are too thin and too lean in methane content to economically
support methane recovery on its own. Since estimates of recovery need to parallel closely the pace of mining and the opening of new mines, there is little lee-
way in making production rate and recovery estimates. However, the actual
rate may vary due to the following:

- The near term production rate and cumulative production may be 20% (10 to 20 Bcf/year) higher than projected if considerable drainage ahead of mining rather than with mining is used.

- The 30 year recovery (used as ultimate recovery for this analysis) could be 20% (or 0.3 Tcf) higher if the rate of growth in mining led to a redoubling of capacity between 1990 and the year 2000. (That is, coal production in the year 2000 would be 4 times current rates rather than the assumed 3 times current rates.)

- It may be possible to place methane recovery facilities in additional mines, accounting for 75% of the methane emission. Should this be done, ultimate recovery would increase by 0.3 Tcf.

- Since the recovery of methane in association with mining depends so greatly on the associated productivity and safety benefits, base production estimates are relatively insensitive to price changes. The major area of price sensitivity centers on the methane recovery efficiency of the captured total gas emissions. As price goes up, coal gas with lower methane concentration can be economically extracted.

METHANE FROM UNMINABLE COAL SEAMS

Resource Base

Gas in Place: In general, coal seams are considered unminable because they are too thin or too deep to be mined economically. While volumetric data on unminable coal is sketchy, at best, several studied (6)(7)(8) do permit gross estimates of gas in place:

- Approximately 290 billion tons of coal are in seams too thin (averaging 15") to be economically mined. This coal contains an estimated 260 cf/ton of methane, or a total of about 70 Tcf.

- An additional 388 billion tons of coal are too deep (3,000 to 6,000 ft) to be mined economically. About 45% of this coal is ranked as bituminous or higher; the balance is subbituminous coals and lignite.

- Gas content of bituminous (and higher) coals averages 480 cf/ton, accounting in total for 80 Tcf.

- Gas content for subbituminous coals is substantially lower than that for bituminous and higher rank coals. Their high moisture content tend to limit their gas-adsorption capacity. Assuming a gas content of 100 cf/ton, one-fifth that of bituminous coals, deep subbituminous coals would contain about 20 Tcf.

In total, the 668 tons of unminable coal contain an estimated 170 Tcf of gas. Previous estimates have used the same gas content for bituminous and subbituminous. This appears to overestimate the gas in place due to the higher moisture content of the lower grade coal. Using the lower gas content for sub-bituminous coal would reduce the initial estimate of 250 to 170 Tcf. Economic recovery of this methane depends on five geological variables:

- Adequate thickness of the coal bed
- Intensity of the natural fracture system
- Uniformity of the bed and seam
- Water content in the coal bed
- Gas content

The methane content in unminable coal beds must be classified as speculative. Major uncertainties surround each of the essential geologic variables. Finding where these five conditions may converge favorably will require considerable basic study of the unminable coal resource.

Because of the controlling effect of the first of these variables, thickness of the coal bed, the analysis assumes the remaining four geologic variables (i.e., fracture intensity, coal seam uniformity, water, and gas content) are favorable. The analysis then establishes the minimum required coalbed thickness as a function of costs and gas price.

Thin Coal Seams: The 70 Tcf of methane held in thin coal seams (averaging 15" thick) cannot be extracted economically for utilization potential under any reasonable set of technological assumptions. At 260 ft^3/ton, an average well drainage area of 72 acres, the drainage area will contain only 40,000 Mcf of methane. The gas in place can be calculated from the following equation:

GIP = (drainage area) x (methane content/cubic foot) x (thickness)

The drainage area equals: acres x 43,560 ft^2

The methane content per cubic foot equals:

260 ft^3/ton x 80 lb/ft^3 ÷ 2,000 lb/ton = 10 ft^3 (0.01 Mcf)/ft^3

Using deviated wells costing $300,000 (2 wells drilled to 800 ft), operating, compressing, separating, and gathering costs of $0.40 Mcf and normal royalty and tax requirements, it is obvious that thin coal seams lack the minimum required coal bed thickness. The calculation of the required thickness is as follows:

Required gas production at $1.75/Mcf to pay back costs equals:

300 MMcf

Required thickness to produce 300 MMcf in 10 years equals:

300 MMcf ÷ [drainage area (ft^2) x gas in place (per ft^3) x 10 yr recovery efficiency]

Required thickness (h) equals:

300 MMcf ÷ [(3 x 10^6 ft^2) x (0.01 Mcf/ft^3) x 30%]

Because thin coal beds are so far from economic viability, they were eliminated from further analysis of methane potential.

Deep Coal Seams: The deep, unminable but thick coalbeds pose a more attractive economic potential. Given the scarcity of published data, much of the above analysis was based on recent test data with state geological societies and Bureau of Mines officials. Thus, only a preliminary appraisal can be made of the five key geological variables that govern the economics of producing methane from deep, unminable coal beds.

Thickness — The distribution of coalbed thickness was based on data in Colorado and the combined data in the three states of Colorado, New Mexico, and Utah. Colorado has about 70% of the bituminous coal and the three states combined have about 90% of the bituminous coal in the Western Basins. In Colorado and these three states, the distribution of thickness is estimated as follows:

Coal Bed Thickness (ft)	. . . . % of the Resources	
	Colorado	Three Western States
0-15	52	61
15-50	36	33
>50	12	6

These data are plotted on a cumulative percentage curve of total resource versus coal bed thickness to provide estimates of coal bed thickness between the three points in the table (Figure 4.4).

Figure 4.4: Estimated Distribution of Total Bituminous Coal Resources by Coal Bed Thickness (8)

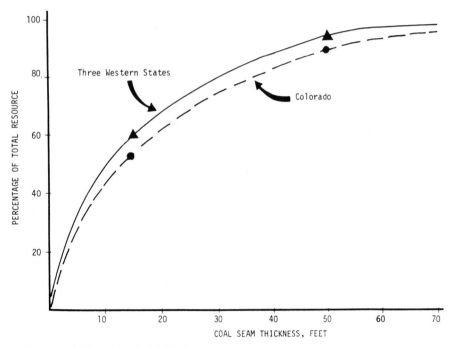

Source: DOE EF-77-C-01-2705

Intensity of Natural Fractures — Unlike the Appalachian Basin that has undergone considerable tectonic upheaval and where the subsurface is highly fractured, the tectonic history of the Western Basins is considered to have been less intense. Discussions with state geological officials and related studies of the

Uinta Basin indicate that low fracture intensity may pose severe geological barriers for recovering methane from this deeper, unminable coal. Preliminary examination of the shallower geologic formations indicates that:

- Only limited fracture systems are evident in the coal, although they do exist in the surrounding sandstone and shale.

- Hydraulically induced fractures tend to migrate vertically to the surface rather than horizontally through the formation.

- The coal is blocky, not friable, with a poorly developed cleat structure.

Whether these conditions hold for the bulk of the deeper bituminous coal will require further resource study.

Uniformity of Deposits — Recent coal mining and methane recovery in the Uinta Basin have identified numerous clay veins that would restrict flow. Also, the depositional history of the basins and geologic horizon in which the deep coals are located has probably created nonuniform deposits; however, further geologic definition is clearly required.

Water Content — High water content in the coal bed can cause major problems for degasification. To produce gas most efficiently, the connate water must be first removed from the coal bed. The general presence of water in Western coal beds (particularly in subbituminous coal beds) may thus pose a serious challenge to economically recovering methane from this resource base.

Gas Content — The gas content of the bituminous and higher grade coals appears high, estimated at 480 cf/ton. In turn, the gas content in subbituminous coals and lignites is low and assumed at one-fifth that for higher grades, due to their higher moisture content. However, even for the higher rank coal, the deposit must have a natural caprock seal to have prevented the escape of the gas over geologic time. Thus, additional research is required to ascertain the true gas content for the deep coals.

Assessment of the Resource Base — The available data on the deeply buried coals are inadequate at present for definitive assessment of these deposits, and thus this resource is classified as speculative. Considerable resource analysis and testing are required before one can begin to make judicious longer term decisions for recovering methane from this resource base.

Applicable Recovery Technology

For efficient recovery, a wellbore will need to communicate with the full natural fracture system in the drainage area. Since for all practical purposes coal is impermeable, the methane must travel through the fracture system or desorb through the structure of the coal in response to a pressure gradient. Three techniques are available for achieving this communication for the deep, unminable coals:

- Vertical wellbores with stimulated fractures
- Large boreholes with horizontal wells
- Deviated wells

Using vertical wellbores with stimulated fractures, under current fracture technology, does not appear to be a technically viable option at this time. The induced fracture will tend to parallel the existing natural fractures and face cleats rather than intersect them, providing only limited connection with the dominant natural fracture system. Additional research may be required to confirm this hypothesis.

Deviated wells, though costly, appear to provide the most technically feasible means for exploiting the deeply buried, unminable coal. However, drilling and controlling deviated wells in coal seams is not a proven technology, particularly since the drilled well will need to intersect the face cleats of the natural fracture system. Although the oil and gas industry has used deviated wells for some time, transferring that technology to the "methane from coal" program will require considerable adaptation in directional drilling and well completion technology, and economic optimization.

Key Economic Issues in Recovery

Only limited information is available on the key geological variables that govern recovery and economics of recovering methane from deep, unminable coal seams. Thus, while the analysis herein served as a first approximation, it is possible to place a range on the economic potential. Assuming that the thickness of the coal seam is the controlling variable and that all other geological features are favorable, economic feasibility will depend on finding coal deposits having sufficient thickness.

Using deviated well costs of $600,000 (2 wells drilled to 4,000 ft, then 1,000 ft into the coal bed) and all other costs as noted above, the minimum 10 year production to yield a 10 year payback can be calculated as a function of price. This, in turn, can be converted into minimum required coal bed thickness.

Required production equals: [(10 yr production) x (1 - royalty and severance taxes) x (gas price)] - [operating costs x 10 yr production] > investment

Actual production equals: drainage area x coal bed thickness x methane content x recovery efficiency (for t = 10 yr)

> Effective drainage area where the pressure is sufficiently low to induce effective desorption is assumed at 72 acres
>
> Methane content is assumed at 480 ft^3/ton or 0.019 Mcf/ft^3
>
> Recovery is calculated by using the recovery efficiency function on Figure 4.5; for time equals 10 years (in seconds) recovery efficiency is 30%
>
> Required production is set equal to actual production and the equation is solved for coalbed thickness

Additional detail is provided in Volume III of Kruuskraa, V.O., Brashear, J.P. and Doscher, T.M., *Enhanced Recovery of Unconventional Natural Gas: The Program–Volume III (of 3 Volumes)*, October 1978. This analysis assumes that deviated wells are used to connect the natural fracture system. Should vertical boreholes with hydraulic stimulation obtain the same recovery efficiency, the minimum required coal seam thickness (or the gas content) could be reduced substantially.

Under the above costs, a diffusion constant of $K = 5 \times 10^{-8}$ cm^2/sec, and assuming a major, natural fracture every 1 ft, one can define the minimum required production (in 10 yr) and, in turn, the minimum coal bed thickness.

Price/Mcf	Production (Mcf)	Minimum Thickness (feet)
$1.75	600,000	34
$3.00	300,000	17
$4.50	188,000	11

Figure 4.5: Relationship of Recovery Efficiency vs Time*

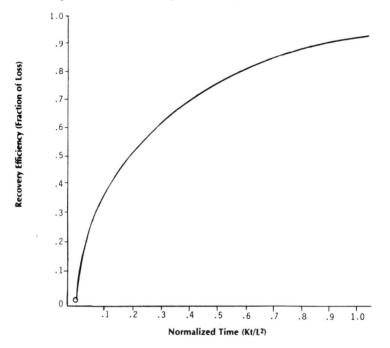

*From fraction of heat loss from a slab (slab thickness 2L) with surface temperature kept at t = 0, as a function of Kt/L^2.

Derived from Carslaw and Jaegger, *Conduction of Heat in Solids*, Oxford, 2nd Ed., 1959, Chap. III, pp 92-102. Equivalent substitute units for any diffusion phenomenon, viz., fraction of adsorbed material lost.

Source: DOE EF-77-C-01-2705

The above analyses, however, make several relatively optimistic assumptions on three variables that affect economic recovery, namely:

- Methane content: assumes 480 cf/ton (bituminous and higher grades).

- Natural fracture system: assumes the fracture system to be extensive enough to provide 30% recovery in 10 years; for this the fractures will need to occur one foot apart

and extend uninterrupted through the drainage area of
the well.

- Diffusion constant: assumes K of 5×10^{-8} cm^2/sec.

The sensitivity of these assumptions on estimates of the total potential is
examined below.

Potential Production from the Western Unminable Coal Seams

Recovery Estimates: The production potential may be estimated by combining the distribution of minimum required economic thickness with the distribution of thickness of the resource. The calculation of the required thickness
is as follows:

Required gas production at $1.75/Mcf to pay back costs equals:

300 MMcf

Required thickness to produce 300 MMcf in 10 years equals:

300 MMcf ÷ [drainage area (ft^2) x gas in place (per ft^3) x 10 yr recovery efficiency]

Required h equals:

300 MMcf ÷ [(3 x 10^6 ft^2) x (0.01 Mcf/ft^3) x 30%

Reading from the Colorado curve in Figure 4.4, the following distribution
can be approximated:

Price/Mcf	Minimum Thickness (feet)	Percentage of Resource
$1.75	34	24
$3.00	17	44
$4.50	11	60

Assuming 30 year recovery efficiency R_{30} of 51% [derived by solving
the Kt/L^2 equation for t = 30 years (in seconds) on Figure 4.5], 480 cf of methane per ton of coal, and 170 billion tons of bituminous and higher grades of
coal, the distribution of economic thicknesses can be converted to estimates of
potential recovery at three prices, rounded to the nearest significant figure. (Because the resource base is considered speculative, the estimates of potential recovery are expressed as ranges.)

Price/Mcf	Recoverable Methane, Tcf
$1.75	0–10
$3.00	0–20
$4.50	0–25

Given the speculative nature of the resource base, no estimates can yet be
made of yearly production rates for recovering methane from the deep, unminable coals.

Sensitivity of the Recovery Estimates to Key Variables: The above recovery estimates have been based on a series of generally optimistic assumptions on
key variables. It is instructive to determine how sensitive these estimates are to
reasonable bounds of variation, particularly on:

- Intensity of the fracture system
- The diffusion constant

- A higher risk premium investment criterion
- Use of vertical hydraulically stimulated wells
- Effective drainage area
- Contribution of fracture porosity gas
- Methane content in subbituminous coals

The results of this sensitivity analysis are shown below:

Intensity of the Fracture System — Should the fractures be less intense, 5 ft rather than 1 ft apart, the recovery potential drops to essentially zero. Assuming that the coal bed is less intensely naturally fractured, one fracture every 5 ft rather than one every 1 ft, the 10 year recovery percentage drops to about 2% and the minimum required coal seam thickness at $3.00/Mcf is over 200 ft.

Diffusion Constant — Should the diffusion constant be lower than assumed, for example, 1×10^{-8} cm^2/sec, the potential from the Western Basins drops to about 1 Tcf. Introducing this lower diffusion constant and keeping all other variables the same, the 10 year recovery efficiency R_{10} becomes about 8% and the required coal seam thickness at $3.00/Mcf is over 60 ft. Since only about 6% of the resource is over 60 ft thick, the potential 30 year recovery drops to less than 1 Tcf, as compared to about 20 Tcf under the assumed higher diffusion constant.

Risk Premium — Should a higher risk premium investment criterion, 20% ROR (in financial terms, a 20% ROR under 6% inflation is equivalent to a real value of 26%; similarly, a 10% ROR is equivalent to 16%), be imposed, the potential from the Western Basins drops to 12 Tcf. Using a 5 year payback (as a proxy for a 20% ROR), and assuming all other parameters stay the same except that t = 5 years, then $Kt/L^2 = 0.035$ and 5 year recovery efficiency R_5 becomes 19%. Required coal seam thickness at $3.00/Mcf becomes 26 ft. Since about 30% of the resource is over 26 ft thick, the potential at $3.00/Mcf drops to about 12 Tcf (as compared to about 20 Tcf under the conventional 10% ROR investment criterion).

Vertical Wells with Artificial Fracturing — If vertically drilled and fractured wells were substituted for the deviated wells, the potential could either increase to 30 Tcf or decrease to 1 Tcf. Assuming that a single vertically drilled and hydraulically stimulated well (costing $300,000) could have the same production as the two deviated wells, the minimum required thickness would drop to 8 ft and the recovery potential at $3.00/Mcf would increase to about 30 Tcf. However, it is likely that the induced fracture would parallel rather than intersect the face cleats. Under this condition, 10 year recovery would drop to 6% (about $\frac{1}{5}$ of that assumed in the Base Case), and even under these lower costs, the potential at $3.00/Mcf would decrease to about 1 Tcf.

Effective Drainage Area — It may be possible to increase the effective drainage area, the area where the coal bed pressure is sufficiently low in the first 10 years, to allow highly efficient desorption. One means for doing this would be to drill, where geologically feasible, the deviated wells further into the coal bed.

Assuming three deviated wells can be effectively drilled 2,000 ft into the coal bed, the drainage area increased fourfold to 288 acres, and 10 year recovery increases threefold. Adjusting for higher investment costs (essentially double those in the Base Case), the analysis shows that the potential at $3.00/Mcf could increase by 5 Tcf.

Contribution of Fracture Porosity Gas — The analysis assumes that the methane is adsorbed within the coal and the fractures are essentially filled with water. Should the fractures (assumed at 5% porosity) be filled with methane, the contribution to ultimate recovery is small, less than 5%.

Subbituminous Coals — Assuming pay thickness and fracture intensity of the subbituminous coals in the sand, the subbituminous coals with a methane content of 100 ft^3/ton could add about 1 Tcf to the above estimates in the $3.00 to $4.50/Mcf price range.

SUMMARY OF R&D PROGRAM AND ITS POTENTIAL

The Research and Development Goals

The geological and technological challenges for recovery of methane from coal seams form the R&D goals of the program.

Methane from Minable Coal Seams—Appalachian Basin:

To ascertain the optimum means for collecting the gas release during coal mining operations.

To establish that methane captured in association with coal mining can provide a reliable economic supply for transmission into natural gas pipelines, or supplemental LNG, or other end uses.

To test and stimulate the installation of methane capture and utilization facilities such that 25 to 38% of the methane released as part of mining operations is captured for use: one-half of all mines would contain methane recovery facilities; the recovery efficiency of the methane at a target mine should reach 50 to 75% of total emissions.

To accumulate additional geological and technical insights into the nature and occurrence of methane in coal seams and in situ diffusion coefficients.

Methane from Unminable Coal Seams—Western Basins:

To identify the geologic characteristics of the thick coal seam basins, particularly the extent of the in situ natural fracture system and the methane content.

To develop and successfully apply deviated drilling technology to coal beds.

To ascertain the economic feasibility of predrainage, given the geology of the Western coal basins and the efficiency and costs of applying the recovery technology.

Production Benefits

The production benefits, stated in terms of ultimate recovery (recovery in 30 years), 1990 cumulative recovery, and 1990 production rate, are as follows:

	. . . . . . Price per Mcf		
	$1.75	$3.00	$4.50
Ultimate recovery, Tcf			
Appalachian Basin	1.1	1.6	1.6
Western Basins	0–10	0–20	0–25
1990 Cumulative production, Tcf			
Appalachian Basin	0.19	0.23	0.23
Western Basins	*	*	*

(continued)

	. . . . . . Price per Mcf		
	$1.75	$3.00	$4.50
1990 Production rate, Tcf			
Appalachian Basin	0.04	0.05	0.05
Western Basins	*	*	*

*Not estimated.

R&D Costs

The research program costs for the two programs in methane recovery from coal are as follows:

	. .5 Year Program Costs. . (MM$)	
	Total	ERDA
Appalachian Basin	16.5	16.5
Western Basins	29.1	24.1

Cost-Effectiveness of the Program

Two cost-effectiveness ratios serve to illuminate the potential benefits of the R&D program:

- Long Term Measure   Ultimate recovery (Mcf) at $3.00 Mcf/total ERDA costs
- Near Term Measure   1990 cumulative recovery (Mcf) at $3.00 Mcf/total ERDA costs

The cost-effectiveness ratios for methane recovery from coal are:

	Long Term Measure (Mcf/total ERDA costs)	Near Term Measure (Mcf/total ERDA costs)
Appalachian Basin	95	14
Western Basins	0–830	—

Summary

These sensitivity analyses show that the methane from minable coal seams can be a steady source of natural gas and will increase as the pace of mining, the efficiency of collection, and prices increase. The methane adsorbed in deep coal seams is a more speculative resource in that less is known of the methane content or the extent of the natural fractures system.

REFERENCES

(1) Irani, M.C., Jansky, J.H., Jeran, P.W. and Hassett, G.L., *Methane Emissions from U.S. Coal Mines in 1975, A Survey*, U.S. Bureau of Mines Information Circular 8133 (1977).
(2) Moore, Jr., T.D. and Zabetakis, M.G., *Effect of a Surface Borehole on Longwall Gob Degasification (Pocahontas No. 3 Coal Bed)*, U.S. Department of the Interior, Report of Investigations 7657 (1972).
(3) Moore, Jr., T.D., Deul, M., and Kissell, F.N., *Longwall Gob Degasification with Surface Ventilation Boreholes Above the Lower Kittanning Coalbed*, U.S. Department of the Interior, Report of Investigations 8195 (1976).
(4) Fields, H.H., Perry, J.H. and Deul, M., *Commercial-Quality Gas from a Multipurpose Borehole Located in the Pittsburgh Coal Bed*, U.S. Department of the Interior, Report of Investigations 8025 (1975).

(5) Fields, H.H., Cervik, J. and Goodman, T.W., *Degasification and Production of Natural Gas from an Air Shaft in the Pittsburgh Coal Bed*, U.S. Department of the Interior, Report of Investigations 8173, (1976).

(6) Averitt, P., *Coal Resources of the United States, January 1, 1974*, Geological Survey Bulletin 1412 (estimates of coal resources) (1975).

(7) Kim, A.G., *Estimating Methane Content of Bituminous Coal Beds from Adsorption Data*, Bureau of Mines, Report of Investigations 8245 (estimates of methane content) (1977).

(8) Booz, Allen and Hamilton, *ERDA's Underground Coal Gasification Program, Volume III-Resources* (estimates of distribution of coal bed thickness and rank) (May 1977).

Methane from Geopressured Aquifers

THE RESOURCE

The information in this and the following three sections is based on *National Gas Survey Report to the Federal Energy Regulatory Commission by the Supply and Technical Advisory Task Force on Nonconventional Natural Gas Resources, Sub-Task Force I—Gas Dissolved in Water,* DOE/FERC-0029, March 1979, prepared by J.W. Harbaugh of Stanford University, Task Force Chairman and Sub-Task Force I, P. Jones of Louisiana State University, T. Jennings of National Gas Survey, P.A. Dennie of Shell Oil Company, C.R. Hocott of University of Texas, P.E. LaMoreaux of P.E. LaMoreaux and Associates, D. Lombard of U.S. Department of Energy and R.H. Wallace, Jr. of U.S. Geological Survey for the U.S. Department of Engery. References for these sections are on p

As reserves of natural gas from conventional sources in the United States continue to decline, natural gas from nonconventional sources will be needed to meet future demands for this important energy resource. Methane gas of bio-genic and thermogenic origin, known to be dissolved in the formation waters of sedimentary basins (1)(2) may offer a viable alternate source of natural gas for the future.

While most of the natural gas produced at the present time represents production from reservoirs where it occurs in the gaseous state or is produced along with crude oil, in which it is very soluble, the total volume occupied by these reservoirs is very small compared to the total volume of water-filled reservoirs. Therefore, even though the concentration of methane dissolved in water-filled reservoirs (aquifers) is low compared to that found dissolved in liquid hydro-carbons or existing as free gas in conventional reservoirs, the large volume of water present outside conventional hydrocarbon producing reservoirs makes the dissolved methane resource base quite significant, assuming saturation or near saturation of this water with methane.

It must be understood that resource base refers only to the amount of energy in place without regard to its recoverability or cost of extraction.

Papadopulos, Wallace, Wesselman, and Taylor (3), in an order of magnitude assessment of the geopressured-geothermal resources of the northern Gulf of Mexico basin, estimated the methane dissolved in water resource base in geopressured Cenozoic sediments in a 145,265 km² area onshore Texas and Louisiana to be equivalent to 23,618 Tcf. This estimate of resource base assumed saturation and included methane dissolved in shale bed waters as well as sand bed waters, although it was recognized that a lesser amount of the resource base would be recoverable from the shale beds.

In an assessment for the National Research Council, Jones (4) estimated the total methane dissolved in water resource base contained in geopressured sand bed reservoirs of the northern Gulf of Mexico basin, both onshore and offshore (an area greater than 150,000 square miles), to be about 49,000 Tcf with a comparable amount present in associated geopressured shales. Jones stated that an appreciable part of the resource base from the shale beds, "might migrate into pressure depleted sand bed reservoirs as production occurs."

Hise (5), however, in a less optimistic paper for the National Research Council, calculated this same resource base to be only 3,000 Tcf for the same area. In each of these assessments, different assumptions were made and different techniques were used. Therefore, meaningful comparisons between them are difficult.

There have been no attempts, heretofore, to estimate the methane dissolved in water resource base in the reservoirs of normally pressured sedimentary sequences. Buckley, Hocott, and Taggert (1), however, gave a clue to the magnitude in their statement that, "The total quantity of gas dissolved in the water of the subsurface formations in this area probably exceeds the known proved gas reserves heretofore discovered in commercial accumulations in the area."

Figure 5.1: Solubility of Methane in Fresh Water as a Function of Pressure and Temperature (7)

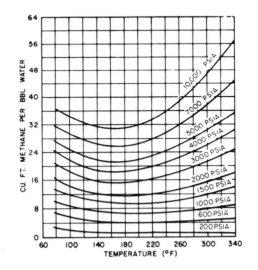

Source: DOE/FERC-0029

If saturation is assumed and reliable data on the temperature, pressure, and salinity of the water is available, the dissolved methane content of the water in many normally pressured and geopressured reservoirs can be estimated from methane solubility studies performed in the laboratory (6)(7)(8). Estimation of methane content of formation water at reservoir temperatures of more than 250°F, pressures of more than 10,000 psi, and salinities of more than 264 g/l sodium chloride, must at present (May 1978) be obtained by extrapolation of the curves of Culberson and McKetta (7) (Figure 5.1) and correction of solubility estimates for water salinity using the experimental work of O'Sullivan and Smith (8).

Methane content estimates based upon extrapolation of solubility data beyond these limits must be considered speculative. Laboratory experiments are underway in the U.S. Geological Survey to determine methane solubility to temperatures of 600°F, pressures of 20,000 psi, and to salinities of 250 g/l sodium chloride.

Geologic Setting

The northern Gulf of Mexico basin occupies the Gulf Coast geosyncline, a subsiding basin of deposition. The geosyncline has been estimated to contain more than 50,000 feet of water filled sediments, mainly sands and clays, deposited in less than about 70 million years, primarily in two centers of deposition (Figure 5.2).

Figure 5.2: Thickness of Cenozoic Sediments (9)

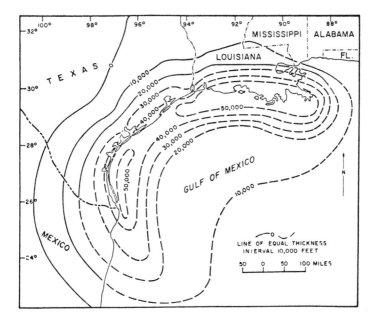

Source: DOE/FERC-0029

These depocenters are aligned with the trend of the geosynclinal axis, and are located adjacent to the south Texas coast and along the Louisiana coastline, both onshore and offshore.

The volume of sediments filling the geosyncline, according to Meyerhoff (10) is about 400,000 mi^3 in southern Louisiana alone, with over half this volume having been deposited in the last 26 million years. The progressive development of the Gulf Coast geosyncline, from Cretaceous time to Pleistocene time, is illustrated by a series of idealized north-south cross sections in Figure 5.3 (11).

Figure 5.3: Stages in the Development of the Gulf Coast Geosyncline (11)*

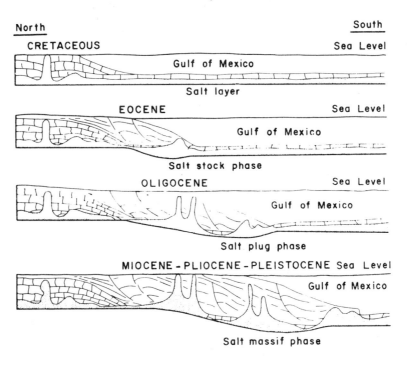

* With salt flowage gulfward under progressive loading, growth of salt stocks and plugs with prograding sedimentation and contemporaneous faulting.

Source: DOE/FERC-0029

Successively younger sequences of deltaic sediment and their prodelta and neritic counterparts were rapidly deposited, gulfward, and deeply buried. Salt tectonics and differential compaction of the clays played key roles in establishing the locus of maximum sediment accumulations, as well as the positions and trends of the major contemporaneous (growth) fault systems (Figure 5.4) that lace the region.

Figure 5.4: Major Regional Fault Zones in Neogene Deposits, Northern Gulf of Mexico Basin (12)

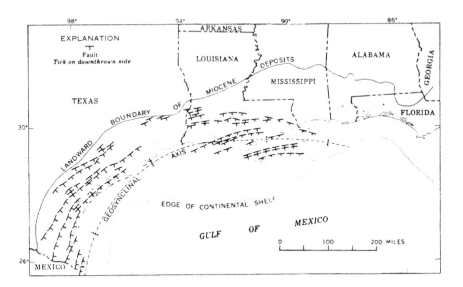

Source: DOE/FERC-0029

If free to drain in response to increasing overburden load and depth of burial, the pore-water from compacting deltaic sands and shales will migrate upward and landward in the system, as shown in the first stage on Figure 5.5. However, as shown in the second stage on Figure 5.5, compartmentalization by faulting (and reduction of porosity in the clay beds), alters and restricts the paths of fluid migration with increasing depth of burial. Therefore, at some critical depth in the Gulf Coast geosyncline, the clay (shale) beds that represent more than 60 percent of the total sedimentary volume, and the sand-bed aquifers will contain more than the normal volume of water for that specific depth of burial.

As a result, the sealed-in fluids assume a part of the weight of the overlying sediments, and are said to be geopressured because of their higher than normal fluid pressures (i.e., in excess of about 0.465 times depth of occurrence). By contrast, the fluids in the sediments of the first stage on Figure 5.5 are said to be hydropressured or normally pressured.

Figure 5.6 shows that geopressures (pressure in excess of about 0.465 pounds per square inch per foot of depth) are encountered in southern Louisiana at depths shallower than 4,000 feet and deeper than 19,000 feet. Maximum depths of occurrence generally coincide with trends of maximum sand accumulation. On the other hand, minimum depths of occurrence generally reflect increased shale accumulation. This would have been expected intuitively because sand beds transmit fluids more easily than shale or clay beds, thus preventing fluid pressure increases.

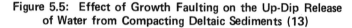

Figure 5.5: Effect of Growth Faulting on the Up-Dip Release of Water from Compacting Deltaic Sediments (13)

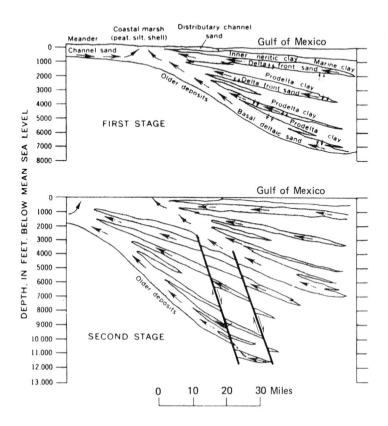

Arrows indicate movement of water released by sediment compaction

Source: DOE/FERC-0029)

Although not readily apparent on Figure 5.6, faulting, salt tectonism, and formation geometry also influence the depth of occurrence of the geopressured zone. Sediments at depths shallower than those shown on Figure 5.6 will be normally pressured.

Hydrogeologic Factors to Be Considered

Exploration for and estimation of the methane dissolved in water resource of the northern Gulf of Mexico basin, as well as other sedimentary basins, involves both regional and site evaluation.

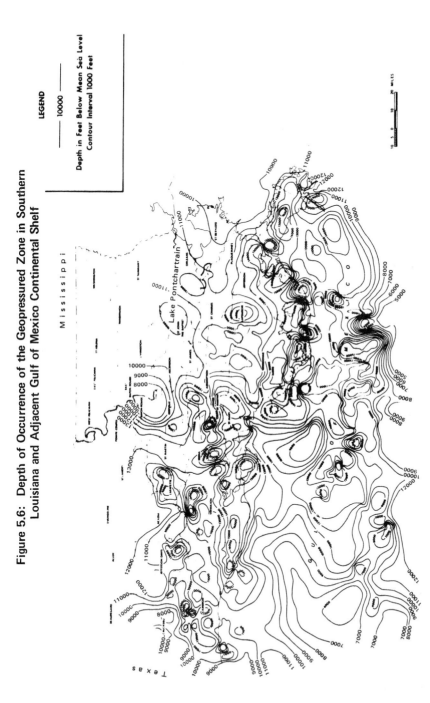

Figure 5.6: Depth of Occurrence of the Geopressured Zone in Southern Louisiana and Adjacent Gulf of Mexico Continental Shelf

Source: DOE/FERC-0029

This is not an easy task because about six hydrogeologic factors must be considered simultaneously in order to evaluate the quantity and quality of the resource. Fortunately, in the Gulf coast, much subsurface information has been collected in the search for conventional oil and gas reservoirs. Key factors that must be evaluated include:

The volume of stacked sand-bed sequences.

The volume of pore-space available in these sand-bed aquifer sequences to contain the fluids.

Reservoir pressures.

Temperature of formation water.

Salinity of formation water.

Ability of the formations to transmit water or gas to the well-bore for extraction (permeability).

A RESOURCE EVALUATION AND ASSESSMENT TECHNIQUE

The method used in this and the following sections was developed for the purpose of assessing the amount of methane dissolved in the formation waters of sand and clay filled sedimentary basins, using the eastern part of the northern Gulf of Mexico basin, onshore and offshore southern Louisiana (Figure 5.7) as a model. This method can be applied to the remainder of this basin as well as to similar basins elsewhere, provided sufficient data (derived from the drilling and production of oil and gas wells) are available to determine temperatures, pressures, salinities, porosities, and sand thicknesses, both areally and vertically within the basin.

As used here, the model is restricted to estimation of only the amount of methane that can be held in solution by formation waters in sand-bed aquifers at reservoir temperatures and pressures with limited knowledge on the effect of salinity. The question of whether these waters are, in fact, saturated with methane under these conditions cannot be answered here, but the assumption is made. Certainly, the potential for saturation increases with depth as temperature and pressure increases (1). This and other problems can only be evaluated after a program of drilling and testing in a variety of hydrogeologic settings has been completed.

Deltaic and associated sediments in the southern Louisiana study area occur as sequences of interbedded sands and shales that form a series of east-west trends of high sand content that merge gulfward into marine shales.

Figure 5.8, constructed from well control shown on Figure 5.7, shows three clearly defined trends of maximum sand deposition in the 11,000 to 12,000 foot depth interval (West Feliciana to Beauregard Parish, Assumption to Cameron Parish, and Plaquemines to Terrebonne Parish), that are separated by two trends of maximum shale deposition.

The geologic age of these trends decreases gulfward as younger deltaic sequences have overridden their predecessors. Deposition of sand beds on the marine components of the previous depositional cycle, coupled with differential compaction and growth faulting, produced these features. Although Figure 5.8 gives the impression of continuity of reservoirs, it must be remembered that the aquifer systems are divided into many compartments of varying sizes by faulting. Nevertheless, these trends of maximum sand deposition are where the maximum resource potential will be concentrated, provided other factors to be considered are also favorable.

Figure 5.7: Location of Study Area and Location of Well Control Used for Estimation of Resource Base

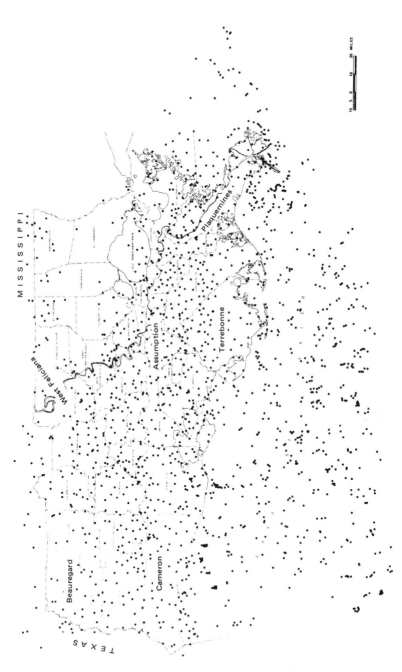

Source: DOE/FERC-0029

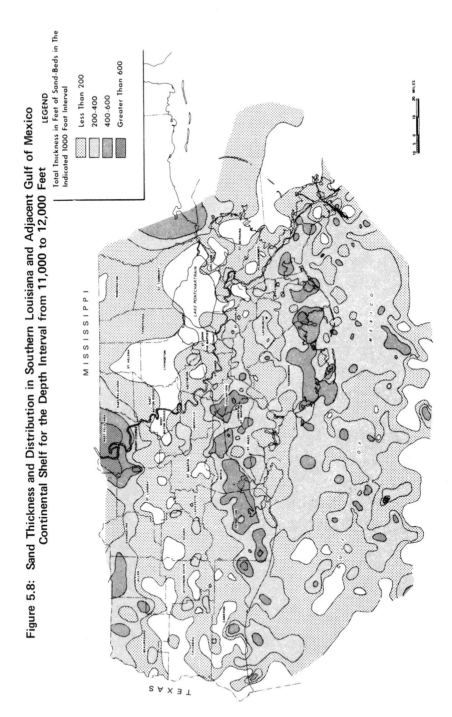

Figure 5.8: Sand Thickness and Distribution in Southern Louisiana and Adjacent Gulf of Mexico Continental Shelf for the Depth Interval from 11,000 to 12,000 Feet

Source: DOE/FERC-0029

As shown on Figure 5.7, the part of the northern Gulf of Mexico basin chosen for development of this resource evaluation and assessment technique is contained largely within the eastern and western geographic boundaries of the state of Louisiana and associated Federal Outer Continental Shelf (OCS) area, and extends roughly from the northward limit of Miocene deposits to the continental shelf edge to the south. This area was selected because of the number and good distribution of deep wells, and the occurrence of hydrocarbon producing reservoirs throughout the region, both areally and vertically. More than 65,000 wells have been drilled in this region.

Net sand data for 1,738 wells (Figure 5.7) provided by a major petroleum company, coupled with reservoir data from 4,790 reservoirs in more than 1,100 gas fields onshore and offshore south Louisiana submitted to the Federal Power Commission by the interstate pipeline companies (Form 15), are the basic data used in this resource evaluation and estimation of the amount of methane gas dissolved in formation water within the area studied.

The first tasks involved extraction and tabular listing of the required reservoir parameters from the FPC data set and geographic location of the gas fields. Each of the reservoirs in the Form 15 data set were cumulated and plotted by 1,000 foot depth intervals.

Both on the distribution and density of control (Table 5.1) a decision was made to limit the assessment to the depth interval 2,000 to 19,000 feet. Reservoir data extracted and tabulated included average top depth, average thickness, average porosity, average temperature (converted from degrees Rankine to degrees Fahrenheit), and initial reservoir pressure. These data were keyed to the American Petroleum Institute (API) unique number for the field's discovery well (or an alternate) and to FPC field name and number to obtain geographic reference. [These data are included in the complete report].

Table 5.1: Number of Reservoirs Listed in the FPC Form 15 Data File by Depth Interval

Depth Interval	Number of Onshore Reservoirs	Number of Offshore Reservoirs	Number of Reservoirs Onshore and Offshore
2,000–3,000	1	11	12
3,000–4,000	4	11	15
4,000–5,000	12	45	57
5,000–6,000	13	88	101
6,000–7,000	22	129	151
7,000–8,000	66	163	229
8,000–9,000	129	222	351
9,000–10,000	154	313	467
10,000–11,000	296	364	660
11,000–12,000	318	336	654
12,000–13,000	341	303	644
13,000–14,000	287	263	550
14,000–15,000	240	114	354
15,000–16,000	205	89	294
16,000–17,000	110	53	163
17,000–18,000	53	16	69
18,000–19,000	17	2	19
Totals	2,268	2,522	4,790*

*Total reservoirs.

Source: DOE/FERC-0029

Fluid pressure gradients (calculated from initial reservoir pressures in the FPC data set) for about 18 percent of the reservoirs examined were hydropressured (about 0.465 psi/ft), 10 percent were superpressured (greater than or equal to 0.7 psi/ft) and 72 percent were intermediate (greater than 0.465 psi/ft but less than 0.7 psi/ft).

This is not indicative of the distribution of the resource base among the three categories, but of the distribution of initial reservoir pressures in nonassoicated gas producing reservoirs in southern Louisiana onshore and offshore. Because more data are available for the intermediate pressured reservoirs, their resource potential can be estimated with greater reliability than can the resource potential of hydropressured and superpressured reservoirs.

Methane Solubility

The data listing also includes a methane solubility calculation of each reservoir derived from the Culberson and McKetta (7) curves (Figure 5.1) (extended) and the average temperature and initial pressure of the given reservoir. Formation waters were assumed to be fresh and saturated with methane for this calculation, although it was recognized that methane solubility is also a function of salinity.

Considering only the relationship of temperature and pressure to methane solubility, Figure 5.1 shows that only 32 scf of methane will be contained in each barrel of water at saturation in a reservoir with a temperature of 200°F and a pressure of 10,000 psi. However, at that same pressure, a temperature increase to 320°F increases the methane content to 52 scf per barrel of water.

In view of the fact that methane solubility increases as temperature increases above about 158°F (70°C) (particularly in combination with increased pressure), four generalized depth of geotherm maps were prepared for this investigation. Figures 5.9, 5.10, 5.11, and 5.12 illustrate the depth of occurrence of temperatures of 158°F, 212°F, 248°F and 302°F respectively. These maps were derived from average reservoir temperature and depth data from the FPC Form 15 file. Each reservoir was arbitrarily assigned the latitude-longitude locations of the discovery well (or an alternate) by API unique number.

Depth to geotherm was then calculated for each reservoir location in each field (considered as a data point) using interpolation between recorded reservoir temperatures. Extrapolation above the uppermost, or, below the deepest sampled interval using that interval's gradient, was also calculated subject to a maximum temperature difference, between the nearest recorded temperature and the desired geotherm temperature of 10°F. To maintain uniqueness for automatic contouring of the geotherms, the temperature was assumed to be monotonically increasing with depth, i.e., only intervals with gradient greater than zero were considered. Equations for the interpolation and extrapolation procedure were derived in the following manner.

For an interval, let T_0 and Z_0 be the temperature and depth of a reservoir, while T and Z are the temperature and depth of a deeper reservoir. Then the interval gradient, GT, is:

$$GT = (T - T_0)/(Z - Z_0)$$

In all cases, to find the depth to geotherm ZT_k:

$$ZT_k = Z_0 + (T - T_0)/GT$$

Figure 5.9: Depth of Occurrence of the 158°F Geothermal Surface in Southern Louisiana and Adjacent Continental Shelf

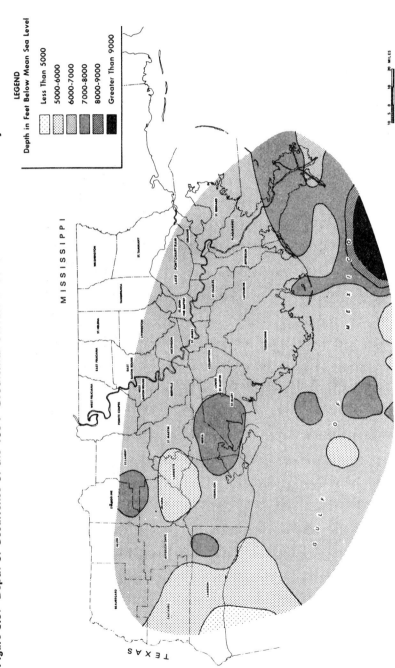

Source: DOE/FERC-0029

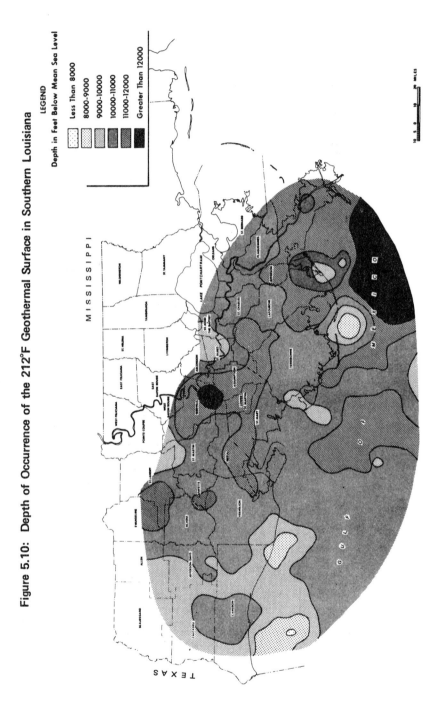

Figure 5.10: Depth of Occurrence of the 212°F Geothermal Surface in Southern Louisiana

Source: DOE/FERC-0029

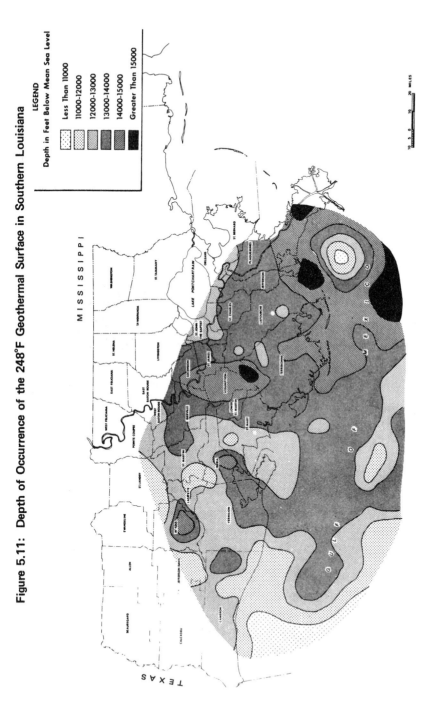

Figure 5.11: Depth of Occurrence of the 248°F Geothermal Surface in Southern Louisiana

Source: DOE/FERC-0029

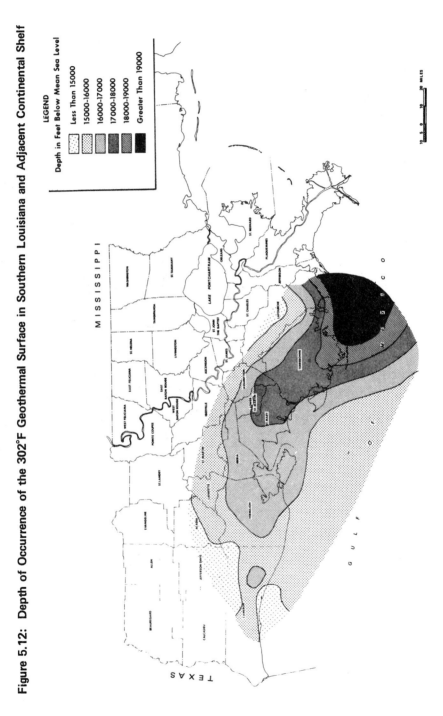

Figure 5.12: Depth of Occurrence of the 302°F Geothermal Surface in Southern Louisiana and Adjacent Continental Shelf

Source: DOE/FERC-0029

The computer-generated file of depths to each of the geotherms is an input to a computer contouring program. For purposes of plotting, interpolations and extrapolations were treated the same. Contour maps generated by automatic plotter were edited to ensure a more realistic presentation of the data.

Comparison of the maps of the depths of occurrence of the four geothermal surfaces (Figures 5.9 through 5.12) to the map of the depth of occurrence of geopressure (Figure 5.6), reveals that: (a) the 158°F geothermal surface (Figure 5.9) occurs in the normally pressured zone with few exceptions; (b) the 212°F geothermal surface (Figure 5.10) also occurs primarily in the normally pressured zone; (c) the 248°F geothermal surface (Figure 5.11) usually occurs within the geopressured zone; and (d) the 302°F geothermal surface (Figure 5.12) occurs almost totally within the geopressured zone.

Therefore, the largest concentrations of dissolved methane gas will usually be found at depths greater than those shown on Figures 5.11 and 5.12, especially in the higher pressured reservoirs.

Broad upwarps and downwarps of geothermal surfaces usually reflect gross lithologic changes. Large masses of interconnected sand beds appear as downwarps because heat is more easily distributed in the system by moving fluids and also because sand is a better conductor of heat than shale. On the other hand, shales restrict fluid movement causing heat buildup in the water of underlying sand beds. That maximum depths of occurrence of geopressure usually coincide with maximum depths of occurrence of geothermal surfaces can be seen by comparing Figures 5.6 and 5.10.

Porosity-Feet

The porosity, or percentage of pore space, of sand and shale beds in the northern Gulf of Mexico basin generally decreases with depth as a result of compaction with burial, but the shale bed porosity decreases more rapidly. The occurrence of geopressure, as shown on Figure 5.13 from Stuart (14), changes this trend. This is significant in that greater pore space than otherwise would be expected is available at depth to contain larger quantities of water containing the dissolved methane.

In this estimation of resource base, only the pore space available in sand beds was considered. If one knows the porosity, the net thickness of the sand deposits, and the areal extent of the sand deposits, the volume of pore space available to contain the water that contains the resource base can be calculated.

In order to calculate the volume of pore space available it is necessary to determine the number of feet of pore space or porosity-feet (porosity multiplied by net sand thickness) available in each control well to be used. For this estimate of resource base, calculation of porosity-feet began with the computation of a weighted porosity value for each 1,000 foot interval of depth from 2,000 to 19,000 feet for the study area. By taking reservoir depth, porosity, and thickness for each reservoir in each field (considered a data point) in the FPC data set, the interval porosity weighted by thickness was computed for each interval.

A weighted porosity was obtained for each of the seventeen 1,000 foot intervals. These weighted porosity values were then multiplied by the recorded sand thickness in the respective 1,000 foot intervals for each of the 1,738 wells in the net sand data file. (The complete report includes the method of computation and the net sand data file). The computer operation involved was programmed to discriminate between no sand values (0) and no value available for a particular well and interval where interval depth exceeded well depth.

Figure 5.13: Relationship of Porosity to Depth for
Sand and Shale (Clay) Beds (14)*

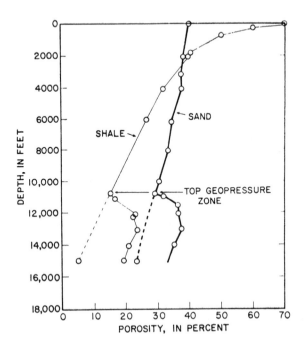

*In the hydropressured and geopressured zones of the northern
Gulf of Mexico basin.

Source: DOE/FERC-0029

The porosity-feet values were plotted by well site location, contoured by machine, and then edited to ensure realistic map presentation.

Figure 5.14 is an example of an edited version of one of the porosity-feet maps obtained. Maximum porosity-feet values, shown on Figure 5.14 for the 11,000 to 12,000 foot depth interval in the south Louisiana area, appear along the same three trends of maximum sand deposition shown on Figure 5.8. The volume of pore space available in any area shown on the porosity-feet map can be estimated by multiplying the contour value by the size of the area.

Estimated Methane Content of Water in Sand-Bed Aquifers

By assuming that the available pore space is filled with water and that the water is saturated with dissolved methane gas for the existing physical and chemical conditions, an estimation of the quantity of gas available as a resource base can be calculated.

Figure 5.14: Porosity-Feet of Sands in Southern Louisiana and Adjacent Continental Shelf—11,000 to 12,000 Feet

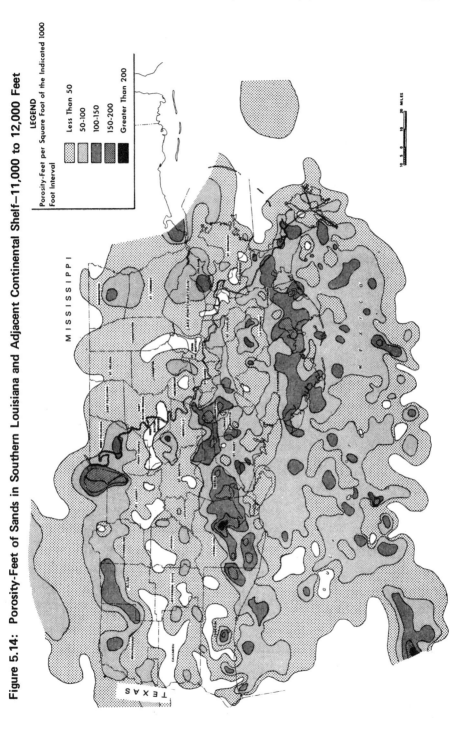

LEGEND

Porosity-Feet per Square Foot of the Indicated 1000 Foot Interval

Less Than 50
50-100
100-150
150-200
Greater Than 200

Weighted methane solubility values were computed from the pressure and temperature values in the FPC reservoir data set for each of the 1,000 foot intervals concurrently with weighted porosity. The weighted solubility values were then multiplied by the porosity-feet values in the respective 1,000 foot intervals for each of the 1,738 wells in the data file.

The computer operation discriminated between no porosity-feet values (0) and no value available as explained previously. The methane content in standard cubic feet per barrel per square foot per 1,000 foot interval thus obtained was then multiplied by a factor of 0.178 to convert to standard cubic feet of methane per square foot per 1,000 foot interval at each well location for machine plotting and contouring.

The computer plotter generated maps (Figure 5.15 is shown as an example) were checked and edited to ensure accuracy of fit of contours to the data points. Uncertainties were resolved by conservative interpretation. The areal extent of each contour value on each of the 17 maps was determined by digital planimetry and multiplied by the average methane content per unit area of that particular area.

Results were summed to yield methane content per contour interval for each 1,000 foot depth interval and then summed to yield total methane content for that 1,000 foot interval (Table 5.2). Methane content by intervals thus obtained were summed to yield total estimated methane content in standard cubic feet for the interval 2,000 to 19,000 feet in the study area, assuming saturated fresh formation waters. The total dissolved methane resource base is, under these conditions, estimated to be 6,143 trillion cubic feet.

Correction for Formation Water Salinity

It is well established that deep subsurface formation water in the northern Gulf of Mexico basin is primarily of sodium chloride type with dissolved solids content ranging from less than 10 g/l to more than 200 g/l. Laboratory and field studies indicate that the solubility of methane gas in water decreases as salinity increases. Therefore, to present a more realistic estimate, an attempt was made to correct for salinity effects by a factor based upon the average salinity per 1,000 foot interval calculated from the spontaneous potential curve of selected geophysical well logs, as sodium chloride, and based upon the work of Wallace, Taylor and Wesselman (15).

The salinity correction factor used (Table 5.2) was derived by plotting the preliminary results from two geopressured reservoir tests in Louisiana and data from the study by Buckley, Hocott and Taggert (1) in relation to experimental results from laboratory tests using distilled water and methane at various temperatures and pressures by Culberson and McKetta (7). The salinity adjustment reduces the total estimated dissolved methane resource base (Table 5.2) to 3,264 Tcf.

Further reduction (amount undetermined) should also be considered to account for those waters expected to be undersaturated in methane at shallower depths, and temperature less than the critical 200° to 230°F required for thermal diagenetic processes in shale beds (16) to provide water carrying dissolved methane of thermogenic origin.

Reliability of Estimates

In view of the quantity and source of the types of data, the assumptions made, and the techniques used, this estimate of the methane dissolved in water resource base for sand bed reservoirs in southern Louisiana and the adjacent Gulf of Mexico continental shelf is probably reliable within less than an order of mag-

Figure 5.15: Dissolved Methane Content of Sand-Bed Aquifer Systems for the
Depth Interval from 17,000 to 18,000 Feet Below Mean Sea Level

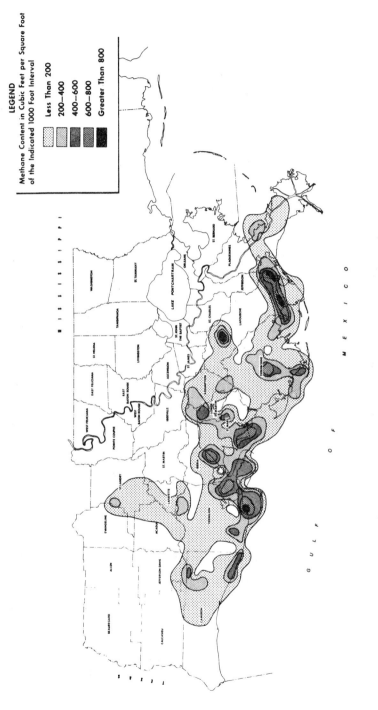

Source: DOE/FERC-0029

Table 5.2: Estimate of Methane Dissolved in the Formation Water of Sand Bed Aquifers of Southern Louisiana and Adjacent Gulf Continental Shelf

Depth Interval Below Mean Sea Level	Area Planimetered (mi²)	Total Estimated Methane* (Tcf)	Average Methane* (Tcf/mi²)	Average Salinity** (mg/ℓ)	Number of Salinity Determinations	Salinity Correction Factor***	Total Estimated Methane After Salinity Adjustment*** (Tcf)
2,000-3,000	59,978	354.25	0.00591	78,000	101	0.610	216.09
3,000-4,000	73,043	572.33	0.00784	102,000	222	0.555	318.65
4,000-5,000	75,529	687.17	0.00909	114,000	257	0.540	371.07
5,000-6,000	59,973	584.58	0.00975	124,000	243	0.525	306.90
6,000-7,000	60,447	603.45	0.00998	131,000	201	0.520	313.95
7,000-8,000	57,034	669.95	0.01170	139,000	233	0.510	341.67
8,000-9,000	58,008	470.06	0.00810	142,000	206	0.505	237.38
9,000-10,000	56,367	418.67	0.00743	146,000	172	0.505	211.43
10,000-11,000	58,023	380.48	0.00656	135,000	128	0.510	194.04
11,000-12,000	54,411	325.69	0.00599	125,000	107	0.525	170.99
12,000-13,000	53,274	294.16	0.00552	115,000	85	0.535	157.38
13,000-14,000	41,254	227.09	0.00550	119,000	69	0.530	120.36
14,000-15,000	30,714	181.33	0.00590	111,000	52	0.540	97.92
15,000-16,000	32,417	165.64	0.00511	120,000	34	0.530	87.79
16,000-17,000	22,919	119.71	0.00522	92,000	27	0.578	69.19
17,000-18,000	9,293	47.98	0.00516	95,000	14	0.571	27.40
18,000-19,000	8,515	40.86	0.00480	108,000	5	0.545	22.27
Total methane for all intervals		6,143.4					3,264.5

*The estimate of total methane and average methane per square mile assumes freshwater at saturation.

**The average salinity values were derived from geophysical log calculation of dissolved solids as sodium chloride.

***The adjustment for salinity is a maximum adjustment derived from the work of Wallace, Taylor and Wesselman (15).

Source: DOE/FERC-0029

nitude. Estimates for the 1,000 foot intervals from 5,000 to 17,000 feet are considered more reliable than the other intervals estimated because more reservoir data were available (Table 5.1). Well control for the determination of net sand thicknesses was considered adequate.

However, the resource base could be better defined by adding net sand data from deep wells drilled since 1972 (the most recent wells in the data set), and by adding future deep well data as it becomes available.

The resource base estimate in this study can easily be revised as new field and laboratory data define more accurately the effects of salinity in combination with pressure and temperature on methane solubility in water. In addition, testing of a number of aquifers is needed before the serious uncertainties surrounding the assumption of saturation can be satisfied. Certainly, saturation would be expected in aquifers in the vicinity of producing hydrocarbon reservoirs (i.e., the intervals of maximum data in Table 5.1), and in deeper aquifers.

Other Hydrogeologic Factors That Could Significantly Increase or Decrease Estimates

Influx of additional methane bearing water from the shale beds, that constitute the largest volume of sediment in the Gulf Coast, to sand bed aquifers during production will probably increase the methane dissolved in water resource base. Many conventional hydrocarbon reservoirs have produced for years without appreciable pressure reduction (17). Shale water drive has been suggested as an explanation.

Short term tests of two geopressured reservoirs in Vermillion Parish, Louisiana (the Edna Delcambre, et al No. 1 well) revealed a produced gas/water ratio about three times the ratio corresponding to saturation of reservoir brine with natural gas at reservoir pressure and temperature. Randolph (18) has hypothesized that this may be due to immobile gas trapped in a few percent of the reservoir pore space. He examined the possibility that expansion of this immobile gas, as reservoir pressure was reduced by water production, resulted in the trapped gas becoming mobile and contributing to production.

His comparison of computer simulation with the preliminary data from the Delcambre well suggests this hypothesis, and that gas in place in highly geopressured reservoirs may be 5 to 10 times as great as estimated herein. However, much more test data and analysis are required before this or any other hypothsis can be accepted as fact.

Buckley, Hocott and Taggert (1) found that shallow Miocene reservoirs were undersaturated in dissolved natural gas almost everywhere, but that the deeper Frio was usually found to be saturated. If these results are typical, the resource base is overestimated in the shallower depth intervals. No attempt was made to determine the probable amount of overestimation, however, because of the many complexities involved.

RECOVERABILITY OF GAS DISSOLVED IN WATER IN GEOPRESSURED AQUIFERS OF THE U.S. GULF OF MEXICO COAST

There are no definitive data on the long term production of natural gas dissolved in water from aquifers on the Gulf Coast known as of March 1979. The production performance of wells drilled into aquifers must be predicted on the basis of theoretical considerations and prior experience with conventional gas wells and water wells.

Although the production rate from any aquifer will inevitably decrease with time, the rate of decline will depend on the complex interaction of a number of physical factors, including expansion of the fluid as pressure falls, exsolution of dissolved gas, sandstone compaction, fluid influx from adjacent shales and the permeabilities of the reservoirs to both fluids and gases.

Any change in reservoir pressure can affect all of these factors. A change in any factor will affect the others. Existing theoretical models, with the aid of computers, are capable of solving such problems, but their predictions are valid only to the extent that all factors are taken into proper account by use of accurate input data (for example the mechanical properties of the sandstone). In general such data are not yet available and the best theoretical predictions that can be made at this time do not enjoy the full confidence of the petroleum engineering profession.

Randolph (19) for instance, has examined the sensitivity of "ultimate production" to several geopressured reservoir parameters, concluding that "today's projections of production contain enormous uncertainties due to lack of quantitative definition of drive mechanisms, pressure and temperature dependence of permeability, relative permeabilities to gas and water, saturation of pores and drainage limitations due to faulting."

Simplistic Model

In the absence of the direct engineering data required as input for sophisticated predictive techniques, perhaps the best approach to estimation of the recoverable fraction of the resource base is to employ a very simple model. Consider a sandstone aquifer at some initial volume and fluid pressure, V_i and P_i. After some period during which water is produced, some depletion of the aquifer has occurred; both the volume and the pressure have been reduced. Let V_f and P_f be the final values. The total volume of water produced in this time is

$$V_i - V_f,$$

and the fraction F of the original resource that has been recovered is

$$F = \frac{V_i - V_f}{V_i} .$$

The pressure has declined by

$$P_i - P_f .$$

The model postulates a simple linear relationship between the pressure decline and the fraction of fluid produced:

$$F = C_e (P_i - P_f)$$

C_e is the effective compressability, representing the fractional change in reservoir pore volume for each unit change in fluid pressure. Given a reservoir at a particular depth, with a known initial pressure in excess of the hydrostatic value, one can postulate that it is produced until the pressure decreases to hydrostatic (see below); thus P_f becomes P_h. By assuming a value for C_e, based on prior experience with oil and gas reservoirs, one can calculate the recoverable fraction:

$$F = C_e (P_i - P_h) .$$

The application of this model to the resource base identified in the previous section will be discussed below. First, however, some of the model's shortcomings should be pointed out. The assumption of a linear relationship between fluid production and pressure decline ignores the complexity of the forces that cause fluid to flow from the reservoir. The value of C_e that represents the behavior of a gas reservoir may not correctly represent the behavior of aquifers with dissolved gas.

Thus, while this model is straightforward and easy to use, it is unlikely to correctly predict the fraction of the resource base that could be produced. Furthermore, the model addresses only the broad issue of resource recovery, without regard to economic factors.

Reservoir Classifications

The resource base identified in the previous section can be categorized according to the fluid pressure gradient in the aquifers as follows:

Hydropressured	~0.465 psi/ft
Geopressured Intermediate	~0.466 to 0.699 psi/ft
Geopressured Superpressured	0.700 psi/ft and higher

Reservoir pressure is calculated by multiplying the pressure gradient by the depth. A hydropressured reservoir at a depth of 5,000 ft would have a pressure of 5,000 x 0.465, or 2,325 psi. This is the pressure at the base of a column of brine 5,000 ft high. If a well were drilled into such a formation, water would rise to the wellhead but no higher. The well would have to be pumped to cause appreciable production of the water, and the contained natural gas. Because there is essentially no documented experience with gas production from this type of reservoir, one cannot make a credible estimate of the fraction of the resource base that could be recovered.

The intermediate reservoirs contain somewhat higher pressures. Consider a 12,000 ft deep aquifer with a pressure gradient of 0.699 psi/ft. The formation pressure would be 12,000 x 0.699, or 8,388 psi. A column of brine extending to the ground surface would exert a pressure of 12,000 x 0.0465 or 5,580 psi. Therefore a well completed in this reservoir would have an excess shut-in wellhead pressure of 8,388 to 5,580, or 2,808 psi.

As long as excess pressure were available to force fluid up the well bore, the well could be produced without pumping. The simple model can be used to estimate the maximum fraction of reservoir fluid produced from the reservoir in question. As an effective compressibility, based on experience with the production of petroleum and gas caps in geopressured reservoirs, a reasonable figure for C_e of 10^{-5} is assumed. Although some authorities have argued that a much higher value of C_e should be used for fluid production from geopressured aquifers, there are no definitive data to support their contention.

Application of the model gives a recoverable fraction of 2,808 x 10^{-5} = 0.028, or 2.8 percent for this intermediate reservoir. Pumping from or reinjection to the aquifer could conceivably increase this percentage but documented relevant studies are not known to exist.

For a particular superpressured formation with a pressure gradient of 0.865 psi per foot and a depth of 15,000 ft, the calculated maximum recoverable fraction is 6 percent, again with the assumption that C_e = 10^{-5}.

Hawkins (20) has employed essentially the same model as used above, but has chosen a value for C_e of twice the value used here. He has discussed the sensitivity of the calculated recoverable fraction of the resource base to the value of C_e assumed, and also has noted that the absence of empirical data on pore volume compressibilities under geopressured conditions and the possible influx of shale bed water into geopressured sandstones contribute a substantial uncertainty to the assignment of a value for C_e.

The magnitude of the resource base identified in the previous section includes methane dissolved in brine without regard to particular pressure gradient regimes. The computer programs employed in this estimation were not written to distinguish between hydropressured, intermediate and superpressured aquifers. Since the recoverable fraction of the resource base is dependent on the pressure gradient the computer analysis would have to be revised and rerun before a quantitative estimate of recoverability could be derived, even by use of the simple model. The substantial uncertainties that exist severely limit the value of such analysis;

Reservoir Size Distribution

Another important factor in the estimation of recoverability is individual reservoir volume. The relationship between recoverability and volume is illustrated by the following example. Consider a reservoir 1 square mile in areal extent, 300 feet thick, with 20% porosity and 40 standard cubic feet of gas dissolved in each barrel of reservoir fluid. The total gas content is:

$$\frac{(5,280)^2 \ ft^2 \ \times \ 300 \ ft \ \times \ 0.20}{5.62 \ ft^3/bbl} \ \times \ 40 \ scf/bbl \ = \ 1.2 \times 10^{10}$$

or 12 billion scf. Assume that 5 percent of this gas— about 600 million scf— is recovered before the returns of production decline below operating costs. Neglecting accumulated operating costs, the returns from this well would be quite insufficient to amortize the costs of drilling at the wellhead price of about $2.50 per Mcf.

Thus if essentially all of the resource base consisted of reservoirs of this size or smaller, the recoverable fraction would be zero at $2.50 per Mcf.

On the other hand, if the reservoir's areal extent were 4 square miles, the thickness were 700 feet, and 10 percent of the in place gas were recovered before returns dropped below operating costs, the amount produced would be over 11 billion cubic feet. A well drilled into such a formation might very well be an economically attractive prospect at a wellhead price of under $2.50 per Mcf. If the bulk of the resource base were represented by reservoirs of this size and larger, the total amount of gas recoverable at $2.50 could be substantial.

Thus, in order to construct a valid supply curve for methane dissolved in geopressured water, it is necessary not only to understand the mechanics of reservoir performance and the engineering properties of the reservoirs, but also to know their size distribution. Some information regarding the size of geopressured reservoirs can be learned by analyzing the data from existing gas wells and available seismic information.

Historically, however, drilling and geophysical measurements have tended to concentrate in areas with a high degree of structural complexity ("on structure") where the probability of encountering pockets of free gas is maximized. The volume of reservoirs that occur "on structure" tend to be small because of the relatively high density of faulting in such areas. Some authorities argue that reservoirs would be larger off structure, where there are fewer existing data. This point is quite controversial, however, and other authorities contend that there is likely to be little difference in reservoir size between areas with a high density of existing wells and other areas.

Although dozens of apparently large "prospect areas" have been identified by government funded projects on the Gulf coast, it is not clear how many of these large areas are subdivided into smaller reservoir units by subsurface growth faulting.

Summary and Conclusion

By making conservative assumptions about the mechanisms that drive the production of fluids from geopressured aquifers, and about production costs, one can readily conclude that only a fraction of one percent of the methane in place is economically recoverable.

An analysis with a more optimistic set of assumptions would suggest a much greater recoverability factor. Thus, estimates of the recoverability factor for methane dissolved in water vary by one to two orders of magnitude. Since the size of the U.S. resource base of methane dissolved in water is also uncertain by more than an order of magnitude, estimates of the recoverable amount (recoverability factor multiplied by resource base) are uncertain within a factor of about 1,000.

It is the conclusion of this sub task force that the present lack of basic technical information related to the amount and recoverability of methane dissolved in water precludes the derivation of a meaningful estimate of the amount recoverable at any particular wellhead price. It is the sub task force's recommendation that the Department of Energy support a vigorous research and development effort to provide the needed technical information.

ENVIRONMENTAL ASPECTS OF LARGE SCALE GAS RECOVERY FROM GEOPRESSURED AND HYDROPRESSURED AQUIFERS

This section is concerned with the various environmental problems that probably will arise with the development of any large scale gas recovery program from either geopressured or hydropressured aquifers along the Miocene trend of southern Louisiana.

Environmental Problems

When consideration is given to large scale gas recovery from these aquifers it probably should be envisioned as a series of smaller scale projects. The focus will be on one of these "smaller projects" which typically could be expected to produce some 50 million cubic feet of methane associated with upwards of one million barrels of brine per day.

The production, processing and disposing of these large volumes of brine from such a project will result in several problems from an environmental standpoint. The two most significant problems will be:

The possible subsidence of the land surface in the immediate vicinity of the producing wells, and

The disposal of the large volume of produced brine.

Lesser environmental problems may be thermal pollution, air pollution, noise, and land use considerations.

With the removal of large volumes of water from either a geopressured or hydropressured aquifer over an extended period of time, there is a strong probability that the area around the project will experience considerable subsidence.

Surface subsidence has been a problem in many areas where large volumes of water (or oil) have been removed. This can be particularly troublesome in low relief, low elevation coastal areas, typical of the Miocene Belt of southern Louisiana. Subsidence of any appreciable degree in these swampy, coastal areas could be extremely bothersome, even in an undeveloped region.

This geographic area (the Miocene trend of southern Louisiana) is thought to be aseismic; however, the removal of large volumes of water, with the resultant drastic reduction of reservoir pressures, conceivably could activate some of the growth faults in these aquifers. Subsequently, it is also possible that such fault movements could result in surface adjustments. Damage due to such movements would probably be limited to the immediate area where the projected fault plane(s) intersect the surface.

Concerning the disposal of large volumes of brine the easiest disposal method would be to discharge to nearby surface waters or into the Gulf of Mexico. This undoubtedly would be the most advantageous from an operating and economic standpoint. This disposal method, however, would raise environmental problems associated with the difference in composition of the produced brine and the receiving water.

There is also the probability that the temperature of the disposed brine will be considerably higher than that of the receiving water, thus resulting in "thermal pollution." Strong objections could be expected from both commercial and sport fishermen, as well as regulatory authorities.

A second and a more environmentally acceptable disposal method would be to inject the brine into shallower, normal pressured aquifers with a series of disposal wells. From an operating and economic standpoint this has great disadvantages. A high capacity disposal well may handle as much as 10,000 barrels per day with suitable high pressure pumps. Thus, a project such as this would require a large network of active disposal wells (plus stand-by or reserve wells) with the associated surface installations, including high pressure pumps which would be required to effect satisfactory injection rates.

Regulatory and Permitting Requirements

These potential environmental problems will involve many Federal and State laws with the resultant difficulties of obtaining permits from the various regulatory agencies. The project discussed herein would be subject to requirements of various sections of at least the following Federal laws:

1. National Environmental Policy Act of 1969 (NEPA)
2. Coastal Zone Management Act of 1972 as Amended in 1976 (if the project is located where it would affect the coastal zone)
3. Clean Water Act of 1977 (Federal Water Pollution Control Act) of 1972, as Amended)
4. Safe Drinking Water Act of 1974
5. Clean Air Act of 1977
6. Resources Conservation and Recovery Act of 1976

The permitting process for a project such as that envisioned will probably be quite time consuming. Because of the research nature and questionable economics of at least the initial projects, it is quite likely that there would be some federal funding. Thus, an Environmental Impact assessment would be required under the National Environmental Policy Act.

Even if a project is completely funded by private sources, the EIS process may be triggered if a permit for a discharge to surface or Gulf waters is needed (see subsequent discussion). Further, if a project is located near Louisiana's Coastal Zone (which these probably will be), "certification of consistency" with the State's approval Coastal Zone Management Program would also be required.

As mentioned previously, two important environmental impacts are probably subsidence and possible displacement along faults caused by large scale water withdrawal. These potential problems are real, but are difficult to quantify. Thus, it is quite likely that project approval would be delayed considerably pending debate and resolution of these issues. Another important related (emotional) environmental matter is the possibility that a large accidental release or spill of hot water or brine could occur and devastate adjacent agriculture lands, and/or disrupt the ecological balance of nearby bays or estuaries.

Such a threat is real, since producing and handling large volumes of hot brine present definite corrosion/erosion problems that could lead to such incidents.

Disposal of the produced water into surface waters or the Gulf would require that a National Pollution Discharge Elimination System (NPDES) permit be obtained in accordance with the Federal Water Pollution Control Act. Such a discharge would be classified as a "new source." This might also activate the EIS process. Unless a state has NPDES permitting authority, for a "new source," an applicant must prepare and submit an environmental assessment to EPA. Based on the information contained in the assessment, the EPA makes a judgement either to prepare an EIS or issue a "Negative Declaration." As discussed above, it would be difficult for the Agency not to prepare an EIS for this type project.

If the produced water contains even minimal (low part per billion) quantities of "toxic pollutants" as defined by the Clean Water Act Amendments of 1977 [Section 301 (B), (C), and (D)], it is quite likely that "special" effluent limitations to control the discharge of these substances would be imposed under and NPDES permit. Because of the large volumes of water that must be handled, removal of toxic pollutants probably would not be economical and reinjection would then be necessary.

If the project is located in the coastal zone, facility siting (land use) will probably require approval under the Coastal Zone Management Act of 1972. This could be a time consuming process. A Corps of Engineers' "dredged or fill material permit" probably will be required under Section 404 of the Clean Water Act for construction of any water cooling or holding basins.

It is quite likely that the produced brine will require subsurface disposal. For such reinjection, permits would be required from the state under the Safe Drinking Water Act of 1974.

Air emissions from the entire facility would also be stringently controlled under state rules and regulations issued pursuant to the Clean Air Act. Of particular concern are the very stringent "Prevention of Significant Deterioration" requirements included in Section 165 of the 1977 Amendments to the Act. If the produced water is disposed of underground, the exhausts associated with the engines needed to provide the tremendous horsepower necessary for reinjection would be a major problem. It is thought unlikely that noxious gases will be produced from the resource base defined in this report.

Last, but not least, there is the Resource Conservation and Recovery Act of 1976 (RCRA), commonly referred to as the Solid Waste Disposal Act. In proposed regulations (40 *CFR* Part 257) published in the *Federal Register* on February 6, 1978 "solid waste" is defined as ". . .any garbage, refuse, sludge from a waste treatment and other discarded materials, including solid, liquid, semi-solid, or contained gaseous material resulting from industrial, commercial, mining, and agriculture operations. . ." Thus, additional environmental constraints not covered by the above mentioned laws may result from this act.

REFERENCES

(1) Buckley, S.E., Hocott, C.R., and Taggart, M.S., Jr., "Distribution of Dissolved Hydrocarbons in Sub-Surface Waters (Gulf Coastal Plain)," *Habitat of Oil--a Symposium*, Tulsa, OK, Am. Assoc. Petroleum Geologists, (1958).

(2) Ritch, H.J. and Smith, J.T., "Evidence for Low Free Gas Saturation in Water Bearing Bright Spot Sands," *Soc. Professional Well Log Analysts, Seventeenth Annual Logging Symposium, Trans.*, June 9-12, (1976).

(3) Papadopulos, S.S., Wallace, R.H., Jr., Wesselman, J.B., and Taylor, R.E., "Assessment of Onshore Geopressured-Geothermal Resources in the Northern Gulf of Mexico Basin," *Assessment of Geothermal Resources of the United States-1975*, U.S. Geol. Survey Cir. 726, (1975).

(4) Jones, P.H., "Natural Gas Resources of the Geopressured Zones in the Northern Gulf of Mexico Basin," *Natural Gas from Unconventional Geologic Sources*, The National Research Council, Board of Mineral Resources, Commission on Natural Resources, National Academy of Sciences, (1976).

(5) Hise, B.R., "Natural Gas from Geopressured Zone," *Natural Gas from Unconventional Sources*, The National Research Council, Board of Mineral Resources, Commission on Natural Resources, National Academy of Science, (1976).

(6) Dodson, C.R. and Standing, M.B., "Pressure-Volume-Temperature and Solubility Relations for Natural-Gas-Water Mixtures, *API Drilling and Production Practices*, (1944).

(7) Culberson, O.L., and McKetta, J.J., Jr., "Phase Equilibria in Hydrocarbon-Water Systems, III—The Solubility of Methane in Water at Pressures to 10,000 psia," *Petroleum Trans., AIME*, vol 192, (1951).

(8) O'Sullivan, T.D. and Smith, N.O., "The Solubility and Partial Molar Volume of Nitrogen and Methane in Water and in Aqueous Sodium Chloride from 55 to 125° and 100 to 600 Atm," *Jour. Phys. Chemistry*, vol 74, No. 7 (1970).

(9) Hardin, G.C., "Notes on Cenozoic Sedimentation in the Gulf Coast Geosyncline, U.S.A.," *Geology of the Gulf Coast and Central Texas*, edited by E.H. Rainwater and R.R. Zingula, Houston, Texas, Houston Geological Society (1962).

(10) Meyerhoff, A.A. (ed), "Geology of Natural Gas in South Louisiana," *Natural Gases of North America: Am. Assoc. Petroleum Geologists, Memoir 9*, vol 1, (1968).

(11) Wilhelm, O. and Ewing, M., "Geology and History of the Gulf of Mexico," *Geol. Soc. America Bull.*, vol 83, (1972).

(12) Murray, G.E., *Geology of the Atlantic and Gulf Coastal Province of North America*, New York, Harper Brothers, (1961).

(13) Jones, P.H., *Hydrology of Neogene Deposits in the Northern Gulf of Mexico Basin*, Louisiana Water Resources Research Inst. Bull. GT-2, (1969).

(14) Stuart, C.A., "Geopressures," *Proceedings of the Second Symposium on Abnormal Subsurface Pressures, Baton Rouge, LA: Louisiana State University, January 30*, (1970).

(15) Wallace, R.H., Jr., Taylor, R.E. and Wesselman, J.B., "Use of Hydrogeologic Mapping Techniques in Identifying Potential Geopressured-Geothermal Reservoirs in the Lower Rio Grande Embayment, Texas," *Third Geopressured-Geothermal Energy Conference, November 16-18, Univ. Southwestern Louisiana, Lafayette, LA: Proceedings* (1977).

(16) Burst, J.F. "Diagenesis of Gulf Coast Clayey Sediments and Its Possible Relation to Petroleum Migration," *Am. Assoc. Petroleum Geologist Bull.*, vol 53, no . 1, (1969).

(17) Wallace, W.E., "Water Production from Abnormally Pressured Gas Reservoirs in South Louisiana," *Gulf Coast Assoc. Geol. Socs. Trans.*, vol 12, (1962).

(18) Randolph, P.L., "Natural Gas Content of Geopressured Aquifers," *Third Geopressured-Geothermal Energy Conference, November 16-18, Univ. Southwestern Louisiana, Lafayette, LA, Proceedings* (1977).

(19) Randolph, P.L., *Natural Gas from Geopressured Aquifers?:* Soc. of Petroleum Engs., *AIME,* preprint No. SPE 6826, (1977).
(20) Hawkins, M.F., Jr., *Investigations on the Geopressure Energy Resource of Southern Louisiana,* Department of Energy final report ORO-4889-14, Louisiana State Univ., Baton Rouge, LA (1977).

AN ALTERNATE ANALYSIS OF THE RESOURCE POTENTIAL

The information in this and the following section is based on *Enhanced Recovery of Unconventional Gas: The Program— Volume II* (of 3 Volumes), October 1978, prepared by V.O. Kuuskraa and J.P. Brashear of Lewin and Associates, Inc., T.M. Doscher of University of Southern California and L.E. Elkins for the U.S. Department of Energy under DOE Contract EF-77-C-01-2705.

A variety of geological assumptions lead to the conclusion that the resource in place is large. The essential question, however, is not the total size of the resource, but the portion that may be technically and economically recoverable.

The results of recent work suggest that, even at prices up to $4.50/Mcf, the technically and economically recoverable methane may be a small fraction of the initial estimates.

Further, geologic and reservoir studies are required before the full potential of this resource base can be confidently assessed. It is possible, however, to establish the minimum geologic conditions that must be found for economic recovery of methane from these aquifers. In addition, the currently limited resource data base can be measured by these standards to illuminate the economic potential. The essential but missing data would provide focus to a reasearch program aimed at ascertaining the total potential.

Nature of the Problem

The basic technology for recovering methane from the geopressured aquifers consists of drilling one or more production wells capable of producing vast quantities of gas bearing water, installing facilities to capture the methane that comes out of solution at atmospheric conditions, and disposing of the water once it has given up its gas.

With improved extraction facilities, it is believed that up to 85% of the gas in solution can be recovered from the produced water. However, only about 2 to 5 percent of the reservoir's water can be produced in 30 years or before exhausting the reservoir's drive mechanism.

The modeling of defined Texas Gulf Coast geopressured aquifers gives recoveries of about 2 to 3 percent. The 5 percent recovery efficiency assumes:

(1) a compression drive coefficient of about 11×10^{-6} psi^{-1};

(2) high permeability (100 md);

(3) thick pay (500 feet); and

(4) high pressure (14,000 psi).

Artifically lifting methane bearing water from the substantial depths of geopressured aquifers was considered to be beyond conceivable economic limits.

Thus, even under optimistic assumptions on reservoir properties and high methane extraction efficiency, less than 5 percent of the gas resource in place is technically recoverable, as shown in Figure 5.16 for three Texas and Louisiana Gulf Coast reservoirs.

Figure 5.16: Analysis of Technical Recovery Efficiency from Geopressured Aquifers

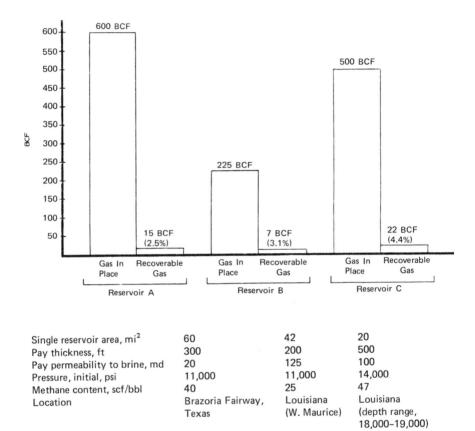

Single reservoir area, mi^2	60	42	20
Pay thickness, ft	300	200	500
Pay permeability to brine, md	20	125	100
Pressure, initial, psi	11,000	11,000	14,000
Methane content, scf/bbl	40	25	47
Location	Brazoria Fairway, Texas	Louisiana (W. Maurice)	Louisiana (depth range, 18,000–19,000)

Source: DOE EF-77-C-01-2705

Moreover, certain technological problems must be solved before even this quantity is recoverable, including:

- Overcoming any well completion or production problems that might impede high rates of production. The poorly consolidated nature of the reservoir sands may pose problems of sand control and may, as the pressure is drawn down, significantly reduce the permeability in the portions of the formation nearest the wellbore.

- Developing high efficiency methane extraction facilities. For optimum economics, particularly for the lower methane concentration brines, the goal of methane extraction operation would be to recover 85% or more of the methane dissolved in the water.

● Disposing of the produced brine in an environmentally safe manner. The brine will need to be injected into the subsurface because of the salinity and residual methane content or reinjected into the original producing formation to maintain the production rate and counteract subsidence.

Economic Feasibility

Minimum economic feasibility is governed by two broad characteristics:

● Large initial investment costs due to the depth of geopressured formations, the size of the casting and extraction facilities required to handle vast quantities of water, and the costs of reinjection wells to dispose of the spent brines.

● Substantial operating and power costs of handling, repressuring, and disposing of the produced water.

Given the cost of applying the technology, to be economic the resource base must support:

● High, sustained rates of water production; although numerous reservoir characteristics contribute to this, the dominant factors are the gross volume of the reservoir (area and thickness, with thickness becoming dominant once a minimum areal extent is reached), and the permeability of the pay.

● High concentrations of methane in the produced water; here the central factors, given the presence of methane, are high pressure, high temperature, and low salinity.

The Contribution of Ongoing R&D: Much of the theoretical analysis and all of the geologic data have accrued from the valuable and quality research efforts that are and have been underway for some time. The R&D program recommendations stemming from this analysis are an extension of this past work.

Minimum Economic Conditions

Basis for the Derivation: Economic recovery of methane from geopressured aquifers depends on meeting two minimum conditions:

● That the economic value of the gas contained in one barrel of produced water at least repay the operating costs of producing and disposing of that barrel of water; and

● That the rate of production be sufficiently high to repay the investment costs of the project.

These two factors, total methane content and production rate, are the two critical economic variables and provide a direct means for representing the numerous geologic and reservoir characteristics that influence these variables. In turn, the minimum required methane content and production rate are based on the operating, investment, and return on capital costs associated with the recovery technology.

The analysis considered the production of methane as the primary purpose and did not consider either the cost or the output value of thermal or hydraulic energy recovery. This was done for two reasons. First, the examination of thermal/hydraulic energy recovery was specifically excluded from the scope of this effort. Second, much of the area defined as being geopressured had temperatures between 200° and 300°F and thus had relatively low potential for thermal energy output. (The theoretical thermal output at 200°F is about ⅕ of the thermal output at 325°F).

Operating Costs — Overall operating costs are composed of four elements:

- General operating and maintenance costs (except for power).
- Capital depreciation on well production and disposal equipment (except for the well).
- Well operating and maintenance costs.
- Power costs involved in water disposal.

Estimates of these costs were developed from theoretical energy balance equations and from field experience with water production and disposal. Costs throughout this section are in constant 1977 dollars.

- General operating and maintenance costs, excluding power, to operate the water handling and methane extraction were estimated at $0.0125 per barrel of water produced. Adding a factor of 20% to cover overhead costs raises this amount to $0.015 per barrel of water.

- Capital depreciation costs were estimated at $0.0125 per barrel of water produced. This cost assumes an initial investment (for separation and handling facilities) of $20.00 per barrel of daily capacity (e.g., a 40,000 barrel per day facility would require $800,000 of investment costs), a 20 year life, and a 15% return on capital. Adding a factor of 10% to cover overhead costs raises this cost to $0.014 per barrel of water.

- Well operating and maintenance costs, to operate and maintain the production well and the four injection wells, were estimated at $100,000 per year plus 20% for overhead costs. At production rates of 35,000 to 40,000 barrels per day, well operating and maintenance costs would be $0.01 per barrel of water.

- Power costs were derived from the energy required to inject water into a two thousand net foot, hundred millidarcy, shallow salt water aquifer. This cost assumes a required average disposal wellhead pressure of 1,000 psi and $0.04 per Kw hr charge for large industrial use electricity. Making power costs of disposal a direct function of gas price provides the following equation: Power costs = [$0.01 + $0.005 (price of natural gas in $/Mcf)] x Disposal pressure (P_i in ksi)

- Combining the four components gives the following equation: total operating costs per barrel of water = $0.04 + [0.01 + 0.005 (price in $/Mcf)] x P_i (ksi)

Since operating costs have such a major and direct effect on the economics of producing methane from geopressured aquifers, these costs were validated by comparison with actual costs of the major water disposal company in east Texas and with independently derived cost estimates by outside consulting engineers. The above costs reconciled closely with these two sources:

Investment Cost Estimate: The investment costs are based on drilling, completing, and equipping a 17,000 foot well, onshore, using 7.0 inch casing. The cost items are provided below:

Drilling costs @ $80 per foot		$1,360,000
Tubing and well equipment		960,000
Reinjection costs Four 5,000 ft wells @ $40 per foot (drilled, completed, and equipped, including pump)		800,000
Dry hole and uneconomic wells cost at 1 in 4*		450,000
	Subtotal	$3,570,000
Overhead and G&A @ 10%		357,000
	TOTAL	$3,927,000

*The essentially exploratory nature of the envisioned drilling program, and the large (areal) step-outs required from one reservoir to another, may lead to large numbers of "dry" (uneconomic) wells. Hence, the dry hole rate could easily be substantially larger, raising the investment costs by $1,000,000 over the numbers shown above.

For the analysis, the investment cost was set at $4,000,000, incurred one year in advance of first year production. These data are consistent with Joint Association Survey cost data and industry estimates.

Decision Criteria — Because of the many major uncertainties still associated with the geology, engineering, and disposal requirements, the simple payback approach was adopted as the financial decision criterion. Two payback rates were used to reflect differing risk premiums:

(1) 10 year payback, before tax (equivalent to a low risk rate of return of about 8 to 10%; and

(2) 5 year payback, before tax (equivalent to a higher risk rate of return of about 20%).

Given the considerable uncertainties and risks inherent in recovering methane from geopressured aquifers, the current investment criterion would be a 5 year payback; successful research and development could, in time, lead to lower risk premiums, as posed in the Advanced Case.

Approach to the Analysis — The analysis followed three steps: (a) calculating the minimum required methane content per barrel of produced water to pay operating and water disposal costs; (b) determining the required production rate/methane content combination required to pay back the initial investment under high and low risk investment criteria; and (c) combining these analyses to compute the overall methane content and flow rate requirements. The analyses assumed a recovery efficiency of 85% of the methane dissolved in the brine, 12.5%

royalty, and 8% severance and other taxes. The analysis was conducted for gas prices of $1.75, $3.00, and $4.50 per Mcf.

Minimum Required Methane Content to Cover Operating Costs: Based on the above operating cost estimates, the following equations define the minimum required methane content to cover operating costs, assuming 1,000 psi injection pressure and 85% methane extraction efficiency:

- 0.8 (price in $/Mcf) $\times$ (recoverable methane per barrel) $\geqslant$
  $0.04 + [$0.01 + 0.005 (price in $/Mcf)] \times P_i$ (ksi)

- Methane content per barrel (in Mcf) $\geqslant$

$$\frac{\$0.073}{\text{price in \$/Mcf}} + 0.007$$

Solving for the three prices of $1.75, $3.00, and $4.50 per Mcf, provides the following minimum required methane content to cover operating costs, as a function of price:

Price per Mcf	Minimum Required Methane Content (ft³/bbl)
$1.75	50
$3.00	30
$4.50	23

The production costs equation is assumed the same across all production rates.

Minimum Required Production Rate and Methane Content to Cover Investment Costs: Using the above investment cost assumptions, the following equation defines the minimum required production rate and total recovery to pay back investment costs:

- Methane content per barrel of water (Mcf) $\geqslant$

$$\frac{\$4,000,000}{(365 \times \text{rate (bpd)} \times \text{years}) \times 0.8 \text{ price/Mcf} \times 0.85}$$

Solving the equation for four production rates of 20,000, 40,000, 60,000, and 80,000 bpd and for the 5 year and 10 year before tax payback criteria, provides the following minimum required combinations of production rate and incremental methane content to cover investment costs, as a function of price:

Price per Mcf	Production Rate (bpd)	Minimum Required Methane Content (ft³/bbl) (10-yr payback)	(5-yr payback)
$1.75	20,000	46	92
	40,000	23	46
	60,000	16	31
	80,000	12	23
$3.00	20,000	27	54
	40,000	14	27
	60,000	9	18
	80,000	7	14
$4.50	20,000	20	40
	40,000	10	20
	60,000	7	13
	80,000	5	10

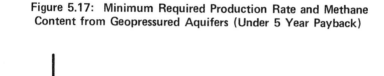

Figure 5.17: Minimum Required Production Rate and Methane Content from Geopressured Aquifers (Under 5 Year Payback)

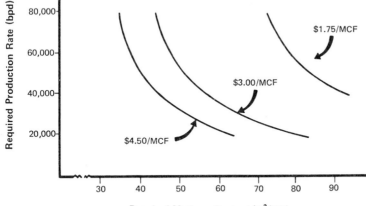

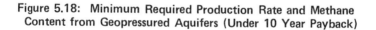

Source: DOE EF-77-C-01-2705

Figure 5.18: Minimum Required Production Rate and Methane Content from Geopressured Aquifers (Under 10 Year Payback)

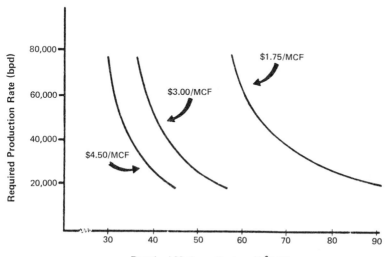

Source: DOE-EF-77-C-01-2705

The lower the production rate, the higher the required methane content. The shorter the payback requirement, the higher the required methane content. Both relationships are linear.

Minimum Required Production Rate/Methane Content Combination to Cover Operating and Investment Costs: Combining the analysis of operating and investment costs establishes the minimum required combination of well production rate and methane content as follows:

Minimum Required Production Rate and Methane Content

Price per Mcf	Production Rate (bpd)	Coverage of Operating Costs	Methane Content (ft^3/bbl)			
			Payback of Investment Costs		Total	
			(10-yr)	(5-yr)	(10-yr)	(5-yr)
$1.75	20,000	50	46	92	96	142
	40,000	50	23	46	73	96
	60,000	50	16	31	66	81
	80,000	50	12	23	62	73
$3.00	20,000	30	27	54	57	84
	40,000	30	14	27	44	57
	60,000	30	9	18	39	48
	80,000	30	7	14	37	44
$4.50	20,000	23	20	40	43	63
	40,000	23	10	20	33	43
	60,000	23	7	13	30	36
	80,000	23	5	10	28	33

Figure 5.17 provides the graph of the minimum required production rate and methane content for the 5 year payback case (the current high risk situation). Figure 5.18 provides the graph for the 10 year payback case and assumes that future resource characterization, technology development, and demonstration will reduce the risk.

Estimating the Resource

This section describes the methods used to estimate methane content and production rates of the identified geopressured resources in Texas and Louisiana and then applies these analytic methods to evaluate their economic potential.

Estimation Methods: *Methane Content* — The volume of methane per barrel of water depends on the presence of methane to be dissolved and the solubility of the methane in water. Solubility, in turn, depends on the interplay of temperature, pressure, and salinity.

The original work by Culberson and McKetta [1] on the saturation values of methane in pure water has been extended to higher pressures and temperatures by Sultanav, Skripka, and Namoit (Figure 5.19). These values were then adjusted using work by Standing and Dodson [2] to account for the salinity of the reservoir water (Figure 5.20).

The resulting estimates are reported later. However, a quick examination of the two curves, in light of the previous analysis, readily shows that an extraordinary combination of temperature, pressure, and low salinity is required for economic feasibility at current and near term gas prices. For example, at the $3.00 per Mcf gas price and assuming a reservoir can support a 40,000 bpd production rate, the minimum methane requirement (under a 5 year payback) could be met only by pressures greater than 16,000 psi, temperature greater than 350°F, and salinity of less than 20,000 ppm.

Figure 5.19: Solubility of Methane in Fresh Water

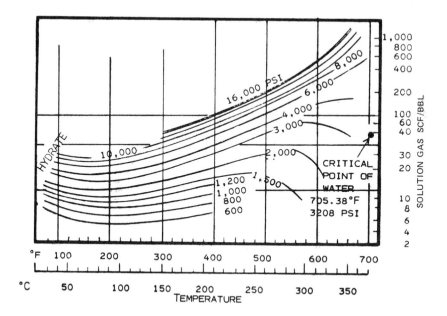

Source: DOE-EF-77-C-01-2705

Figure 5.20: Effect of Salinity on Gas In Solution

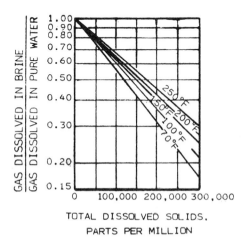

Source: DOE-EF-77-C-01-2705

(Because the three variables interact to determine methane content, less than ideal conditions on one variable may be compensated by improved conditions on the others.)

Production Rate: *Analysis of Production Rate – Sample Reservoir –* The rate of production from geopressured aquifers is determined by the interplay of numerous reservoir and fluid properties. Within the range of conditions expected to be encountered in the geopressured aquifers, however, many of these have relatively small effect. In general, production rate is governed by six critical factors:

- Compressibility
- Permeability
- Net pay thickness
- Area
- Porosity
- Pressure

The effect of these variables on the production rate is analyzed below.

(1) Compressibility – The compressibility coefficient (c) is defined as the fractional change in the pore volume per unit change in pressure. It reflects the combined effects of water compression, rock matrix compression, and compaction of the formation.

While the compression coefficient for water is well known (at approximately 3×10^{-6} psi^{-1}), little data exists regarding the compression coefficient for the matrix rock or the compaction coefficient for geopressured reservoirs, thus introducing an additional bound of uncertainly in estimates of potential production rate.

One company, after having studied the question for two years, estimates that the rock matrix compression and compaction of the formation is approximately 7.6×10^{-6} psi^{-1}, for unconsolidated sandstone, and 3×10^{-6} psi^{-1} for semi-consolidated sandstone. When the water and rock (formation) coefficients are combined, one arrives at 7×10^{-6} psi^{-1} for the compression drive coefficient.

It may, however, be possible to have higher compression coefficients, estimated by some as high as 40×10^{-6} psi^{-1}. Should these higher compression coefficients be found with no counterbalancing effects, it could be possible, theoretically, to recover considerably larger portions of the resource and to produce substantially smaller reservoirs. The effect of variation in this parameter has been shown to have direct, linear relationshp to recovery efficiency. However, the counterbalancing effects of high compressibility coefficients could be severe:

- Permeability near the wellbore will be dramatically lowered as the compressibility increases.
- Risks of environmentally serious subsidence increase, probably necessitating reinjection of the produced water into the geopressured aquifer.
- High compression coefficients that imply "soft" reservoirs could, because of high sand flow, be rejected as technologically infeasible to complete or produce.

Because of the vast unknowns concerning compressibility, a relatively optimistic coefficient of 11×10^{-6} psi^{-1} with no counterbalancing effects were assumed throughout the analysis.

(2) Permeability, Thickness, and Area — Assuming a fixed compressibility coefficient and the other baseline reservoir data on the sample reservoir (Table 5.3), the production from geopressured aquifers is governed by permeability, thickness, and areal extent.

For analytic purposes, constant rates were estimated at permeabilities of 5, 10, 20, 50 and 100 md; net pay of 100, 200, 300, and 500 feet; and areas of 5, 10, 20, and 50 and 100 square miles. The results of these calculations are displayed in Table 5.3.

Figures 5.21 and 5.22 show the relationship of production rate to net pay and permeability for a small (5 square miles) and a large (50 square miles) reservoir.

(3) Porosity and Pressure — The final two parameters, porosity and pressure, were analyzed for their effect on production rate estimates according to the flow analysis performed by Randolph (3), as shown on Figures 5.23 and 5.24.

(4) Influence of Reservoir Parameters on Production Rate — The analysis shows that economic exploitation will require reservoirs of substantial size. Of the three parameters varied in the analysis, over the likely range of parameters considered, formation thickness (net pay) shows the most marked effect on production rates, followed closely by areal extent and permeability. If the lower methane content reservoirs (with gas contents of 40 cf/bbl or less) are ever to be produced economically, at affordable gas prices, reservoirs of over 5 cubic miles, equivalent to reservoirs approaching Prudhoe Bay in size, (50 square miles x 500 feet in thickness) will be required to provide the minimum required flow rates to pay back investment. The investigation, as reported below, showed that such large reservoirs are episodic and difficult to find.

Engineering Calculations of the Production Rate — The Basic Formula: The basic mechanics of fluid flow through the aquifer formed the basis for calculating the production rates of the identified geopressured reservoirs. Two formulas were used to calculate fluid flow:

- The infinite reservoir solution (or the exponential integral solution) was used to describe pressure behavior around a single well producing at constant rate from an infinite reservoir.

- The pseduo-steady state formula was used to describe the pressure behavior of a well producing from a bounded reservoir after the pressure disturbance of the well has reached the boundaries of the reservoir.

Potential of Producing Geopressured Aquifers at Gas/Water Ratios in Excess of Solution Value: Recently, the possibility that geopressured aquifers might be produced at gas/water ratios in excess of the ratio that exists in the reservoir has been posed.

This hypothesis may stem from experience with undersaturated crude oil reservoirs that produced oil and gas at solution ratios between initial pressure and bubble point pressure. Production during this initial period is due to liquid expansion such as that presumed to occur in the production of geopressured aquifers. However, below the bubble point pressure, gas/oil ratios increase quite rapidly for undersaturated crude oil reservoirs. This is due to the release of gas from solution within the bulk of the reservoir and the subsequent and rapid migration of this gas to the wellbore because of its far greater mobility than that of the crude oil.

Table 5.3: Ten Year Average Water Production From Geopressured Aquifiers as a Function of Net Pay, Permeability and Reservoir Area (Mbpd)

Area (mi²)	Net Pay (ft)	Permeability, md				
		5	10	20	50	100
5	100	4	5	6	7	7
	200	8	10	12	13	14
	300	11	15	17	19	19
	500	18	23	26	28	29
10	100	5	8	10	12	13
	200	10	15	19	22	24
	300	15	21	26	30	32
	500	23	31	36	40	42
20	100	6	10	15	20	22
	200	12	19	26	33	36
	300	17	26	34	41	44
	500	26	37	44	49	52
50	100	7	12	20	30	36
	200	13	22	33	44	48
	300	19	30	41	50	54
	500	28	40	50	57	59
100	100	7	13	22	36	43
	200	13	23	35	48	54
	300	19	31	43	54	58
	500	29	42	52	59	62

▨ Current Technology.

Initial Pressure:	11,000 psi
Depth:	13,000 ft
Porosity:	0.216
Viscosity:	0.199 cp
Wellbore Radius:	0.275 ft
Minimum Wellhead Pressure:	500 psi
Fluid Head:	$0.465 d = 6,045$ psi
Friction Loss (psi):	$7.477 \times 10^{-11} \, dq^2 = 9.7201 \times 10^{-7} \, q^2$
Compressibility:	11×10^{-6} psi^{-1}
Shape Factor:	30.8828 (for one well)
Time:	10 yr for economic payback calculations—30 yr for ultimate recovery

Note: q is constant rate of production in barrels per day, more fully defined in DOE EF-77-C-01-2705, Volume III.

Source: DOE EF-77-C-01-2705

Figure 5.21: Ten Year Average Production Rate As A Function of Net Pay, Permeability, and Area (5 Square Miles)

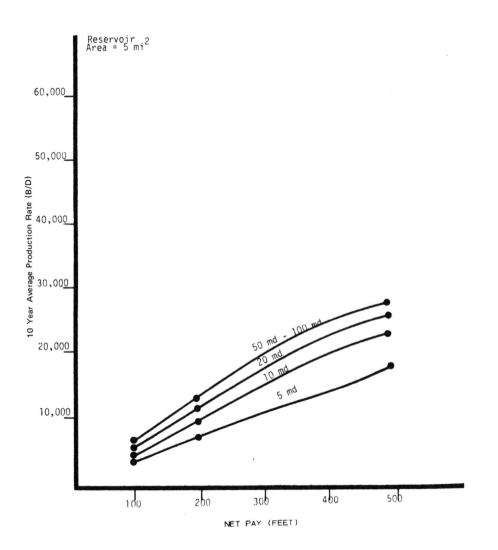

Source: DOE EF-77-C-01-2705

Figure 5.22: Ten Year Average Production Rate As A Function of
Net Pay, Permeability, and Area (50 Square Miles)

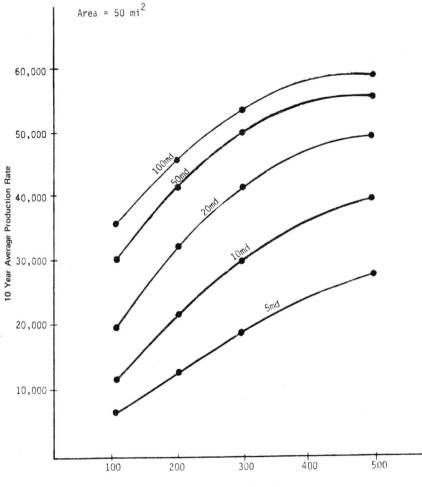

Source: DOE EF-77-C-01-2705

Figure 5.23: Effect of Porosity Upon Production History (3)

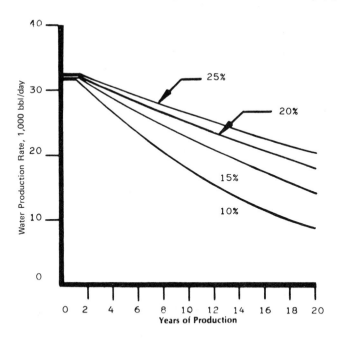

Source: DOE-EF-77-C-01-2705

Before this released gas can flow it must first build up in saturation to some critical value. Most estimates of critical gas saturation are based on laboratory flow studies, and are usually in the range of 2 to 3% pore volume. These may be high estimates, because of problems in calculating material balances on small volumetric changes and the actual observation of very small flow rates in laboratory studies.

The key question becomes what is critical gas saturation for gas flow to occur. For crude oil reservoirs with solution gas ratios in the hundred and even as high as 1,000 to 2,000 ft³/bbl, there is no problem in accounting for the release of sufficient gas to reach a critical gas saturation.

However, in the case of geopressured brines containing 40 ft³/bbl or less, the assurance of critical gas saturation becomes more problematic. Should the reservoir pressure of a geopressured aquifer decrease from 11,000 to 6,000 psi (approximate economic limit), the equilibrium gas saturation decreases from some 40 ft³/bbl to 30 ft³/bbl. At 6,000 psi, the released 10 ft³/bbl will occupy a reservoir volume of only 0.7%. This saturation is significantly less than any reported or presumed values of a critical gas saturation.

One assessment of the way in which gas bearing geopressured aquifers came into existence is to assume that gas was developed in bounding shales and then entered the brines which saturated the adjacent porous sands. Under such conditions, the gas upon reaching a critical gas saturation would begin to migrate and collect under conventional trapping conditions.

Figure 5.24: Effect of Reservoir Pressure Upon Production History (3)

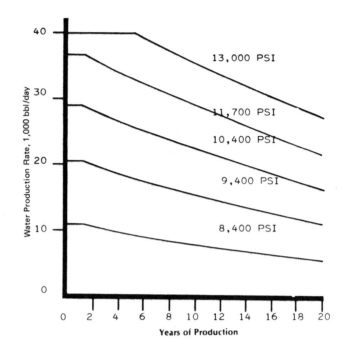

Source: DOE EF-77-C-01-2705

Thus, in the body of a geopressured aquifer, no more than critical gas saturation should be expected to be encountered except where the reservoir configuration provided a conventional gas trap. If only critical gas saturation is encountered, then the additional small release of gas on pressure drawdown will not contribute significantly to gas production at ratios above solution values.

On the other hand, where critical gas saturation was earlier reached and gas generation continued, the accumulation of free gas must be sought in conventional traps by relatively conventional exploration techniques. The industry has been and continues to explore for geopressured gas accumulations. The development of such reservoirs, because of the high gas bearing capacity of a geopressured pore, is a very rewarding and profitable operation.

However, as noted above, the total volume of such accumulations tend to be quite small. These accumulations of free gas in anticlinal, stratigraphic, or fault traps are not considered part of the potential of gas supply from geopressured aquifers, even should geopressured aquifers be adjacent or the source of such gas.

General Resource Base — Recent Findings

Areal Extent of Geopressured Aquifers: Generalizations about the likely size of geopressured aquifers have proven unreliable since local geology, especially tectonic activity and local geological deposition, determine the size of the reservoirs. Several recent studies illustrate this point.

The geologic studies by Wilson and Osborne (4) indicate that the likely maximum sizes of individual sand bodies in selected delta facies is on the order of 3 to 4 cubic miles and the median volume would be somewhat less than 1 cubic mile. This is prior to the secondary modifications that result in faulted reservoirs. They find that the occurrence of multiple, "stacked" reservoirs amenable to multiple completions has low likelihood based on the depositional history of the aquifers.

Bebout and his associates at the University of Texas have carried out some exemplary studies on regional and site specific sedimentology (5). What appears to them to be a prime target as a geopressured aquifer occurs in the Austin Bayou prospect. Three individual sandstone bodies lie between 13,500 and 16,500 feet within an area of about 60 square miles. The largest of the three is a wedge shaped body 150 feet thick; the three sand bodies together total 300 feet. Thus, the total volume of the multi-story package is about 3.4 cubic miles.

While geological analysis has inferred that there are some 800 to 900 net, but dispersed, feet of sands at the prospect location, only about 300 feet of pay appear to have adequate permeability to ensure economic flow. Recent data indicate that the 60 square miles may be comprised of four or more separate sand bodies, thus a well at the targeted location is expected to drain sands within a 15 to 16 square mile area.

Recently, Wallace (USGS)(6) has begun to identify the vertical extent and composition of Louisiana reservoirs. However, even though these formations are known to be highly faulted, little is known of the areal extent of these individual fault blocks.

To gain some insight into sizes of potentially similar structures, the areal distribution of the fault blocks that comprise the 31 largest oil fields in the Gulf of Mexico were examined (Table 5.4).

Table 5.4: Distribution of Fault Block Sizes, Gulf of Mexico

Size Distribution (mi²)	Number of Reservoirs	Total Area (mi²)	Percentage
0-1	292	146	40
1-2	56	84	23
2-3	15	37	10
3-4	7	24	6
4-5	1	4	1
5-10	8	60	16
>10	1	16	4
Total	380	371	100

Source: DOE EF-77-C-01-2705

Based on 380 fault blocks in these 31 fields, covering 371 square miles, only one fault block (covering 16 square miles) met the minimum size requirement of 10 square miles that could, when combined with the thick pay and high permeability, support a production rate of 40,000 bpd.

Should the same conditions hold for the geopressured water bearing aquifers as for oil bearing fault blocks, only about 4% of southern Louisiana's surface area would contain aquifers of sufficient size to provide economically attractive production rates.

The possibility that the geopressured aquifers and oil and gas traps may have a different distribution parallels the assumption that the productive geopressured aquifers will be found in massive deposition occurring in the salt withdrawal synclinal areas. Oil and gas traps, of course, are found at structure locations. However, for the synclinal reservoirs to be large the assumption must also be made that these areas are not as frequently faulted.

Other Key Reservoir Parameters: After areal extent, the next two most important variables are the thickness of the producible net pay and its permeability.

The pay thickness determines reservoir size and influences the production rate (kh) and the slope of the production decline curve. For example, based on the above analysis, a reservoir with 500 feet of pay (when combined with an area of 50 square miles and permeability of 20 md) will support a 50,000 bpd rate for 10 years; the same reservoir with 200 feet of pay will support a rate of 33,000 bpd for 10 years and one that steadily declines after that time.

Recent geologic data (4) indicate that pay thickness may be considerably lower than initially assumed. For example, the recovery estimates of 250 Tcf [Dorfman (7)] assumed 1,500 feet of pay. Recent data indicate high porosity, adequate permeability methane bearing pays are in the 300 foot range, about 20% of the initial estimates of pay thickness.

Finally, the analysis shows that decreases in permeability can have significant effect on initial production rates. While the initial estimates of recovery were based on assumed permeabilities of 100 md, recent data gathered from geopressured reservoirs (5) reveal that absolute permeabilities to gas range from 1.5 md to 50 md in Texas Gulf Coast (although they may be higher in Louisiana), and are probably substantially less to hot brines.

While the work to date has shed considerable light on the geology and reservoir properties associated with geopressured aquifers, much remains unknown.

The remainder of this section combines the general analysis of geopressured aquifers with detailed geological studies in defining the potential of the "identified to date" resource base in Texas and Louisiana.

Geopressured Aquifers of the Texas Gulf Coast

Recently completed work by D.B. Bebout of the Bureau of Economic Geology (University of Texas at Austin) has identified prospective geothermal/geopressured reservoirs in the Texas Gulf Coast-Frio Formation.

Five major geothermal/geopressured fairways were identified by Bebout in the Frio Formation (Tertiary) of the Texas Gulf Coast. The geologic data (Table 5.5) and the analytic methods discussed above were used to estimate the methane content of the water and calculate the average production rates for the five identified prospects, as follows:

	Estimated Methane Content (ft^3/bbl)	Estimated Production Rate (average bpd, 10-yr)
Hidalgo Fairway	45	7,000
Armstrong Fairway	35	34,000
Corpus Christi Fairway	40	9,000
Matagorda Fairway	55	2,000

(Continued)

	Estimated Methane Content (ft³/bbl)	Estimated Production Rate (average bpd, 10-yr)
Brazoria Fairway		
Austin Bayou	45	51,000*
Other	40	41,000

*Assumes a sand body of 60 square miles; subsequent analysis indicates that the 60 square mile area may be composed of numerous discrete sand bodies and that a well in the Austin Bayou prospect may drain sands only with a 16 square mile area. Should the more recent data hold, the estimated 10-year production rate would decrease to about 35,000 bpd, and the Austin Bayou prospect would be economic only at a higher, $4.50/Mcf, gas price.

Table 5.5: Reservoir Properties of Prospective Fairways in Texas

Prospective Fairway	Tempera-ture (°F)	Pressure (psi)*	Dissolved Solids (ppm)**	Permea-bility (md)	Net Pay (ft)	Area of Individual Sand Body (mi²)	Area 1 Extent (mi²)
1. Hidalgo	300*	11,000	20,000	1.5	300	50	500
2. Armstrong	250	10,000	20,000	20	300	50	50
3. Corpus Christi	300	10,000	20,000	5	350	4	200
4. Matagorda	350	12,000	20,000	35	30	4	100
5. Brazoria							
Austin Bayou	325	12,000	60,000	50	300	60	60
Other	300	11,000	60,000	20	300	50	140

*Assuming a pressures gradient of 0.85 psi/1,000 ft.   **Estimated for Fairways 1-4.

Source: DOE-EF-77-C-01-2705

When these five prospects were compared to the minimum economic conditions for the (low risk) payback period of ten years, the following results were drawn (also shown in Figure 5.25):

	Gas in Place	Technically Recoverable (Tcf)	Economic Status
Hildalgo Fairway	6.7	0.03	Not economic at $4.50/Mcf
Armstrong Fairway	0.6	0.01	Not economic at $4.50/Mcf
Corpus Christi Fairway	2.2	0.17	Not economic at $4.50/Mcf
Matagorda Fairway	0.2	0.02	Not economic at $4.50/Mcf
Brazoria Fairway			
Austin Bayou	0.8	0.02	Economic at $3.00/Mcf*
Other	1.7	0.05	Economic at $4.50/Mcf
Total	12.2	0.30	

*Assumes a sand body of 60 square miles; subsequent analysis indicates that the 60 square mile area may be composed of numerous discrete sand bodies and that a well in the Austin Bayou prospect may drain sands only with a 16 square mile area. Should this more recent data hold, the estimated 10-year production rate would decrease to about 35,000 bpd, and the Austin Bayou prospect would be economic only at a higher, $4.50/Mcf, gas price.

Figure 5.25: Economic Analysis of Texas Gulf Coast Fairways

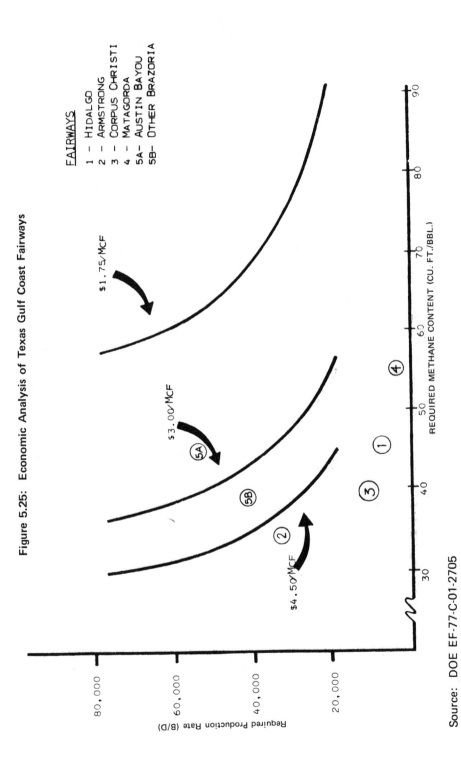

FAIRWAYS

1 – HIDALGO
2 – ARMSTRONG
3 – CORPUS CHRISTI
4 – MATAGORDA
5A– AUSTIN BAYOU
5B– OTHER BRAZORIA

Source: DOE EF-77-C-01-2705

Production from the economic fairways was estimated under 10 year and 30 year constant production rates. Figure 5.25 shows that at $3.00 per Mcf only the Austin Bayou appears economic; at $4.50 per Mcf, the Other Brazoria Fairways become economic. The economic recovery from these fairways was extrapolated to the total Texas Gulf Coast area to the 300°F isotherm on the assumption that these fairways are representative of the total area.

If one assumes that the five identified fairways that jointly cover about 1,000 square miles are representative of the total 5,000 square mile Texas Gulf Coast from the shoreline to the 300°F isotherm of the lower Frio formation, one can extrapolate these findings to Texas. Figure 5.26 shows the boundaries and area to which the results have been extrapolated. The result of this extrapolation is as follows (in Tcf):

	Identified Texas Prospects	Total Texas Gulf Coast
	. . . . (Tcf)	
Total gas in place	12	60
Recoverable gas in place	0.30	2
Economic recovery at:		
$1.75/Mcf	—	—
$3.00/Mcf	0.02	0.1
$4.50/Mcf	0.07	0.4

Geopressured Aquifers of the Louisiana Gulf Coast

Analysis of Geopressured Aquifers—LSU: Two recent studies by Hawkins (8) of Louisiana State University and Wallace of the U.S. Geological Survey have gathered valuable geological data on the geopressured reservoirs of Louisiana.

Eight large geopressured aquifers were located and defined by Hawkins in the Louisiana Gulf Coast area. In addition, 55 prospective areas were identified from sand count and pressure maps, however, further geological study is required to ascribe any volumes to these prospective areas.

Based on the available reservoir data on the eight identified areas, the natural gas content of the geopressured water is estimated by Hawkins at 13.6 Tcf. However, Hawkins qualifies this gas in place estimate with the following statement:

> The reader is cautioned to realize that these figures tell only the estimated total energy in place and that at this point very little is known about (a) the rate at which geopressured energy could be produced, and (b) the fraction of the total geopressured energy resource which occurs in aquifers which are large enough to warrant drilling even one well which would produce for say ten years at a sufficiently high volume rate to be of economic interest. Indeed, it is to be inferred from the general geology of southern Louisiana, that much of this geopressured water is contained in aquifers far too small to justify the drilling and completion of a single well, much less a cluster of wells.

The methane content for these eight areas is estimated in the table which follows.

Figure 5.26: Potential Areal Extent of Geopressured Aquifers—
Gulf Coast of Texas

From: D.G. Bebout

Source: DOE EF-77-C-01-2705

Prospective Area	Temp. (°F)	Pressure (psi)	Dissolved Solids (ppm)	Methane Content (ft³/bbl)
		...... Reservoir Properties		
Newton	190	7,000	80,000	14
S. Midland	250	10,000	80,000	21
N. Lake Arthur	240	9,000	80,000	20
W. Lockport	220	10,000	80,000	19
W. Maurice	250	11,000	80,000	25
S. White Lake	280	9,000	80,000	25
Big Mouth Bayou	220	8,000	80,000	17
SE Peron Island	270	12,000	80,000	28

*Estimated.

Under a minimum required methane content of 35 to 45 cubic feet per barrel (assuming a 10 year production rate of 40,000 bpd), none of the eight prospective areas are economic at $4.50/Mcf.

Analysis of Geopressured Aquifers—USGS: A second effort currently underway by USGS (Wallace) involves analysis of over 1,000 wells covering about 75,000 square miles from onshore and offshore Louisiana.

The analysis had identified about 6,000 Tcf of gas in place for the full depth interval of 2,000 to 19,000 feet. Approximately half of the area was onshore (half offshore), and the geopressured zone began at about 10,500 feet of depth. Thus, about 800 Tcf of gas in place is in reservoirs that are both geopressured and onshore.

Using data on temperature, pressures, and salinities from USGS, the following methane concentrations were determined for the geopressured interval 10,500 to 19,000 feet:

Geopressured Interval (ft)	Temp. (°F)	Pressure (psi)	Salinity (ppm)	Methane Content (ft³/bbl)	GIP Onshore (Tcf)
		.... Reservoir Properties			
10,500–11,000	210	8,000	110,000	16	95
11,000–12,000	220	9,000	100,000	18	160
12,000–13,000	230	9,000	90,000	19	150
13,000–14,000	240	10,000	90,000	21	115
14,000–15,000	250	11,000	80,000	26	90
15,000–16,000	260	12,000	70,000	31	85
16,000–17,000	270	13,000	50,000	37	60
17,000–18,000	280	13,000	50,000	42	25
18,000–19,000	290	14,000	50,000	47	20
Total					800

Since limited conclusive data are available on areal size or permeability of the south Louisiana geopressured aquifers, for economic and recovery purposes it was assumed that each interval could support a ten year production rate of 40,000 bpd, a reasonably optimistic assumption. (Such a rate could be provided by large reservoir volumes with high permeability, see Table 5.3).

Plotting the methane content and assumed production rates on Figure 5.27 shows that:

- The 18,000 to 19,000 foot interval could be economic at $3.00/Mcf if wells in this interval can sustain a ten year production rate of 40,000 bpd.

Figure 5.27: Economic Analysis of Louisiana Geopressured Aquifers

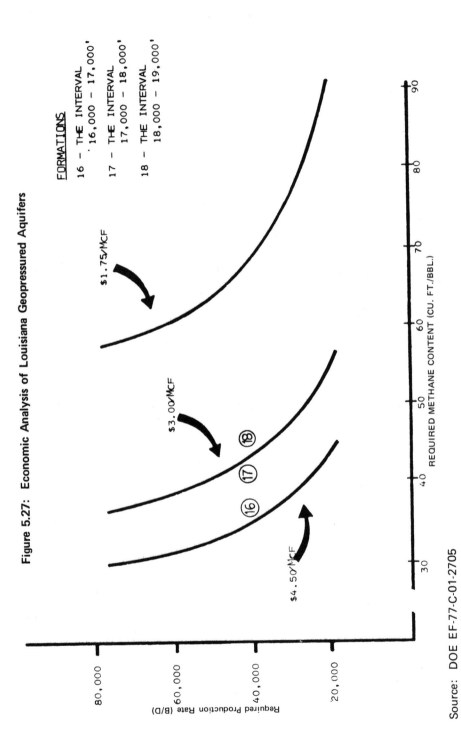

Source: DOE EF-77-C-01-2705

- The 16,000 to 19,000 foot interval could be economic at $4.50/Mcf if wells in this interval can sustain a ten year production rate of 40,000 bpd.

Based on the gas in place in each of the economic depth intervals, and assuming a five percent recovery efficiency these results were extrapolated for the 25,000 square miles of the Louisiana onshore target area, as follows (in Tcf):

- Total gas in place   800
- Recoverable gas in place   40
- Economic recovery at:

    $1.75/Mcf   –

    $3.00/Mcf   1

    $4.50/Mcf   5

The assumption of a five percent recovery efficiency with a sustained ten year production rate of 40,000 bpd requires thick (500 feet+) and highly permeable (100 md+) reservoirs.

Summary and Sensitivity Analyses of Economically Recoverable Methane from Geopressured Aquifers

Summary of Potential: The analysis above shows that the economic potential of currently identified Texas and Louisiana reservoirs ranges from 1 to 5 Tcf, at gas prices of up to $4.50/Mcf.

	Texas	Louisiana
	. . . . (Tcf)	
Gas in place	60	800
Recoverable gas in place	2	40
Economically recoverable at:		
$1.75/Mcf	–	–
$3.00/Mcf	0.1	1.0
$4.50/Mcf	0.4	5.0

Sensitivity of the Recovery Estimates to Key Variables: These recovery estimates have been based on a series of key assumptions about economics, geology, and reservoir performance. It is useful to determine how sensitive these estimates are to reasonable bounds of variation, particularly:

- Investment costs for the well
- Operating costs
- Risk premium investment criteria
- New producing horizons/supplemental free gas
- Free gas saturation
- Thickness of pay
- Areal extent
- Rejection pressures

The results of this sensitivity analyses are shown below.

Investment Costs — The recovery estimates of 1 to 5 Tcf are insensitive to ±$1,000,000 changes in investment costs. Reducing or increasing investment costs by $1,000,000 per well changes the minimum methane content requirements (at the 40,000 bpd level) by only ±2 to 3 cubic feet per barrel and would not have an appreciable effect on the economically recoverable portion of the resource base.

Operating Costs — The recovery estimates of 1 to 5 Tcf are highly sensitive to changes in the operating and energy costs of reinjection. The operating costs used herein are tied to the selling price of the produced gas, and to the energy required for disposing of the produced water; for example, if the average disposal pressure was 2,000 psi rather than 1,000 psi, the increase in operating costs would reduce the economic potential of the identified Texas and Louisiana reservoirs to essentially zero at prices up to $4.50/Mcf.

It may be possible to design large reliable gas fired compressors for reinjecting the reservoir brines. Should this be feasible, the power costs, and thus the operating costs, could decrease to $0.05 + [0.002(price per Mcf) x P_i (ksi)]. Under this assumption (and reinjection pressures of 1,000 psi), the 1 to 5 Tcf potential would increase to a range of 2 to 9 Tcf gas prices of $3.00 and $4.50 per Mcf, respectively.

Investment Criterion — Requiring a higher risk premium reflected in a 20% ROR (or a five year payback) would reduce the economically recoverable estimates of 1 to 5 Tcf to about 1 Tcf. When the production rates and methane content of the Texas and Louisiana reservoirs are analyzed using the five year payback curves (as shown previously in Figure 5.16), the economic potential becomes as follows:

	Texas	Louisiana
	. . . . (Tcf)	
Economic at:		
$1.75/Mcf	—	—
$3.00/Mcf	—	—
$4.50/Mcf	0.03	1

New Horizons — New producing horizons may serve to increase the economic potential. It may be that other formations, such as the Wilcox and Vicksburg in Texas, and the Upper and Lower Cretaceous in Louisiana, contain sufficiently geologically favorable aquifers of high methane content to be economically recoverable. Should this be vertified by further R&D, the potential of the geopressured aquifers would increase.

Free Gas Saturation — The potential from geopressured aquifers could be substantially enhanced should free gas in the pores or in an overlying gas cap be consistently found in conjunction with large geopressured aquifers. However, it is generally believed that geopressured free gas accumulations occurring in conventional traps will be sought and developed as an intrinsic part of conventional exploration and production activities.

Additional analysis and resource definition will be required to define any parallels between the occurrence of geopressured methane bearing aquifers and geopressured reservoirs.

Thickness — Should the formations prove to be twice as thick as have been found in Texas and be able to produce at 60,000 bpd in Louisiana, economically recoverable methane would increase by 1 Tcf. In Texas, given the areal, permeability, and methane content data on the five identified fairways, only the Armstrong and Brazoria are economic even at pays up to 600 feet. However, since the volume of two economic fairways doubles, the total recoverable from Texas would double. In Louisiana, increasing the productive capacity sufficient to raise the 10 year production rate to 60,000 bpd would not economically bring on the next interval of 15,000 to 16,000 feet, and thus would not add to currently identified recovery.

Areal Extent — Should the five identified fairways in Texas be the most favorable and the fault blocks in Louisiana be small (comparable to the offshore oil fields), the economically recoverable potential drops to below 1 Tcf. The five identified Texas fairways, on their own, will produce 0.01 Tcf (at $4.50 per Mcf). If the fault blocks in Louisiana are small with only 4% being 10 square miles or larger, the 5 Tcf in Louisiana now estimated to be economically recoverable at $4.50 per Mcf may drop to 0.2 Tcf. Thus, together these two areas would provide about 0.3 Tcf.

Reinjection Into the Producing Aquifer — Should the produced water need to be reinjected into the producing reservoir, the economics of recovery would be severely impaired. For injection into the original reservoir, at 6,000 psi, operating costs could reach $0.15 to $0.20 per barrel, depending on fuel costs. The minimum required methane content, even at $4.50 per Mcf, would be over 70 cubic feet/barrel, and would make all of the above identified prospects uneconomic.

Summary of the R&D Strategy and Its Potential

The Research and Development Goals:

- Establish the economic sensitivity of production for a wide range of parameters to define the critical bounds.

- Estimate the anticipated distribution of reservoir sizes from geological and analog studies.

- Confirm by drilling and completion the extent of reservoir continuity.

- Determine the water salinities, pressure gradients, and water temperatures in these reservoirs, as well as the quantities of methane dissolved per barrel of water.

- Determine the optimum manner for the disposal or reinjection of the produced fluids.

- Define any potential well completion and production problems due to sand production and corrosion, and secure technical contributions in these problem areas from industry and service companies.

- Develop highly efficient methane extraction facilities.

Production Benefits: The production benefits are only stated in terms of ultimate recovery. Given the speculative nature of the resource base, no estimates are made for cumulative recovery or production rates.

	. . . . Price per Mcf		
	$1.75	$3.00	$4.50
Ultimate Recovery, Tcf	—	1	5

(Future work may identify new producing horizons to increase this potential.)

R&D Costs: The research program costs for the methane recovery from geopressured aquifers programs are as follows:

	. . .Program Costs (MM$). . .	
	5 Year	Total
Total	36.2	42.6
ERDA	36.2	42.6

Cost Effectiveness: The long term cost effectiveness measures of yield per ERDA dollar (Ultimate Recovery in Mcf/Total ERDA 5 Year Costs) at three prices are:

	. . . . Price per Mcf		
	$1.75	$3.00	$4.50
Ultimate Recovery, Tcf	—	30	140

Summary: From these summary sensitivity analyses, it is evident that considerable uncertainty exists regarding the potential of geopressured aquifers for contributing to the nation's energy supply. This uncertainty can only be resolved by using R&D to define the geological value and extent of this resource base. It is clear, however, that the early estimates of hundreds of years of supply are grossly exaggerated.

References

(1) Culberson, O.L. and McKetta, J.J., "Phase Equilibria in Hydrocarbon Water Systems III— The Solubility of Methane in Water at Pressures to 10,000 psia," *Trans. Amer. Inst. Mining Eng.,* v. 192, (1951).

(2) Dodson, C.R. and Standing, M.B., *Drilling and Production Practices,* American Petroleum Institute, (1974).

(3) Randolph, P.L., "Natural Gas for Geopressured Aquifers?", SPE Paper 6826 presented at the 52nd Annual Fall Technical Conference and Exhibition of the Society of Petroleum Engineers of AIME, Denver, Colorado (1977).

(4) Wilson, T. and Osborne R., *The Likely Sizes and Reservoir Parameters of Geopressured Aquifers,* University of Southern California, (1977).

(5) BeBout, D.G., Loucks, R.G., and Gregory, A.R., *Geopressured Geothermal Fairway Evaluation and Test Well Site Location - Frio Formation, Texas Gulf Coast,* prepared for the U.S. Energy Research and Development Administration, (May 1977).

(6) Unpublished work on over 1,000 well logs and records in Southern Louisiana.

(7) Dorfman, M.H., "Potential Reserves of Natural Gas in the United States Gulf Coast Geopressured Zones," and Jones, P.H., "Natural Gas Resources of the Geopressured Zones in the Northern Gulf of Mexico Basin," *Natural Gas from Unconventional Geologic Sources,* Washington, D.C.: National Academy of Sciences, (1976).

(8) Hawkins, Murray F., *Investigation on the Geopressured Energy Resource of South Louisiana,* ORO-4889-14, Louisiana State University, (April 15, 1977).

Fracture Fluid-Reservoir Interactions
in Tight Gas Formations
A Literature Review

The information in this chapter is based on *Review of Fracture Fluid-Reservoir Interactions in Tight Gas Formations,* June 1979, DOE BETC/IC-79/3, prepared by B.A. Baker and H.B. Carroll, Jr. of Bartlesville Energy Technology Center for the U.S. Department of Energy.

The discovery of large easily-produced natural gas reservoirs in the continental U.S. is now a thing of the past. The only large gas reservoirs available are locked in low-permeability, low-porosity gas sands such as the Uinta Basin in Utah and the Piceance Basin in Colorado. Currently, the best known method for stimulating these tight gas formations seems to be large-scale massive hydraulic fracturing (MHF). The problems associated with massive hydraulic fracturing are many, but so are the benefits. Therefore, a comprehensive review of these problems and some of their solutions, as set forth in various publications, would be a useful endeavor. The information contained in such a review should be helpful in solving the remaining problems.

Many of the mechanical techniques of hydraulic fracturing were solved long ago; however, many remaining problems are associated with the interactions of the fracture fluids and their components with the formations being fractured. Since these interactions can be physical as well as chemical, the studies of the physical properties of the rock matrix in which the fracture is propagated and of the proppants used in that fracture are also important.

The main problems in producing gas from low-permeability reservoirs by hydraulic fracturing are establishing a sustained and adequate flow rate and maintaining it until most of the gas has been recovered. It is not enough just to create a good fracture system in the formation, but it must be fractured in such a way that the flow path is kept open during the production phase. Variables that affect the fracture conductivity are fracture length and width, proppant placement and properties, particle movement in the reservoir formation, reaction products in the reservoir, and residues left in the gels and fracturing fluids after the fracturing treatments.

WELL STIMULATION IN TIGHT GAS RESERVOIRS

In the past, the stimulation of wells in oil and gas reservoirs has been accomplished by such varied methods as wellbore heating, explosive shooting, steam and solvent injection, acidizing, and hydraulic fracturing. Not all of the methods used are applicable to tight gas formations; however, the last two methods have been widely used for this purpose. Currently, hydraulic fracturing in the form of massive hydraulic fracturing has just about replaced acidizing as a well stimulation method in low-permeability gas formations.

Acidizing

Before 1950, acidizing was the primary well stimulation method used by the oil and gas industry (38). Actually, the treatment of oil wells with acid was tried as early as 1895, but the process was not used much during the next 30 years. However, in 1932 the Pure Oil Company and the Dow Chemical Company used hydrochloric acid to successfully stimulate several wells in Michigan (38). As early as 1933, the Haliburton Company pumped a mixture of hydrochloric and hydrofluoric acids into a well in Texas (38) with discouraging results. The commercial application of this process was first introduced to the industry by the Dowell Company in 1939 (38) in the form of a product called "Mud Acid." However, little use was made of it for the next 20 years.

Because acidizing was usually ineffective in sandstone formations since they are essentially nonreactive, other methods had to be developed. However, an exception to this trend was reported in 1976 by Gidley, Ryan, and Mayhill (19). They were successful in acidizing sandstone formations in the field where they used an acid-mutual solvent treatment which was actually developed and reported at an earlier date (17). The results of this field test differed from those of McCune, Ault, and Dunlap (40), who reported on a laboratory study that implied sands with permeabilities less than 100 md could be acidized more successfully than sands with greater permeabilities. However, Gidley, Ryan, and Mayhill (19) actually reported a 100% success in 5 treatments with the highest permeabilities.

Some current work on fracture acidizing (16) shows that acid can now be placed selectively in a vertical fracture by controlling the density and viscosity of the preflush. Such practices as preflushes and overflushes are commonly used to cool the fracture and lower fluid loss of the acid solution which follows them (16).

The actual design of acid fracturing treatments is discussed in a paper by Williams and Nierode (68), published in 1972. In it, they develop a mathematical model for predicting acid penetration into a reservoir formation which allows for the effect of fracture geometry, acid penetration, acid concentration, formation temperature, and rock type.

In 1975, Keeney and Frost (32) published the results of an investigation of the problems of recovering the spent acid used during the acid fracturing of gas formations. More specifically, they investigated the routine use of alcohols to reduce the water saturation in the reservoir matrix surrounding the wellbore. Their results indicated that such problems could be minimized by limiting the alcohol concentration to 32 vol % and the treatment temperatures to 175°F for sandstone and 200°F for limestone formations.

Hydraulic Fracturing

As a result of the search for more effective formation stimulation methods, hydraulic fracturing was first introduced to the oil and gas industry in late 1949 by J.B. Clark (5). Pumping rates of these early hydraulic fracturing attempts were only 2 to 5 bbl/min. Pumping rates have steadily increased since then until some as high as 400 bbl/min have been achieved. Rates of 40 to 60 bbl/min are now commonplace.

Hubert and Willis (30) in 1957 studied the fracturing of rocks from a theoretical standpoint. They concluded that in tectonically-relaxed formations (characterized by normal faults), the fractures formed should be vertical, and they should be formed when the injection pressures are less than the overburden pressures. Furthermore, in tectonically-compressed areas, the fractures formed should be horizontal, and they should occur at pressures equal to or greater than the total overburden pressures. Both laboratory data and field observations supported these conclusions.

Most engineers now concede that the majority of fractures in deep wells are generally vertical, and those in shallow wells are horizontal. However, the geology of the well site can sometimes alter these conclusions.

Massive Hydraulic Fracturing

Hydraulic fracturing as a well stimulation method has now evolved into what is termed "massive hydraulic fracturing" or simply MHF. The only real difference between MHF and normal hydraulic fracturing is the size of the operation. MHF is at least an order of magnitude larger in scope.

A typical MHF job involves more than 100,000 gal of frac fluid and over 200,000 lb of a proppant, such as sand (18). The sand used for this purpose is stored on the site and is usually delivered to the blender via conveyor belt. The frac fluid or base fluid may also be stored at the site in large metal tanks or plastic-lined ponds dug into the earth. Once it mixed, the fracturing fluid is pumped into the formation at rates varying from 10 to 60 bbl/min, although much higher rates are used at times. Massive pumping equipment must be used which is capable of operating at very high pressures for long periods of time.

Batch mixing was once used exclusively, but it is now being supplemented by continuous mixing. This method became necessary when the faster-acting gelation processes came into use. In actual practice, certain of the chemical components are added on-the-fly to the liquid being pumped downhole.

Anderson and Baker (2) in 1974 observed in their paper that fast hydrating (fine grind) polymers are normally used in continuous-mix operations, while the slower acting polymers are used in batch-mix operations.

Such hydraulic fracturing treatments are designed to produce fractures in tight formations that extend out from the wellbore 1,000 to 2,000 ft at typical depths of 5,000 to 12,000 ft below the surface. At such depths, only vertical fractures are created (18).

The calculated fracture widths created in tight gas sands are normally quite small, from 0.05" to 0.15". However, wider fractures have actually been initiated. The sand used as the proppant is relatively fine, and is usually no larger than 20 to 40 mesh.

It is interesting to note that it normally requires from 30 to 60 days of production following the above treatment to clean up the formation into which a fracturing fluid has been injected. During this time, 50 to 80% of the fracturing fluid is usually recovered from the tight formation.

Propagation and Design of Fractures

The actual generation and propagation of a fracture is usually a simple process of pumping an essentially incompressible fluid downhole and back into the producing formation through perforations in the casing at the desired depth. However, such fluids as foams and emulsions are also used. The process of creating a fracture requires adequate high-pressure pumping equipment to generate enough pressure to overcome the rock strength. Even after that, it is not easy to predict the length, height, width, and direction of the fracture. However, such authors as Kiel (34) have listed equations from earlier investigators which can be used to predict the fracture geometry.

The first attempt to analyze the factors which affect fracture extension was published by Howard and Fast (29) in 1957. They showed that the accurate knowledge of the fluid-loss properties of the fracturing fluids is essential to the effective design of a fracture treatment. They proved that reducing the fluid loss out of a fracture has the same effect on fracture area as increasing the pumping rate. They also defined fracture-fluid coefficients which are still used in many predictive models.

Hall and Dollarhide (23) in 1964 and Williams (67) in 1970 showed that dynamic fluid tests represented the actual fluid-loss conditions in a fracture much better than static tests. Thus, a more accurate basis was established for predicting the extension of fractures.

In 1969, Tinsley, Williams, Tiner and Malone (58) used an electrolytic analog model to study the movement of fluids in a reservoir. Data thus collected were used to develop curves to determine theoretical production increases considering fracture height as a variable. Other variables in the model were fracture length, reservoir height, drainage radius, fracture flow capacity, formation permeability, and wellbore radius.

In 1977, Salz (50) investigated the relationship between pore pressure and fracture propagation in the geopressured Vicksburg formation in South Texas. He developed an empirical relationship from this data to show the decrease in fracture propagation pressure with reduced reservoir pore pressure. However, the relationship was for this specific tight, low-permeability sandstone formation. The porosity of the Vicksburg formation ranged from 12 to 18%, and the permeability was calculated to be less than 1.0 md.

In a paper presented by Hannah, Harrington, and Anderson (25), a mathematical model was developed to predict the location of selected fluid segments in fractures in terms of distances from the wellbore. When combined with a temperature profile along the length of the fracture, this model becomes a very useful scheduling tool for placing polymers, breakers, proppants, etc., in a fracture at a desired location. It can also be used to select the minimum viscosities of the fluids being pumped and thus helps determine approximate uniform break times.

WELL PERFORMANCE IN TIGHT GAS RESERVOIRS

One of the first papers which discusses the effect of vertical fractures on well performance was written by McGuire and Sikora (41) in 1960. They related well productivity increases to fracture lengths and conductivity. In their paper, they also define relative conductivity as the ability of the fracture to conduct fluid relative to that of the formation.

Many other papers have been written since then on the same subject. One of the important ones was written by Wattenbarger and Ramey (62) in 1969, who published a mathematical finite-difference model which they developed to represent the flow of real gases in vertically-fractured wells. The model was designed to simulate well test conditions in that it accounts for wellbore gas storage and turbulent gas flow. The method of analysis was substantiated with examples of actual field data from two fractured gas wells.

In 1975, Tannich (55) published a paper in which he discussed the problems of fracturing-fluid build-up in gas wells. He observed that the productivity of a gas well is restricted by the presence of the fracturing fluid in the formation near the wellbore. The author also describes a numerical simulator clean-up model which was designed to determine the pressure distribution in the reservoir just outside of the invaded zone.

In 1976, Cinco-Ley, Samaniego, and Dominquez (4) wrote a paper describing a numerical simulator for the transient pressure behavior of a well with finite conductivity vertical fracture. However, this technique is limited to instances where the compressibility is constant and small.

Also in 1976, Holditch and Morse (28) reported on the reservoir conditions needed for non-Darcy gas flow in a fracture. A single-phase, two-dimensional, finite-difference reservoir simulator was used to design a numerical model for this purpose. Laboratory data and empirical data from a paper by Cooke (7) were used to check out the accuracy of the model. The results agreed very closely.

In 1977, Agarwal, Carter, and Pollock (1) presented a paper in which they reported on a study using an MHF simulator (two-dimensional, single-phase, finite-difference model) to evaluate fracture treatments mathematically. They developed a set of what they called "type curves" that were for constant well rates and for constant well pressures. The concept of finite capacity fractures was used in the development of these curves because the conventional pressure transient methods, based on infinite capacity fractures, do not adequately represent the behavior of finite capacity fractures. The curves compared very well with those generated by Cinco-Ley, Samaniego, and Dominquez (4). Also when the curves were applied to actual MHF field examples, they matched the data fairly well.

Effect of Stress-Sensitive Permeability

In 1971, Variogs, Hearn, Dareing, and Rhoades (61) published a paper in which they presented laboratory data on a variety of cores, in the form of plots of permeability ratio versus confining pressure. They also designed a mathematical model to show this effect on the gas flow. The original permeabilities of the cores used in this study varied from 0.04 to 191 md. Confining pressures ranged up to 20,000 psi. Their major conclusions were that very tight cores are affected by stress more than the permeable ones and that permeability reduction due to stress can significantly affect the production characteristics of wells in tight gas reservoirs. In general, most of the permeability reduction occurred within the first 4,000 psi of confining pressure.

In 1972, Thomas and Ward (56) published a paper which showed similar results for a series of tight gas cores with permeabilities less than 0.2 md. Most of the permeability reduction in their study occurred within the first 3,000 psi of confining pressure.

In 1977, Evers and Soeiinah (12) published a paper in which they attempted to simplify the works of Variogs and Rhoades (60) in making well performance predictions in stress-sensitive gas reservoirs. They eliminated the need for core data and for a numerical model. They considered their results to be comparable with those of Variogs and Rhoades.

Fracture Conductivity

In 1975, Cooke (8) published a paper in which he discussed the effect on fracture conductivity of degraded polymer left in fracturing fluids after treatment. These residues as well as those of the fluid-loss additives used in the fracturing fluids can substantially reduce the sand permeability.

Cooke's paper gives the results of testing several types of fracturing fluids to see how they affect the fracture conductivity. These include guar, cellulose derivatives, polyacrylamide polymers, and polymer emulsions. Then, a predictive method was developed to describe the results obtained.

The experimental procedure consisted of (1) measuring the amount of residue per volume of fracturing fluid for several types of fluids; (2) conducting filtration tests on these fluids; (3) measuring the permeabilities of stressed sands containing high concentrations of degraded fracturing fluids; and (4) making conductivity measurements of simulated fractures from which fracturing fluid leak-off occurred.

In the above permeability measurements of stressed sands, the stresses used were 3,000; 5,000; 7,000; and 8,000 psi, respectively.

WELL TESTING AND ANALYSIS IN TIGHT GAS RESERVOIRS

In a paper by Gochner and Slater (22) in 1977, a single well gas simulator model (mathematical) was used to characterize the properties of several gas wells in tight reservoirs. However, the simulator model is only briefly described, but the field examples are explained in detail. Two cases were particularly interesting. One was a long-term isochronal test where none of the gas flow could be stabilized. The other case was one where the results of the conventional analysis were suspected to be in error. The authors showed that their model was a useful tool for the analyses of gas well tests, particularly where unstabilized data could not be interpreted by conventional methods.

In that same year, Firoozabadi and Katz (15) presented a paper in which they showed a correlation for the velocity coefficient of the squared term in the expanded gas flow equation. This was done to explain the high-velocity gas flow in a porous media; for example, near the wellbore.

Also in 1977, a paper by Cinco-Ley and Samaniego (3) presented type curves which were used for the analyses of the pressure interference test data from wells with fully penetrating vertical fractures (finite conductivity fractures). In their case, they were trying to determine the orientation of the fractures.

FRACTURING FLUIDS

Along with the development of fracturing techniques and equipment, there has been a corresponding development of fracturing fluids. The first fracturing (frac) fluids used in such well treatments were napalm gels (38). They had a rather limited use, and were soon replaced by lease oil or water pumped at high rates. As higher and higher pumping rates were attempted, the use of friction-loss reducers (38) and fluid-loss additives or control agents began (38). This trend was soon reversed; however, as more and more studies were initiated to learn the true relationships of the properties of these fluids to fracture widths, lengths, and their role in proppant placement. Efforts in the last few years have been concentrated on very viscous fracturing fluids which can be pumped at slower rates and which can be made to degrade thermally or chemically after they have been forced back into the reservoir formation (26). Thus, recovery of the fracture fluid components is enhanced. Chemicals used for this purpose are called gel breakers.

Some of the desirable properties of fracture fluids are listed in a paper by Howard and Fast (29) in 1970. These include friction reduction, fluid-loss control, proppant transport ability, and ease of recovery from the formation.

In 1973, Holditch and Ely (27) published a list of their requirements for a fracture fluid which would be desirable for stimulating deep, high-temperature wells. These were as follows:

It must have low friction-pressure properties;

It should be temperature stable;

It should be nondamaging to the formation;

It should be compatible with the clay-stabilization salts used;

Its viscosity in the fracture should be independent of its surface viscosity;

It must have good proppant support properties; and

It must degrade to the viscosity of the base fluid once it has been injected into the formation.

Today, these properties are still important.

Some of the materials used to prepare these fracturing fluids are natural guar gels, cross-linked guar gums, viscous oil-external emulsions, cellulose polymers, and gelled oils. These materials and other commonly used frac fluids are discussed by White and Free (65). Their discussion includes water- and oil-base fluids, polymer emulsions, foams, and liquified gases.

Base Fluids

The type of formation or sand being fractured along with the fluids present in the reservoir must be considered when deciding on the kind of frac fluid to use. The properties of these frac fluids are largely determined by the base fluid used to prepare them (57). Such base fluids as fresh water or water with sodium or potassium chloride added, alcohols or mixtures of alcohols and water, acid solutions, and hydrocarbons or oils are generally used (65). However, the base fluid for the foams is nitrogen gas (60 to 80%). The remaining components are water and surfactants (65). The base fluid for polymer emulsions probably would be the oil used, since these fluids are essentially composed of two parts oil to one part water.

Alcohols have been effective base fluids in preparing frac fluids for low permeability reservoirs, as have mixtures of alcohols and water. The alcohol most commonly used is methanol. Alcohols, in general, possess lower surface tensions and higher vapor pressures than water. Both are desirable properties in a frac fluid. The actual gelling of the alcohol is done with a synthetic polymer which leaves little or no residue in the formation.

Another group of base fluids widely used to prepare frac fluids are the hydrocarbons or oils. This includes such fluids as kerosene and diesel fuel.

Even acids have been used as base fluids in acid-oil emulsions, combining both the acidizing and the fracturing processes into one fluid. One of the most commonly used is hydrochloric acid.

Water-Base Fracturing Fluids

Water-base frac fluids are commonly used because they can be easily adapted to such a wide range of reservoir conditions. They are usually less expensive and are safer to handle. They are easily thickened with natural polymers (guar gum) or chemically modified natural polymers (derivatized guar gum or cellulose). White and Free (65) discussed these fluids in a paper published in 1976. In that paper, they stated that water gels for water-base frac fluids represent two-thirds of the fracture treatments performed in the field in the U.S. The other one-third is composed of oil-water emulsions and gelled oils.

Alcohol-Base Fracturing Fluids

Frac fluids made with alcohol-base gels are particularly useful in fracturing water-sensitive gas sands. Tight gas-bearing formations (low permeability) are prime candidates for fracturing with alcohol-base frac fluids. Thus, frac fluids containing methanol were developed (20). Such fluids have been successfully used in low-porosity, low-permeability gas reservoirs (57) where formation damage cannot be tolerated. However, these fluids are costly at times, hazardous to handle, and require the use of costly specialized equipment.

Alcohol/Water-Base Fracturing Fluids

Alcohol/water-base gels are economic alternatives to gelled alcohol (57). When used to prepare frac fluids, alcohol/water solutions can contain up to 40% alcohol (methanol). As with alcohol-base gels, these mixtures increase the viscosity of the frac fluid, in this case as much as two-fold. They also help to protect the water-sensitive formations. In addition, fracture fluid recovery from the formation is increased over that of water-base frac fluids which are normally difficult and slow to clean up (57).

Hydrocarbon- or Oil-Base Fracturing Fluids

Hydrocarbon- or oil-base frac fluids are especially useful for fracturing water-sensitive gas formations which are easily damaged by water. These also include low-permeability oil zones and gas zones which produce small amounts of condensate. The main advantages of oil-base frac fluids are their low friction properties, low damage to the clays in the formation, and their saturation changes which approach ideal behavior in many cases. Major limitations of this type of frac fluid are cost, availability of the oil used, and the fire hazard involved (65).

Acid-Base Fracturing Fluids

This type of fracturing fluid must be used selectively because acids can easily react with many of the mineral components present in the reservoir matrix. Other reactions are also possible. As an example, alcohols are sometimes used in acidic stimulating fluids to aid in the recovery of spent acids (65). In the case of organic acids, such as acetic or formic acid, the formation of esters is possible. This simply uses up the acid present. However, organic chlorides can form when hydrochloric acid reacts with an alcohol. These reactions can be minimized by limiting alcohol concentration to 32 vol %, and by limiting the treatment temperature to 175°F for sandstones and 200°F for limestone formations (65).

If the formation being fractured is a limestone or a dolomite, acid-base fracturing fluids will increase the permeability of the reservoir as material is actually dissolved (57). Of course, the dissolving action can also loosen some of the material in the reservoir, and the movement of these residues can cause plugging of the formation.

Polymer Emulsions

Polymer emulsions have been widely accepted as fracturing fluids, particularly in massive hydraulic fracturing operations. Many papers have been written on their development. For example, in 1970, Kiel (34) reported on what he called "A New Hydraulic Fracturing Process". It used a fracturing fluid composed of 1 part brine, plus 0.1% surfactant, and 2 parts viscous oil blended together to form an oil-in-water dispersion or emulsion. The resulting high-viscosity fracturing fluid provided good proppant support, and it reduced fluid loss and produced wide fractures and improved fracture conductivity. Results supporting the use of this new frac fluid on more than a dozen wells are also listed in Kiel's paper.

Then in 1972, Kiel and Weaver (36) published another paper describing an emulsion fracturing system. In 1973, Kiel (35) followed up these two papers with a patent on such a process.

Further development of the hydraulic fracturing system is described by Sinclair, Terry, and Kiel (52), who tested it under laboratory and field conditions. Fluids used in this system were viscous emulsions prepared from lease crude, refined oil, condensate, or LPG as the internal phase; and water, brine, or acid containing a water-soluble polymer and a surfactant as the external phase. For field applications, the preferred emulsion contained from 60 to 75% oil, by volume, and from 25 to 40% water or brine containing 1 to 2 pounds of guar per barrel of liquid.

A surface-active emulsifier is added to the aqueous phase in the amount of about 0.5 wt %. The surfactants used are sodium thallate (for fresh water) and a quaternary amine (for brine). Sometimes a fluid loss additive (FLA) like silica flour is also used. Breaking of the emulsion is accomplished primarily by adsorption of the emulsifier on the matrix material of the reservoir. Other investigators, such as White and Free (65), define a polymer emulsion as a fluid having 2 parts oil to 1 part gelled water.

The advantages of such emulsions include being a good transport media for the proppants used and possessing excellent matrix fluid efficiency. The oil to water ratio must be watched closely during mixing, however, to optimize viscosity and minimize friction pressures. The disadvantages are their cost, availability of the oil used, poor clean-up characteristics, high surface pressures, and fire hazard.

Some laboratory investigations (13) show that the damage to the formation permeability and to the fracture conductivity during fracturing is low. More problems appeared to occur in the low-temperature wells than in the high-temperature ones.

Foams

The use of foam fracturing fluids appears to be restricted to shallow, low-pressure, low-permeability gas wells. The foams are composed of 60 to 80% nitrogen and 20 to 40% water with surfactant. The water is sometimes gelled.

In many respects, foams approach the ideal fluid behavior. They minimize water damage to the formation, have excellent sand (proppant) transport capabilities, and are easily mixed (65). A limiting factor to their use, however, may be the local availability of nitrogen. Also, clean-up can be a problem when the foam fails to break satisfactorily.

A study of the factors which affect the dynamic fluid leak-off during foam fracturing was presented in a paper at the 52nd annual SPE Fall Meeting by King (37). The parameters affecting foam leak-off of foams containing 70, 80, and 90 vol % nitrogen gas through a variety of rock core wafers with permeabilities from less than 1 to over 5 md were studied. The author also describes the use of new liquid fluid-loss additives whose advantage over the long-chain polymers is to reduce the residues left in the formation after fracturing.

Liquified Gases

In 1972, Hurst (31) reported on the use of liquified gases as fracturing fluids for dry gas reservoirs. The gases used were selected to remain in the liquid state during the fracturing operation, but vaporize afterward at conditions of reservoir temperature and pressure. This prevents water damage to the reservoir because of liquid retention. However, liquified gases are far from ideal in proppant transport, fluid-loss efficiency, and cost. They also require special equipment and well-trained personnel to mix them properly (65).

GELLING AGENTS

Gelling agents (gels) are mixed with base fluids to prepare the fracture fluids. Depending upon reservoir conditions, the frac fluids are specially designed to minimize formation damage during and after the fracturing treatment. Four basic gel systems are used to prepare these fracturing fluids. These are discussed by White and Free (65) and by Tiner, Stahl, and Malone (57). These systems are water, alcohol, alcohol-water, and oil or hydrocarbon base gels.

Water-Base Gels

The most widely used gelling agents for water-base frac fluids have traditionally been the naturally-occurring gums, such as the guar gums. In fact, as noted by Anderson and Baker (2) in 1974, the oldest form of polymer-thickened aqueous fracture fluid was an acid solution containing starch or gum karaya. The natural gums are economical, possess good fluid-loss control, and can be easily crosslinked. The latter property is necessary in order to increase viscosity and enhance the proppant-carrying ability of the frac fluid. In fact, such fluids have achieved what is considered to be perfect proppant transport.

In a continuing effort to improve the characteristics of the natural gums, various companies have chemically altered their structure. For example, Githens and Burnham in 1977 described a derivatized guar gelling agent which resulted in an 85% reduction of the inherent residue over that of conventional guar gum. These authors state that the three major classes of polymers used in fracturing fluids are guar, guar derivatives, and cellulose derivatives.

In 1973, Kiel (35) reported that the temperature stability of common guar was improved by using the hydrolyzed form.

In 1974, Tuttle and Barkman (59) described what they called new non-damaging and acid-degradable drilling and completion fluids. In their investigation, they examined one refined natural polymer, hydroxyethylcellulose (HEC) and two synthetic polymers, polyacrylamide and polyoxyethylene. They compared the core permeability damages of these two polymers with those of guar gum and showed the differences. Then, they described three high-viscosity completion fluids formulated to use calcium carbonate (chalk) for fluid-loss control.

Although the use of crosslinked gels is widespread, very little information has actually been published on their effect on formation permeability. However, in 1975 an investigator named Pence (48) did report that crosslinked guar gel gave an 11 to 15% reduction in permeability when used in the Mesa Verde formation.

Also in 1975, White and Means (66) described a relatively new aqueous thickener with low particle residue. It was a polysaccharide derivative (PSD), and it was applicable in both limestone and sandstone formations. The fluid-loss properties of PSD are compared to those of guar and cellulose.

In 1976, Hannah and Baker (24) reported on a new aqueous crosslinked gel with good proppant support which was nondamaging to the formation. The polymer used was carboxymethylcellulose which was hydrated before pumping into the formation. The system was designed for continuous mixing only because of the high viscosity of the gelled fluid.

Unlike natural gums, synthetic polymers do not have inherent solid particles present. Thus, they are very useful whenever low particle residues in the formation are desirable.

Acid-Base Gels

In general, the same chemical compounds which gel water will gel acids (acid and water). However, the use of pH control agents to adjust the hydration rate is not possible with the cellulose derivatives (20); instead, natural guar gum and derivatized guar polymer are usually used for this purpose.

Alcohol-Base Gels

In 1974, Tiner, Stahl, and Malone (57) discussed alcohol base gels and their use in fracturing tight, gas-producing formations. In their paper, they indicate that the alcohol (usually methanol) is gelled with synthetic polymers which are relatively free of residues. Field results, in terms of production, of fracturing treatments with such fluids are given on a before-and-after basis. The exact gelling agents used are not indicated, although, such polymers as PSD, derivatized cellulose, and derivatized guar gum could be used (66)(20). The limiting factors in the widespread use of alcohol solutions seem to be the availability of alcohol and its cost.

Alcohol/Water-Base Gels

The above authors also discussed alcohol/water-base gels which are applicable to fracturing the same low-porosity, low-permeability gas reservoirs as are the alcohol-base fluids. The advantage of using a mixture of alcohol and water and a gelling agent is an economic one. As before, the gelling agents used are those with little or no inherent residues. Mixtures with up to 40% alcohol can be effectively used.

Hydrocarbon- or Oil-Base Gels

Fracturing fluids made from the hydrocarbon-base gelling agents such as the alkyl quaternary surfactants are particularly good in water-sensitive formations, even those once thought to be compatible with water. These gelling agents are best suited for gelling high-gravity crudes, distillates or condensates, or diesel. Tiner, Stahl, and Malone (57) are in agreement on these statements. They give field results to show the increase in production which occurred when oil-base gels were used to fracture appropriate gas formations.

Another relatively new gelling agent for oil uses a phosphate ester.

Gel Breakers

Obviously, it is important to have gelling agents whose properties are designed for a particular situation involving a specific type of formation. Just as important in some cases, but not so obvious, is the need to provide some way to destroy these same desirable characteristics so that the gas can be produced after the formation has been fractured. Gel breakers are needed in these instances. An explanation of the mechanism is given by Holcomb and Smith (26).

Because guar gum solutions are so widely used for a variety of purposes, a breaker for these types of gels is needed. The most commonly used breakers for guar gums are the cellulose enzymes (10). Other substances used are acids and oxidizing agents (20).

Enzymes do not affect the synthetic cellulose derivatives very much; thus, other substances such as acids and oxidizing agents are used as breakers for these types of gels. Compounds such as ammonium persulfate and potassium persulfate are commonly used (10).

pH CONTROL AGENTS

The use of chemicals to control the pH of fracturing fluids and thus control their hydration rate is a common practice (20). For example, a 3 to 5% HCl solution could be used (26).

FLUID-LOSS ADDITIVES

The use of additives (chemical agents) to control the fluid loss to the formation (leak-off) during the fracturing treatment has been practiced for several years (49). All of these additives introduce insoluble residues into the reservoir matrix which may reduce the flow capacity of the propped fracture. One such common fluid-loss additive (FLA) is silica flour. However, there are others that are classified as low or nondamaging (57), and which usually dissolve slowly in the fracturing fluid.

A few papers have been written specifically on the subject of fluid leak-off. One such paper presented in 1976 by Sinha (53) gave the results of an investigation of the differences in static and dynamic behavior of fracturing fluids, using two different types of apparatus and two fluid systems (diesel plus a fluid-loss additive and gelled diesel). Although Sinha's results are a little hard to interpret, he did observe large differences in the results obtained with the two different methods, as others have done. He actually showed that lower fluid leak-off occurred under dynamic conditions than under static conditions.

More recently, a paper presented by King (37) discussed the fluid leak-off during foam fracturing. Usually, foams do not contain solid materials, but sometimes natural polymers and synthetic polymers have been used to control fluid leak-off. Unfortunately, they have certain inherent faults, such as leaving residues in the formation. New liquid additives were developed to solve some of these problems. They affect the bubble size, shape, and interaction in such a way as to make it more difficult to push the bubbles through the reservoir matrix.

FRICTION-LOSS REDUCERS

Use of friction-loss reducers in hydraulic fracturing processes has come about because of the higher injection rates needed to create fractures and to propagate them through the formation. These chemical additives improve fracturing efficiencies by reducing the friction loss during the pumping operation.

In 1961, Ousterhout and Hall (47) published a paper in which they reported on flow tests of thickened aqueous solutions in field size pipe, using a Triplex pump mounted on a commercial fracturing unit. They used a natural gum polymer and a synthetic polymer in these experiments. Their results were reported as plots of pressure gradients versus flow rates for given pipe sizes.

Then in 1964, White (64) published a paper which included a survey of five different types of friction reducers. These were guar, anionic synthetic polymers, nonionic synthetic polymers, synthetic polymer solutions, and in situ soap gels. He also included plots of flow rates, and cost versus friction pressure or pressure reduction ratio for these reducers, and cost and concentration versus viscosity and pressure reduction.

In a more recent paper, Woodroof and Anderson (69) discussed the reduction of formation damage from use of synthetic polymer friction reducers. The application, in this case, was for acid stimulation of the formation. In an earlier paper, White (64) discussed several types of polymers, such as the natural polymers, the nonionic synthetic polymers, the anionic synthetic polymers, the cationic synthetic polymers, and the synthetic liquid polymers. The authors also include data on a stability study of some 16 different friction reducers.

A friction-loop apparatus was designed and used by these authors for the above tests. The results are given in 6 tables and 14 figures. The figures include several plots of flow rate versus friction pressure for different polymers.

FRACTURE PROPPANTS

Even though a fracture has been successfully propagated, it does not mean the stimulation process is completed. To be of use the fracture must be kept open during the production phase. This involves the use of efficiently designed proppants. There are many kinds of proppants available, but the most common material used is sand. It is divided into several size ranges by sieving. These ranges usually include 10 to 20, 20 to 40, 40 to 60, 60 to 80, and 100 mesh sands. Other materials are also used, such as walnut shells, plastic, sintered bauxite or aluminum, and oxide ceramic. Additional proppant materials are being developed as the need arises. In fact, as early as 1961, aluminum particles were used as proppants, and were discussed by Kern, Perkins, and Wyant (33).

More recently, in 1977, Neel, Parmley, and Colpoys (44) presented a paper in which they described the use of oxide ceramic proppants for the treatment of deep well fractures. This material was developed as an alternative to sandglass bead mixtures. The new material is particularly resistant to crushing, and was specifically designed for multilayer pack distribution in fracture-closure pressure applications.

Proppant Transport

In a paper by Clark, Harkin, Wahl, and Sievert (6), a large vertical transparent proppant transport model is described. It is used to study the proppant-carrying abilities of both Newtonian and non-Newtonian fluids through a vertical fracture.

Parameters such as proppant size and concentration, fracture width, and fluid velocity were found to be related in this model. In the actual experiments 20, 30 or 40 lb guar gel (per 1,000 gal of water) was used as the non-Newtonian fluid, and glycerol was used as the Newtonian fluid. Particle movement within the physical model was traced with a camera and then transformed to numeric form with a Tektronix graphics tablet. A computer was then used to process the data, that is, to calculate horizontal sand transport velocities from the measured proppant settling velocities encountered during flow and no-flow conditions.

The authors concluded that measured proppant settling velocities during flow conditions were as much as three times the single particle fall rate. They also observed particle coalescence (clustering) after flow was stopped.

In 1977, Novotny (46) presented a paper describing a computer program designed to model the transport of proppant material while fracturing. It used proppant settling velocities in the model as well as fracture closure data to represent the flow behavior, which other models did not.

The actual flow behavior of fracture fluid-proppant slurry is a complex process, best described in a computer program. As slurry flows through a formation, it is heated and the formation rock is cooled. Proppant concentration increases as the slurry loses fluid to the reservoir matrix. As a result, fluid velocity is also decreased. Another factor which affects fluid velocity is the decreasing width of the fracture as it gets farther away from the wellbore.

The mathematics of the computer model used are discussed in detail in the paper. The author then gives four computer-simulated treatment examples, using the model. These treatments were made with an emulsion fracturing fluid composed of 33% brine and 67% oil in which the concentration of the gelling agent varied from 0.5 to 2.0 lb/bbl of brine. The injection rate used was 10 bbl/min. Both 10 to 20 and 20 to 40 mesh sand were used as proppant material in amounts of 2 lb/gal. Fluid loss additive was also used in the amount of 20 pounds per 1,000 gallons of brine.

The results of the four treatments are give in a series of plots relating proppant settling velocities with fluid shear rate, proppant diameter, Reynolds number, and proppant concentration.

The important conclusions reached by the author were: (1) proppant settling during fracture closure affects the distribution of the proppant in the fracture and may even determine the success or failure of the treatment; (2) the resulting stimulation of the formation is a function of the position of the propped portion of the fracture relative to the permeable reservoir rock.

Proppant Concentration

The advantages of high proppant concentration in fracture stimulation were discussed by Coulter and Wells (10) in 1972. In their paper, they provided a table in which example design calculations for two types of fluids were given. One type is a conventional low-viscosity fluid in which the proppant settles out during treatment. The other type is an extremely viscous fluid in which the proppant remains suspended during and after treatment.

The authors concluded that increasing the proppant concentration in the fracture causes an increase in fracture flow capacity, lessens sand flow capacity, and increases the tolerance of the proppant system to the presence of fines. These high flow capacity proppant systems may also bring about longer periods of increased production.

In 1973, Holditch and Ely (27) describe the stimulation of deep, high-temperature wells in low-porosity, low-permeability formations where high proppant concentrations were used to create and pack a wide fracture.

In this paper, the authors present results of stimulating 17 wells in two different fields in the Vicksburg formation and 11 wells in the Devonian, Wilcox, and Hosston formations. A unique viscous fracturing fluid and high concentrations of proppants were used in these stimulations. The permeabilities of the 17 wells in the Vicksburg formation vary from 0.03 to 0.82 md, depending upon the type of stimulation treatment used. In one of the tables, the comparison of low-viscosity treatments is made with those of high-viscosity treatments. The Vicksburg formation is typical of the deep, high-temperature reservoirs being fractured today.

The desirable qualities of a fracturing fluid (gel) for such stimulations are also listed. These are as follows:

- Low friction pressure properties;
- Temperature stability up to 450°F;
- Low residue in fluid after breaking;
- Compatibility with salts used for clay stabilization;
- Viscosity in fracture must be independent of surface viscosity;
- Good proppant support characteristics; and
- Ability to degrade to viscosity of base fluid at end of fracturing treatment.

Such a fluid was developed, and the results of its use with high-proppant concentrations are given in a series of tables.

In 1977, Novotny (46) related proppant concentration to the settling velocities of proppant-fracture fluid slurries and to the Reynolds number. This has already been discussed in the previous section.

Proppant Conductivity

In 1973, Cooke (7) published a paper in which he discusses the conductivity of brittle proppants in multilayers in vertical fractures. His work differs from previous investigations of fracture conductivity in that he considers the effect of environment (the fluid in place in the reservoir and reservoir temperature) and the flow rate of the fluid being pumped. The multiple layers of proppants in a fracture occur when the viscosity of the fracturing fluid is low enough to let the proppant settle to the bottom of the vertical fracture during the injection process.

Proppant permeability versus stress was obtained at different stress levels by flowing the liquid (usually hot brine) through a layer of proppant sandwiched between two metal platens attached to a 50 ton hydraulic press. Thus, pressure or stress could be applied and maintained during flow conditions.

The results of these experiments for different fluids and different proppants are give in a series of stress-permeability plots. Another series of plots is also given in which turbulence factors and permeabilities are related for different proppant-liquid systems.

An important conclusion reached by the author was that the measured turbulence factors for the fracture sand (proppant) under stress were 2 to 50 times higher than predicted by extrapolation of a literature correlation for consolidated sandstone.

RESERVOIR MATRIX CHARACTERISTICS

An understanding of the physical parameters of the gas-bearing reservoir itself is necessary before any kind of stimulation process can be successfully designed. These parameters vary from rock properties, such as compressive and tensile strength and hardness, to pore properties, such as pore size and distribution. They also include fracture properties like fracture length and width and chemical properties of the formation such as clay and mineral content.

Formation Damage

The term "formation damage" usually refers to reduction in permeability. In a publication by Simon and Derby (51), some of these parameters are discussed, and photomicrographs of pore openings showing various deposited minerals and clay platelets are shown before and after treatment with a gelled 3% hydrochloric acid solution.

The authors also discuss threshold pressure and its effect on flow of gas through the reservoir. A table is included which compares threshold pressures and effective interfacial tensions. A discussion of the distribution of clays in tight gas formations is also included.

Formation damage (permeability reduction) can be caused by residue deposits from the fracturing fluids being used (57).

Effect of Overburden Pressure

A reduction in reservoir or formation permeability can also be brought about by mechanical stress caused by overburden pressure on the formation. This was indicated by Variogs and Rhoades (60) and by Thomas and Ward (56). Both of these papers have been discussed earlier in the section on "Well Performance."

Particle Movement

Even before well stimulation methods such as hydraulic fracturing were introduced to improve oil and gas production in problem wells, studies were undertaken to investigate pore plugging from drilling fluid (mud) invasion. These early attempts at least pointed the way for such investigators as Nowak and Kreuger (45) in 1951 and Kreuger and Vogel (39) in 1954, who studied underneath-the-bit and above-the-bit conditions, respectively, with inert cores in the restored-state condition. Glenn and Slusser (21) in 1957, extended this work to show that submicron particles from drilling fluids penetrated at least 2 to 5 cm into the pore spaces. They also observed that certain particle-size/pore-size relationships existed which allowed the particles to move freely through the reservoir rock.

Donaldson, Baker and Carroll (11) showed that particle distributions could be deliberately chosen, based on given pore-size distributions, and pumped unimpeded through the reservoir rock. They postulated that this invasion of particles into the pore spaces usually formed an internal mud cake and thus blocked or drastically reduced the flow thereafter. Glenn and Slusser (21) found that permeability damage could be minimized by using saline and oil filtrates. Multivalent salts were found to be more effective than sodium chloride for this purpose. Although these studies actually concerned drilling fluids or muds, their observations are just as valid for fracturing fluids that are pumped into the formation.

In an early laboratory study, Kern, Perkins, and Wyant (33) proved that the existing theory of the sand transport mechanism was wrong. They showed that sand actually settles to the bottom of a vertical fracture at normal flow rates.

Pore Structure

As already mentioned, Simon and Derby (51) published photomicrographs of pore openings in Morrow sandstone cores. Other authors also used the microscope to gain insight into the pore structure of reservoir rocks. One such publication was by Weinbrandt and Fatt (63) who used a scanning electron microscope (SEM) to study Berea and Boise sandstone. They had to use special mounting, coating, and polishing techniques in order to get realistic results. Two types of cores were used to cover a wide range of permeability and a modest range of porosity.

With SEM, they measured the tube interconnection parameter, β, which is defined as the number of flow channels connected to each flow channel. If the pore-size distribution is known from capillary pressure data, then β can be calculated. Values thus determined agreed with the results of other investigators (14) (42).

Swanson (54) presented a paper at the 52nd Annual SPE Fall Meeting in which he described an investigation using the scanning electron microscope in combination with injection of Wood's metal into rock cores to study the concepts of nonwetting phase distribution and of fluid flow in rocks.

A technique of stepwise bleeding of the vacuum on the core was used to control the nonwetting phase saturation. Thus, the applied capillary pressure is used to control the nonwetting phase saturation. This method was used on several different kinds of cores with permeabilities varying from 0.35 to 1,200 md. Photomicrographs were taken of these cores at different fluid saturations. However, not enough cores were studied to draw any significant conclusions. The method does appear to be one which could be useful in determining the fluid distributions in a specific reservoir which is to be subjected to a particular producing scheme.

Types of Clays

In a paper presented at the 25th Annual Technical Meeting of the Petroleum Society of CIM, Tiner, Stahl and Malone (57) listed the most common clays found in gas- and oil-producing formations. These are illite, montmorillonite, mixed layer (illite-montmorillonite and chlorite-montmorillonite), kaolinite, and chlorite.

The authors also mention that the best way to determine the type of clay minerals present in the formation is by x-ray analysis.

Effect of Dispersed Clays

In 1977, Neasham (43) presented a paper at the 52nd Annual SPE Fall Meeting in which he discussed the effect of clay dispersal in the pore spaces of sandstone cores. In this paper, Neasham describes three basic types of dispersed clay in sandstones and lists several geological-petrophysical properties measured in the laboratory which are associated with these three types of clay.

SEM analyses were performed on some 14 selected sandstone cores, and they were classified into three categories, depending upon the type of clay they contained. These categories were: discrete-particle clays, pore-lining clays, and pore-bridging clays.

Example photomicrographs of cores illustrating these types of clays are included in this paper. Significant observations are made concerning different clay and mineral deposits in and across the pore openings and their effect on fluid flow.

Some of the specific geological and petrophysical properties associated with these three types of clay are porosity, air permeability, pore-size distribution, air-mercury capillary pressure curves, relative permeability, and clay mineral content.

To help prevent some of the problems described in the above paper and to improve well performance, many types of clay control (stabilization) agents have been developed and tried. A discussion of four such agents was presented in a paper by Copeland, Coulter, and Harrisberger (9).

In their investigation, these authors mixed clay stabilizers with a carrier fluid and injected the mixture into four different test mediums (unconsolidated sands) to simulate downhole treatment of a clay-bearing formation. The results were then evaluated and reported. The zirconium salt was the most efficient clay stabilizer studied.

REFERENCES

(1) Agarwal, R.G., Carter, R.D. and Pollock C.B., "Evaluation and Prediction of Performance of Low Permeability Gas Wells Stimulated by Massive Hydraulic Fracturing", Pres. at SPE 52nd Annual Fall Meeting, Denver, CO, October 9-12, 1977, Preprint Paper No. 6838, 9 pp.

(2) Anderson, R.W. and Baker, J.R., "Use of Guar Gum and Synthetic Cellulose in Oilfield Stimulation Fluids", Pres. at SPE 49th Annual Fall Meeting, Houston, TX, October 6-9, 1974, Preprint Paper No. 5005, 7 pp.

(3) Cinco-Ley, H. and Samaniego, V., "Determination of the Orientation of a Finite Conductivity Vertical Fracture by Transient Pressure Analysis", Pres. at SPE 52nd Annual Fall Meeting, Denver, CO, October 9-12, 1977, Preprint Paper No. 6750, 5 pp.

(4) Cinco-Ley, H., Samaniego, V. and Dominquez, A., "Transient Pressure Behavior for a Well with a Finite Conductivity Vertical Fracture", Pres. at SPE 51st Annual Fall Meeting, New Orleans, LA, October 3-6, 1976, Preprint Paper No. 6014.

(5) Clark, J.B., "A Hydraulic Process for Increasing the Productivity of Wells", *Trans. AIME* 186, pp. 1-8 (1949).

(6) Clark, P.E., Harkin, M.H., Wahl, H.A. and Sievert, J.A., "Design of a Large Vertical Prop Transport Model", Pres. at SPE 52nd Annual Fall Meeting, Denver, CO, October 9-12, 1977, Preprint Paper No. 6814, 6 pp.

(7) Cooke, C.E., Jr., "Conductivity of Fractured Proppants in Multiple Layers", *J. Petrol. Technol.,* vol. 25, pp. 1101-1107 (September 1973).

(8) Cooke, C.E., Jr., "Effect of Fracturing Fluids on Fracture Conductivity", *J. Petrol. Technol,* vol. 27, pp. 1273-1282 (October 1975).

(9) Copeland, C.T., Coulter, A.W. and Harrisberger, W.H., "Designed Application of Clay Stabilizer Improves Performance", Pres. at SPE 52nd Annual Fall Meeting, Denver, CO, October 9-12, 1977, Preprint Paper No. 6759, 8 pp.

(10) Coulter, G.R. and Wells, R.D., "The Advantages of High Proppant Concentration in Fracture Stimulation", *J. Petrol. Technol.,* vol. 24, pp. 643-650 (June 1972).

(11) Donaldson, E.C., Baker, B.A. and Carroll, H.B., "Particle Transport in Sandstones", Pres. at SPE 52nd Annual Fall Meeting, Denver, CO, October 9-12, 1977, Preprint Paper No. 6905, 9 pp.

(12) Evers, J.F. and Soeiinah, E., "Transient Tests and Long-Range Performance Predictions in Stress-Sensitive Gas Reservoirs", *J. Petrol. Technol.,* vol. 29, pp. 1025-1030 (August 1977).

(13) Fast, C.R., Holman, G. and Calvin, R.J., "A Study of the Application of MHF to the Tight Muddy 'J' Formation, Wattenberg Field", Pres. at SPE 50th Annual Fall Meeting, Dallas, TX, September 28 through October 1, 1975, Preprint Paper No. 5624, 2 pp.

(14) Fatt, I., "The Network of Porous Media. Parts I, II and III.", *Trans. AIME* 207, pp. 144-181, (1956).

(15) Firoozabadi, A. and Katz, D.L., "An Analysis of High Velocity Gas Flow Through Porous Media", Pres. at SPE 52nd Annual Fall Meeting, Denver, CO, October 9–12, 1977, Preprint Paper No. 6827, 5 pp.

(16) Fredrickson, S.E. and Broaddus, G.C., "Selective Placement of Fluids in a Fracture by Controlling Density and Viscosity", J. Petrol. Technol., vol. 28, pp. 597–602 (May 1976).

(17) Gidley, J.L., "Stimulation of Sandstone Formations with the Acid-Mutual Solvent Method", J. Petrol. Technol., vol. 23, pp. 551–558 (May 1971).

(18) Gidley, J.L., Mutti, D.H., Nierode, D.E. and Kehn, D.M., "Tight Gas Stimulation by Massive Hydraulic Fracturing–A Study of Well Performance", Pres. at SPE 52nd Annual Fall Meeting, Denver, CO, October 9–12, 1977, Preprint Paper No. 6867, 5 pp.

(19) Gidley, J.L., Ryan, J.C. and Mayhill, T.D., "Study of the Field Application of Sandstone Acidizing", J. Petrol. Technol., vol. 28, pp. 1289–1294 (November 1976).

(20) Githens, C.J. and Burnham, J.W., "Chemically Modified Natural Gum for Use in Well Stimulation", Soc. of Petrol. Eng. J., 10 pp (February 1977).

(21) Glenn, E.E. and Slusser, M.L., "Factors Affecting Well Productivity–II Drilling Fluid Particle Invasion into Porous Media", J. Petrol. Technol., pp. 132–139 (May 1957).

(22) Gochner, J.R. and Slater, G.E., "Well Test Analysis in Tight Gas Reservoirs", Pres. at SPE 52nd Annual Fall Meeting, Denver, CO, October 9–12, 1977, Preprint Paper No. 6842.

(23) Hall, C.D., Jr. and Dollarhide, F.E., "Effects of Fracturing Fluid Velocity on Fluid-Loss Agent Performance", J. Petrol. Technol., vol. 16, No. 5, pp. 555–560 (May 1964).

(24) Hannah, R.R. and Baker, J.R., "A New Nondamaging, Aqueous Cross-Linked Gel with Improved Fracturing Properties and Perfect Proppant Support", Proc. of 23rd Annual Southwestern Petroleum Short Course Assoc. Mtg., Lubbock, Tex., pp. 123–128.

(25) Hannah, R.R., Harrington, L.J. and Anderson, R.W., "Stimulation Design Applications of a Technique to Locate Successive Fluid Segments in Fractures", Pres. at SPE 52nd Annual Fall Meeting, Denver, CO, October 9–12, 1977, Preprint Paper No. 6815, 5 pp.

(26) Holcomb, D.L. and Smith, M.O., "The Use of Low-Concentration Crosslined Hydroxy-alkyl Polymer System as a Highly Efficient Fracturing Fluid", Proc. of 22nd Annual Southwestern Petroleum Short Course Assoc. Mtg., pp. 129–134.

(27) Holditch, S.A. and Ely, J., "Successful Stimulation of Deep Wells Using High Proppant Concentrations", J. Petrol. Technol., vol. 25, pp. 959–964 (August 1973).

(28) Holditch, S.A. and Morse, R.A., "The Effects of Non-Darcy Flow on the Behavior of Hydraulically Fractured Gas Wells", J. Petrol. Technol., vol. 28, pp. 1169–1179 (October 1976).

(29) Howard, G.C. and Fast, C.R., "Optimum Fluid Characteristics for Fracture Extension", API Drill. and Prod. Practice 261 (1957).

(30) Hubert, M.K. and Willis, D.G., "Mechanics of Hydraulic Fracturing", J. Petrol. Technol., pp. 153–166 (June 1957).

(31) Hurst, R.E., "Use of Liquified Gaseous Fracture Fluids for Dry Gas Reservoirs", Pres. at SPE 47th Annual Fall Meeting, San Antonio, TX, October 8–11, 1972, Preprint Paper No. 4116, 2 pp.

(32) Keeney, B.R. and Frost, J.G., "Guidelines Regarding the Use of Alcohols in Acidic Stimulation Fluids", J. Petrol. Technol., vol. 27, pp. 552–554 (May 1975).

(33) Kern, L.R., Perkins, T.K. and Wyant, R.E., "The Mechanics of Sand Movement in Fracturing", J. Petrol. Technol., pp. 403–405 (July 1959).

(34) Kiel, O.M., "A New Hydraulic Fracturing Process", J. Petrol. Technol., vol. 22, pp. 89–96 (January 1970).

(35) Kiel, O.M., Method of Fracturing Subterranean Formations Using Oil-in-Water Emulsions, U.S. Patent 3,710,865 (Exxon Production Research Co.) (January 1973).

(36) Kiel, O.M. and Weaver, R.H., "Emulsion Fracturing System", Oil and Gas J., pp. 72–73 (February 21, 1972).

(37) King, G.E., "Factors Affecting Dynamic Fluid Lead-off with Foam Fracturing Fluids", Pres. at SPE 52nd Annual Fall Meeting, Denver, CO, October 9–12, 1977, Preprint Paper No. 6817, 4 pp.

(38) Kreuger, R.F., "Advances in Well Completion and Stimulation During JPT's First Quarter Century", *J. Petrol. Technol.,* vol. 25, pp. 1447-1462 (December 1973).

(39) Kreuger, R.F. and Vogel, L.C., "Damage to Sandstone Cores by Particles from Drilling Fluids", *API Drill. and Prod. Prac.,* 158 (1954).

(40) McCune, C.C., Ault, J.W. and Dunlap, R.G., "Reservoir Properties Affecting Matrix Acid Stimulation of Sandstones", *J. Petrol. Technol.,* vol. 27, pp. 633-640 (May 1975).

(41) McGuire, W.J. and Sikora, V.J., "The Effect of Vertical Fractures on Well Productivity", *J. Petrol. Technol.,* pp. 72-74 (October 1960).

(42) Millington, R.J., "Gas Diffusion in Porous Media", *Science,* p. 100 (1959).

(43) Neasham, J.W., "The Morphology of Dispersed Clay in Sandstone Reservoirs and its Effect on Sandstone Shaliness, Pore Space and Fluid Flow Properties", Pres. at SPE 52nd Annual Fall Meeting, Denver, CO, October 9-12, 1977, Preprint Paper No. 6858, 3 pp.

(44) Neel, E.A., Parmley, J.L. and Colpoys, P.J., "Oxide Ceramic Proppants for Treatment of Deep Well Fractures", Pres. at SPE 52nd Annual Fall Meeting, Denver, CO, October 9-12, 1977, Preprint Paper No. 6816, 3 pp.

(45) Nowak, T.J. and Kreuger, R.F., "The Effect of Mud Filtrates and Mud Particles Upon the Permeabilities of Cores", *API Drill. and Prod. Prac.,* 164 (1951).

(46) Novotny, E.J., "Proppant Transport", Pres. at SPE 52nd Annual Fall Meeting, Denver, CO, October 9-12, 1977, Preprint Paper No. 6813, 8 pp.

(47) Ousterhout, R.S. and Hall, C.D., "Reduction of Friction Loss in Fracturing Operations", *J. Petrol. Technol,* vol. 13, No. 5, pp. 217-222 (March 1961).

(48) Pence, S.A., "Evaluating Formation Damage in Low-Permeability Sandstone", Pres. at SPE 50th Annual Fall Meeting, Dallas, TX, September 28 through October 1, 1975, Preprint Paper No. 5638.

(49) Pye, D.S. and Smith, W.A., "Fluid-Loss Additives Seriously Reduce Fracture Proppant Conductivity and Formation Permeability", Pres. at SPE 48th Annual Fall Meeting, Las Vegas, NV, September 30 through October 3, 1973, Preprint Paper No. 4680.

(50) Salz, L.B., "Relationship Between Fracture Propagation Pressure and Pore Pressure", Pres. at SPE 52nd Annual Fall Meeting, Denver, CO, October 9-12, 1977, Preprint Paper No. 6870, 3 pp.

(51) Simon, D.E. and Derby, J.A., "Formation Damage is Evaluated by New Instruments, New Methods", *Oil and Gas J.,* pp. 209-214 (September 20, 1976).

(52) Sinclair, A.R., Terry, W.M. and Kiel, O.M., "Polymer Emulsion Fracturing", *J. Petrol. Technol.,* vol 26, pp. 731-738 (July 1974).

(53) Sinha, B.K., "Fluid Lead-Off Under Dynamic and Static Conditions Utilizing the Same Equipment", Prep. for SPE 51st Annual Fall Meeting, New Orleans, LA, October 3-6, 1976, Preprint Paper No. 6126, 5 pp.

(54) Swanson, B.F., "Visualizing Pores and Non-Wetting Phase in Porous Rock", Pres. at 52nd Annual Fall Meeting, Denver, CO, October 9-12, 1977, Preprint Paper No. 6857.

(55) Tannich, J.D., "Liquid Removal from Hydraulically Fractured Gas Wells", *J. Petrol. Technol.,* vol. 27, pp. 1309-1317 (November 1975).

(56) Thomas, R.D. and Ward, D.C., "Effect of Overburden Pressure and Water Saturation on Gas Permeability of Tight Sandstone Cores", *J. Petrol. Technol.,* vol. 24, pp. 120-124 (February 1972).

(57) Tiner, R.L., Stahl, E.J. and Malone, W.T., "Developments in Fluids to Reduce Potential Damage from Fracturing Treatments", Prep. for Pres. at 25th SPE Annual Technical Meeting of Petrol. Soc. of CIM, Calgary, Canada, May 8-10, 1974, Preprint Paper No. 374022, 9 pp.

(58) Tinsley, J.M., Williams, J.R., Jr., Tiner, R.L. and Malone, W.T., "Vertical Fracture Height—Its Effect on Steady State Production Increase", *J. Petrol. Technol.,* vol. 21, pp. 633-638 (May 1969).

(59) Tuttle, R.N. and Barkman, J.H., "New Nondamaging and Acid-Degradable Drilling and Completion Fluids", *J. Petrol. Technol.,* vol. 26, pp. 1221-1226 (November 1974).

(60) Variogs, J. and Rhoades, V.W., "Pressure Transient Tests in Formations having Stress-Sensitive Permeability", *J. Petrol. Technol.,* vol. 25, pp. 965-970 (August 1973).

(61) Variogs, J., Hearn, C.L., Dareing, D.W. and Rhoades, V.W., "Effect of Rock Stress on Gas Production from Low-Permeability Reservoirs", *J. Petrol. Technol.,* vol. 25, pp. 1161–1167 (September 1971).

(62) Wattenbarger, R.A. and Ramey, H.J., Jr., "Well Test Interpretation of Vertically Fractured Gas Wells", *J. Petrol. Technol.,* vol. 21, pp. 625–632 (May 1969).

(63) Weinbrandt, R.M. and Fatt, I., "A Scanning Electron Microscope Study of the Pore Structure of Sandstones", *J. Petrol. Technol.,* vol. 21, pp. 543–548 (May 1969).

(64) White, G.L., "Friction Pressure Reducers in Well Stimulation", *J. Petrol. Technol.,* vol. 16, No. 8, pp. 865–868 (August 1964).

(65) White, G.L. and Free, D.L., "Properties of Various Frac Fluids as Compared to the Ideal Fluid", *Proc. AGA Symposium on Stimulation of Low Permeability Reservoirs,* pp. 1–14 (February 16–17, 1976).

(66) White, J.L. and Means, J.O., "Polysaccharide Derivatives Provide High Viscosity and Low Friction at Low Surface Fluid Temperatures", *J. Petrol. Technol.,* vol. 27, pp. 1067–1073 (September 1975).

(67) Williams, B.B., "Fluid Loss from Hydraulically Induced Fractures," *J. Petrol. Technol.,* vol. 22, pp. 882–888 (July 1970).

(68) Williams, B.B. and Nierode, D.E., "Design of Acid Fracturing Treatments", *J. Petrol. Technol.,* vol. 24, pp. 849–859 (July 1972).

(69) Woodroof, R.A, Jr. and Anderson, R.W., "Synthetic Polymer Friction Reducers Can Cause Formation Damage", Pres. at SPE 52nd Annual Fall Meeting, Denver, CO, October 9–12, 1977, Preprint Paper No. 6812, 6 pp.

Technology Studies—Tight Gas Basins

The information in this chapter is based on the following sources: *Fourth DOE Symposium on Enhanced Pit and Gas Recovery and Improved Drilling Methods, Volume 2—Gas and Drilling, Tulsa, Oklahoma, August 29-31, 1978,* DOE CONF-7808 25-P3, edited by B. Linville of Bartlesville Energy Technology Center. This symposium will be referred to as Symposium I. *Fifth DOE Symposium on Enhanced Oil and Gas Recovery and Improved Drilling Technology, Volume 3—Gas and Drilling, Tulsa, Oklahoma, August 22-24, 1979.* DOE CONF-790805-P3, edited by B. Linville of Bartlesville Energy Technology Center. This symposium will be referred to as Symposium II. References in this chapter are at the end of each section.

DEMONSTRATION OF MHF, PICEANCE BASIN

The information in this section is based on "Demonstration of Massive Hydraulic Fracturing, Piceance Basin, Rio Blanco County, Colorado" by J.L. Fitch and W.L. Medlin of Mobil Research and Development Corporation (Symposium I) and "Demonstration of Massive Hydraulic Fracturing, Piceance Basin, Rio Blanco County, Colorado" by J.L. Fitch, W.L. Medlin and M.K. Strubhar of Mobil Research and Development (Symposium II).

Mobil is using massive hydraulic fracturing in an attempt to stimulate gas production to commercially attractive rates from the low permeability gas sands of the Piceance Basin. Mobil's first massive frac test in the basin ended in failure because of extremely low reservoir permeability.

Location of the present test is in the Piceance Creek Gas field, Sect. 13, T2S, R97W, Rio Blanco County, Colorado. The well, Piceance Creek Unit No. F31-13G, was drilled in 1977 to a total depth of 10,800 ft. Drilling, coring, logging and some testing of the well were reported at last year's Symposium (I).

Overview: During 1977 the well was drilled and prepared for testing and fracturing by setting 7" casing to 10,800 ft. Top of the cement is at ±8,100 ft. Logs were used to select a number of intervals with potential for massive fracturing. During 1977 five of these intervals were perforated and tested by running short flow tests followed by pressure buildups. Figure 7.1 shows the distribution of these intervals in the well.

Figure 7.1: Distribution of Intervals

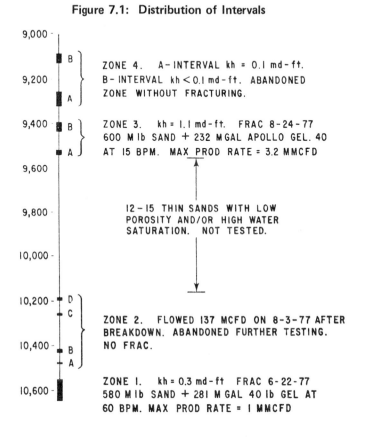

Source: Symposium I

Two of these intervals were judged to have sufficient permeability-thickness to justify fracturing. These intervals were fractured and flow tested. During the period from January 1 through March 31 all of the intervals open to flow were produced to the gas sales line. Testing of additional intervals above 9,086 ft. is planned for the remainder of 1978.

Testing and Fracturing of Intervals—Zones 1-4

Testing and Fracturing Gross Interval 10,549-10,680 ft: This 131 ft interval, containing about 60 ft of net sand, was judged to be one of the most promising zones in the well on the basis of log analysis. A formation breakdown treatment was carried out by pumping 5,000 gallons of 2% KCl water at 8 bbl/min with an excess number of ball sealers. Ball off was not achieved but noise logs run during

flow-back indicated that most of the zone was contributing to flow. After a
3-day flow-back period, at a maximum rate of 400 Mcfd, the zone was shut in
for pressure buildup. Figure 7.2 shows the buildup data which gave a kh of
0.3 md-ft. This kh is somewhat below the preselected 0.5 md-ft cutoff. Nevertheless, a decision was made to carry out a massive frac treatment in this interval.

Figure 7.2: Buildup Data—Zone 1

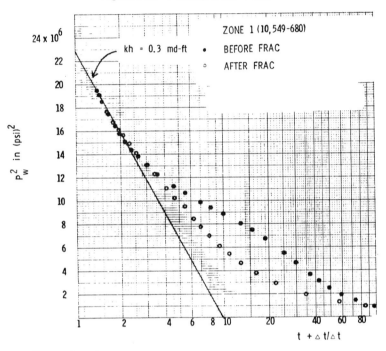

Source: Symposium I

     A massive frac treatment was designed to carry 600,000 lbs of 20-40
sand in a 40 lb/1,000 gallon cellulose gel at a pump rate of 60 bbl/min. The
basic concept of this design was to generate a fracture of minimum practical
width with thin fluid sand transport. This scheme was expected to provide a
settled bed of sand over most of the fractured height extending over roughly
half of the generated frac length. During the treatment there was an unexpected
increase in treating pressure which forced shut-in after 580,000 lbs of sand had
been put away. The premature shut-in occurred before all sand had been flushed
to the perfs. Flowback was started within 15 min after shut-in. There was no
significant restriction to flow due to sand remaining in the wellbore and only a
modest amount of sand was produced during the first hours of flow.
     The maximum production rate after fracturing was 1 MMcfd after 3 days of
flow. This rate had declined to 800 Mcfd after 7 days when the zone was shut
in for pressure buildup. Figure 7.2 shows the results of this after-frac buildup
test. The points at long time fall on the same final slope as the prefrac points.
This match indicates that the massive frac treatment did not open up a significant

amount of additional productive interval. On the other hand, the peak flow rates before and after fracturing do not seem consistent with the productivity increase expected for a 580,000 lb treatment. This result, together with the observed treating pressure increase, suggests fracturing out of zone into non-productive formation. Following the after-frac pressure buildup, Zone 1 was flowed for 6 days at rates declining from 2.8 to 0.83 MMcfd. The well was then shut in and killed with KCl water in preparation for testing Zone 2. A retrievable bridge plug was set above Zone 1 for isolation.

Testing Gross Interval 10,186-10,476 ft: This interval, denoted as Zone 2, contained four thin sands, labeled A-D in Figure 7.1. Thicknesses were 4, 14, 6 and 18 ft, respectively. Sands A and B were perforated with 3 and 10 holes, respectively, and broken down in one treatment using 5,000 gallons of 2% KCl water pumped at 8 bbl/min with excess ball sealers. A retrievable bridge plug was set above these perforations. Sands C and D were then perforated with 4 and 12 holes, respectively, and broken down in the same way. Good ball action was observed in both breakdown treatments. The retrievable bridge plug was moved below sand D and the entire Zone 2 interval opened to flow.

After some swabbing, gas flow was initiated but the maximum rate was only 137 Mcfd. Noise-temperature logs run during the flow period showed that all four sands were contributing to flow. Based on the experience of Zone 1 it seemed unlikely that the kh for Zone 2 would exceed 0.1 md-ft. Therefore, the well was killed with KCl water and a retrievable bridge plug set above Zone 2 for isolation.

Testing and Fracturing Gross Interval 9,392-9,538 ft: Based on conventional log analysis, the interval between 9,550 and 10,150 did not contain sands with promising characteristics for massive frac applications. As indicated in Figure 7.1 there were 12 to 15 thin sands in this interval with either low porosity or high water saturation. The next sands selected for testing were at 9,517 to 9,538 and 9,392 to 9,432.

These sands were included together in Zone 3 and are labeled A and B, respectively, in Figure 7.1. The A sand was perforated with 16 holes and the B sand with 18 holes. Both sands were broken down together with 12,000 gallons of 2% KCl water pumped at 12 bbl/min with 84 ball sealers. Good ball sealer action was observed. Maximum production rate during flowback was 930 Mcfd declining to 700 Mcfd in 3 days when the zone was shut in for buildup. Figure 7.3 shows results of this buildup test which gave a kh of 1.0 md-ft. Since this value was well in excess of the 0.5 md-ft cutoff, plans were made to carry out a massive frac treatment in Zone 3.

Based on the experience of Zone 1, the massive frac treatment was designed for low rate to avoid large treating pressure increases. To insure adequate sand transport under these conditions a crosslinked water base fluid system was chosen. Final design called for 600,000 lb of 20-40 sand to be carried in 230,000 gallons of 40 pounds per 1,000 gallon crosslinked guar at a rate of 15 bbl/min down 2⅞" tubing.

The treatment was carried out according to design with no significant mechanical problems. Treating pressure rose by 1,500 psi during the treatment but most of the increase was limited to the last 10% of the pumping time. The well was shut in for 5 hr after the treatment to allow complete breaking of the cross-linked gel. During shut-in a temperature log was run to evaluate fracture height. Sand fill was found at 9,500 ft just above the A sand perfs. Above this depth the temperature log indicated a well defined, upper fracture reasonably well contained within the B sand interval.

Figure 7.3: Buildup Data—Zone 3

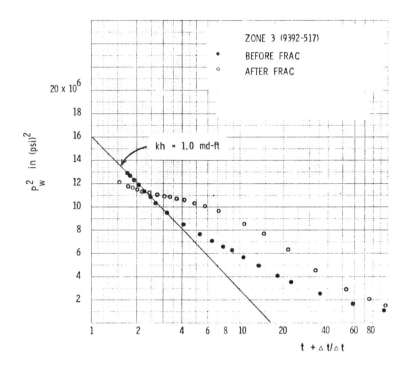

Source: Symposium I

During initial flow-back following massive frac treatment the maximum rate was 1.2 MMcfd over a 3-day period. The well was then killed with KCl water to clean out sand fill. Mechanical problems were encountered in unseating the packer on which the tubing string was set. During unseating operations the well was bled down to reduce surface pressure. Solid material containing drilling mud components and formation sand was produced, followed by a sharp increase in production rate. The zone was again opened to flow with a maximum rate of 3.2 MMcfd. This rate declined to 2.6 MMcfd after 3 days when the zone was shut in to resume work on the stuck packer.

A two-week operation was required to unseat and remove the packer. Sand fill was then circulated out of the wellbore. These operations resulted in a loss of approximately 5,500 bbl of KCl water in the fractured interval. To remove part of this load water, the zone was again opened to flow for a 6 day period. Maximum gas rate was 1.1 MMcfd with no decline and water recovery was approximately 2,000 bbl.

Gamma and noise/temperature logs were run during this flow period. The gamma log provided fracture height evaluation on the basis of radioactive tracer added to the frac sand during the massive frac treatment. It showed reasonably good containment of two individual fractures within the A and B sands. The noise and temperature logs showed nearly all of the gas flow to be about equally divided between the upper 5 feet of both the A and B sand intervals. This flow pattern was consistent with an expected high water saturation in the lower part of each fracture.

Following the 6 day flow period Zone 3 was shut in for pressure buildup. Figure 7.3 shows the results of this after-frac buildup test. At long time, the after-frac pressure data do not coincide even approximately with the before-frac data. The comparison is difficult to interpret. It is not clear that the after-frac curve has reached its final slope. If it has, the after-frac condition corresponds to an increased kh and a reduced reservoir pressure.

Alternatively, if it is assumed that the after-frac buildup curve will turn up at longer times to match the before-frac reservoir pressure, the kh value would be significantly reduced after fracturing. A clear interpretation of these results seems impossible. However, the mismatch in buildup data, together with the anomalous flow-back behavior indicate that the condition of the fractured zone is less than ideal.

Following the buildup test, Zone 3 was again opened to flow for a 6 day period. The maximum gas rate was 1.9 MMcfd declining to 1.6 MMcfd in 6 days.

Testing Gross Interval 9,254-9,320 ft: This interval consists of an almost continuous sand sequence containing about 55 net feet. On the basis of log analysis it appeared to have about the same quality as the Zone 3 sands. After perforating with 33 holes the interval was broken down with 5,000 gallons of KCl water pumped at 12 bbl/min with excess ball sealers. Good ball action was obtained with complete ball off after about 50 ball sealers had reached the perforations. The interval was immediately opened to flow, reaching a maximum rate of 400 Mcfd and declining to 125 Mcfd after 3 days. Noise-temperature logs showed most of the gas being produced from the lower part of the interval.

Figure 7.4: Buildup Data—Zone 4

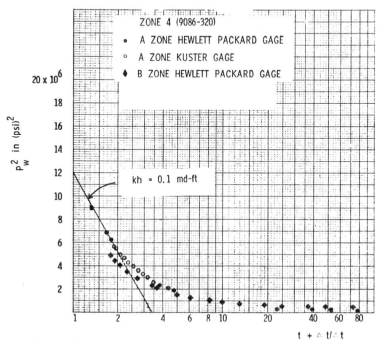

Source: Symposium I

After a 3 day flow period the zone was shut in for pressure buildup. Figure 7.4 shows the buildup data for this interval designated as the A sand of Zone 4. The Hewlett Packard downhole tool normally used for buildup measurements malfunctioned for a time during this buildup and a downhole Kuster gage was used as a backup device. Agreement between the two was quite satisfactory as indicated in Figure 7.4. The computed kh for this interval was only 0.1 md-ft with an extrapolated reservoir pressure of 3,460 psi. These disappointing results led to a decision to test the next thick overlying sand to obtain additional kh.

Testing Gross Interval 9,086-9,124 ft: This interval contained an almost continuous sand of about 35 ft net thickness. Based on log analysis its reservoir quality again appeared to be comparable to that of the Zone 3 sands. The interval was perforated with 32 holes and broken down in the usual way. The maximum flow rate following breakdown was 400 Mcfd declining to 125 Mcfd in 3 days. The zone was then shut in for pressure buildup. Figure 7.4 shows the buildup data labeled as the B sand of Zone 4. The corresponding kh was estimated to be no more than 0.05 md-ft. Thus the combination of A and B intervals of Zone 4 in Figure 7.4 gave a kh of no more than 0.15 md-ft, well below that desired for a massive frac test.

Completion of the Zone 4 work had carried the project to mid-November. It was decided to discontinue further work until the spring of 1978 and produce the well to a sales line during the winter months.

Production Testing of Fractured Intervals

Upon completion of Zone 4 testing the wellbore contained two retrievable bridge plugs and some sand fill. These had been removed and 2⅜ inch production tubing installed by December 13 at which time the well was flowed to the pit for three days. A sales line connection did not become available until mid-January. Beginning on January 18 flow through the sales line was started through a ¾ inch wellhead choke.

Figure 7.5 shows the production history of the well through March 30, 1978.

Figure 7.5: Production History

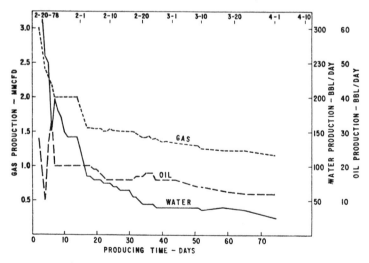

Source: Symposium I

Gas production through this date has been as expected, based on the after-frac production and buildup tests. Water rate has declined to an almost insignificant rate in terms of estimated load water still to be recovered. Figure 7.5 suggests that the well would now average a little less than 1.0 MMcfd gas and 10 bpd condensate during the first year of production.

Testing and Fracturing Procedure—Zones 5-9

Figure 7.6 shows remaining potential massive frac sands in relation to those previously tested and fractured in 1977. These sands were grouped, as shown, into 5 zones numbered 5 through 9.

Figure 7.6: Piceance Creek 31-13 Well

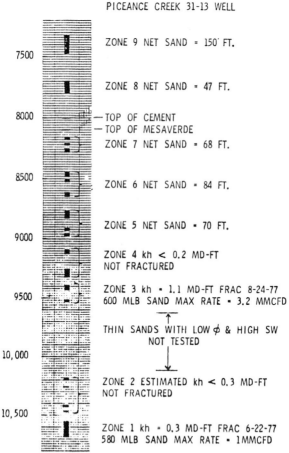

PICEANCE CREEK 31-13 WELL

ZONE 9 NET SAND = 150' FT.

ZONE 8 NET SAND = 47 FT.

— TOP OF CEMENT
— TOP OF MESAVERDE
ZONE 7 NET SAND = 68 FT.

ZONE 6 NET SAND = 84 FT.

ZONE 5 NET SAND = 70 FT.

ZONE 4 kh < 0.2 MD-FT
NOT FRACTURED

ZONE 3 kh = 1.1 MD-FT FRAC 8-24-77
600 MLB SAND MAX RATE = 3.2 MMCFD

THIN SANDS WITH LOW ϕ & HIGH SW
NOT TESTED

ZONE 2 ESTIMATED kh < 0.3 MD-FT
NOT FRACTURED

ZONE 1 kh = 0.3 MD-FT FRAC 6-22-77
580 MLB SAND MAX RATE = 1 MMCFD

7500
8000
8500
9000
9500
10,000
10,500

Source: Symposium II

All were chosen on the basis of log analysis as before. The uppermost zones, 8 and 9, overlie the Mesaverde formation whose top is at 8,100 ft. They presumably correlate with the Ft. Union sands in the CER Geonuclear RB-MHF-3 massive frac well some 20 miles to the southwest (2). Both are above the top of the cement and required remedial cement squeezes before testing. Zone 9 is locally identified with the Ohio Creek formation. This sand body is known to be less lenticular than Mesaverde sands as judged from well to well correlations in the Piceance Creek field.

Gross Interval 8,765-8,972 (Zone 5): This 207 ft interval contains three sands ranging in thickness from 10 to 20 ft with considerable separation. Perforations were restricted to sand intervals only with 40 holes covering about 70 net feet of sand. The perforated intervals were broken down with 10,000 gallons of 2% KCl water pumped at 10 to 12 bbl/min with ball sealers. Good ball sealer action was observed. After breakdown the well was flowed to the pit with gas to surface in 1½ hr. The gas rate declined from 800 to 480 Mcfd in 48 hr after which the well was shut in for pressure buildup. The pressure buildup (PBU) data are shown in Figure 7.7 on an m(p) plot.

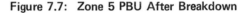

Figure 7.7: Zone 5 PBU After Breakdown

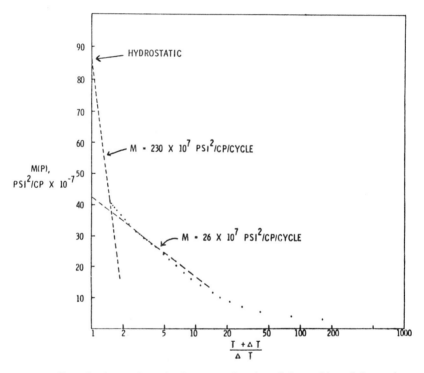

Normalized stress-intensity factor as a function of the position of the crack tip to the interface. The Young's modulus to the right of the interface is twice the modulus to the left, $E_2 = 2E_1$; crack is approaching the interface from the left.

Source: Symposium II

Two straight line segments can be defined as shown. Their slopes correspond to kh values of 1.0 and 0.25 md-ft, respectively. The higher value was taken to be more consistent with the preshut-in flow rates and a massive frac treatment was carried out in zone 5.

The planned frac treatment called for 245,000 gallons of crosslinked gel to be pumped at 15 bbl/min with 600,000 lb of 20-40 sand in concentrations increasing from 2 to 4 pounds per gallon (ppg). Provisions were made to add radioactive tracer to the last 100,000 lb of sand. The treatment was pumped down the annulus between the 7 inch casing and 2⅜" tubing which was hung open-ended to 8,650 ft. Provisions were made to inject 25 ball sealers midway through the treatment for fluid diversion if needed. The actual treatment was not carried to completion because of a sand-out after 388,000 lb of sand had been placed. The sand-out was the direct result of a severe pressure increase throughout the treatment. Figure 7.8 shows a plot of treating pressure and pump rate vs time.

Figure 7.8: Zone 5 MHF Treatment 5-10-78

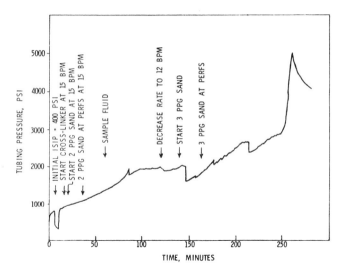

Normalized stress-intensity factor as a function of the position of the crack tip the interface. The Young's modulus to the left of the interface is twice the modulus to the right, $E_1 = 2E_2$; crack is approaching the interface from the left.

Source: Symposium II

An initial shut-in following a 3,000 gallon prepad gave an ISIP of 400 psi. Before sand-out the extrapolated ISIP was about 2,400 psi. The pressure rose at roughly a constant rate during the first 90 min and then remained nearly constant for the next hour. This was followed by a period of sharp pressure increases and breaks with an average rate of pressure increase about equal to that of the first period. After a total increase of 2,000 psi sand-out occurred accompanied by an abrupt pressure increase to the wellhead limit.

The pressure increase was thought to be associated with rock properties rather than sand transport problems. The best evidence of this is to be found in the first 36 min of Figure 7.8. The early constant rate of pressure increase is clearly established well before the first sand has reached the perforations.

After fracturing, zone 5 was flowed for 12 days for evaluation. Cleanup was slow and the earliest measured gas rate was only 330 Mcfd, 4 days after fracturing. The gas rate increased gradually to 430 Mcfd during the next 4 days and then began to decline slightly during the final 4 days. All rates were depressed by frac water production. The zone was clearly stimulated since the rate of decline was much less after fracturing than before. However, fracture effecttiveness was clearly much less than expected. Pressure drawdown analysis of after-frac production gave a kh of approximately 1 md-ft.

Gross Interval 8,443-8,650 (Zone 6): This interval is quite similar to zone 5. It covers 207 ft, with three main sands of about equal thickness and log quality. After isolating the zone with a bridge plug the sand sections were perforated with 47 holes spread over 84 net feet. The perforations were broken down in the usual way with 2% KCl water. Good ball sealer action was observed. Production after breakdown was much like that of zone 5, declining from 850 Mcfd a few hours after breakdown to 300 Mcfd 3 days later. The zone was then shut in for pressure buildup which gave the results plotted in Figure 7.9.

Figure 7.9: Zone 6 PBU After Breakdown

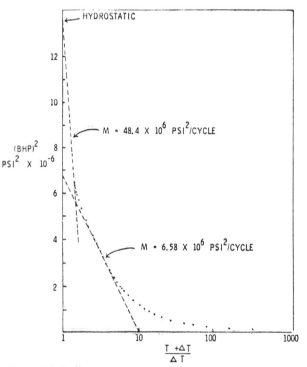

Source: Symposium II

By more optimistic evaluation, zone 6 had adequate kh to justify massive frac treatment. However, its similarity to zone 5 coupled with the massive frac experience in that zone prompted testing of additional sand intervals to be added to zone 6 before fracturing.

Gross Interval 8,173-8,372 (Zone 7): This 199 ft interval contained 3 sands which were perforated with 40 holes covering 68 net feet. After the usual breakdown treatment with 10,000 gallons of 2% KCl water, the zone was flowed four days before shut-in. The gas rate averaged about 200 Mcfd during this test period. Steady flow could not be maintained as the well flowed by heads and produced some formation water. The pressure buildup data are shown in Figure 7.10.

Figure 7.10: Zone 7 PBU After Breakdown

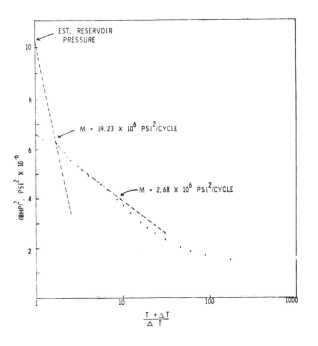

Source: Symposium II

The usual intermediate and final slopes are again apparent and give kh values of 1.6 and 0.3 md-ft, respectively.

Fracture of Zones 6 and 7: It was decided to combine zones 6 and 7, covering the gross interval 8,173 to 8,650 ft in a single massive frac treatment. It was assumed that the large to severe pressure increases observed in deeper zones would occur here and provide a fluid diversion mechanism to assure fracturing of all perforated intervals. The treatment design called for 900,000 lb of 20-40 mesh sand to be carried in 346,000 gallons of crosslinked gel with 5% hydrocarbon phase added to reduce leak off. Pump rate was 15 bbl/min. This treatment ended prematurely with sand-out after 660,000 lb of sand had been placed. The cause of sand-out was again a severe treating pressure increase as shown by treatment record of Figure 7.11. This record is remarkably similar to the one for zone 5.

Figure 7.11: Zones 6 and 7 MHF Treatment

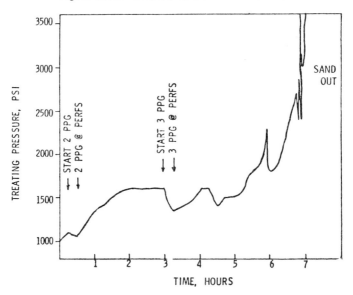

TIME, HOURS

Source: Symposium II

There is an early near-linear increase in pressure during the first 2 hours. Following is a 3 hour period of almost constant pressure marked with occasional sharp breaks and slower recoveries. This is followed by a final rapid increase leading to sand-out.

After fracturing, zones 6 and 7 were produced for 19 days for evaluation. Cleanup was slow and intermittent swabbing was required to produce sustained flow. The gas rate was first measured 7 days after fracturing at 500 Mcfd. This rate had declined to 320 Mcfd 12 days later when production was shut in. Although the slower decline rate showed that these zones had been stimulated the after-frac production was far below expectations.

Gross Interval 7,704-7,796: This 92 ft interval overlies the Mesaverde formation and is the rough time equivalent of the uppermost Ft. Union sand in the CER Geonuclear RB-MHF-3 well (2). This sand in the CER well was considered to be less lenticular and more uniformly distributed than the Mesaverde sands. Furthermore, one is able to recognize a tentative correlation between zone 8 and a similar sand in Mobil's PCU F54-13G well 2,000 ft away. All of these factors point to a more continuous lithology for zone 8.

Remedial cementing was required for zone 8 before perforating. For this purpose the zone was isolated with a bridge plug and the casing was perforated at 7,818 to 7,820 ft. Circulation to surface was attempted without success. Additional perforations were then placed at 7,650 to 7,652 ft and circulation across the zone was established, followed by placement of 150 sacks of cement. Circulation was lost during cementing but a bond log showed good bonding to at least 7,670 ft.

Following cement cure 27 perforations were placed over the 47 net feet of sand. The usual breakdown treatment was then carried out with 6,000 gallons of 2% KCl pumped at 10 to 12 bbl/min with good ball sealer action. During a

3 day flow test, gas production showed little decline, going from 580 to 440 Mcfd. However, water production was significant, averaging 500 bpd well after all load water had been recovered. Noise/temperature logs showed that most of the water production was from the lower part of the zone. Zone 8 has been the only interval in the well with significant water production.

The pressure buildup plot shown in Figure 7.12 was obtained during shut-in following the 3 day flow test.

Figure 7.12: Zone 8 PBU After Breakdown

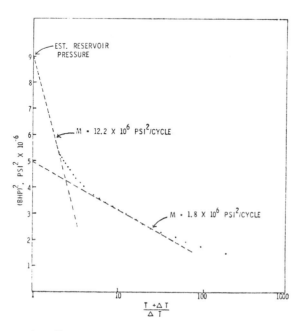

Source: Symposium II

The characteristic intermediate and final slopes can again be identified and give kh values of 1.2 and 0.5 md-ft, respectively. Since there were no additional sand intervals below the thick Ohio Creek sand and since a more continuous lithology was expected for zone 8 it was decided to carry out a massive frac treatment here.

The planned treatment called for 600,000 lb of 20-40 sand to be carried in 230,000 gallons of crosslinked fluid containing 5% hydrocarbon phase and pumped at 15 bbl/min. This treatment, like the two previous ones, ended prematurely with sand-out due to severe treating pressure increase. Figure 7.13 shows the treating pressure record. Comparison with Figures 7.9 and 7.11 show the zone 8 treatment to be the most extreme case of pressure increase yet experienced.

After fracturing, zone 8 was flowed for 10 days. Cleanup was slow and measured gas rates were only obtained over the last 5 days during which they increased from 150 to 180 Mcfd. Sand fill was found to be covering the lower part of the interval, apparently restricting flow severely.

Figure 7.13: Zone 8 MHF Treatment 9-7-78

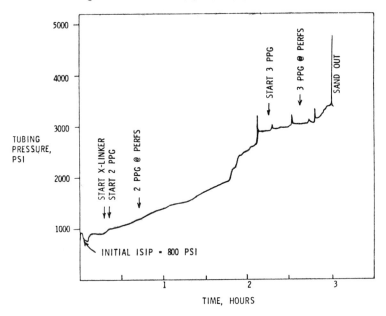

Source: Symposium II

Time did not permit removal of the sand and further testing. This brief flow test was considered sufficient to show that fracture effectiveness was far below expectations.

Gross Interval 7,324-7,476: This interval covers the thick Ohio Creek sand which is clearly less lenticular than the Mesaverde sands. Log analysis gives low porosities over most of the interval with an average of 8 to 9%. Remedial cementing would also be required before testing and fracturing. Thus, massive frac potential was not particularly encouraging. However, the thickness of this sand, its greater continuity of deposition and the need for additional productive interval clearly outweighed other considerations.

Remedial cementing was carried out as in zone 8. Cement was circulated from perforations below to perforations above the zone. An additional cement squeeze treatment was required in the upper perforations when they failed to hold pressure. A bond log run after this operation showed poor bonding around the lower cementing perforations. Therefore, a squeeze treatment was carried out through these perforations also. A subsequent bond log showed acceptable bond quality.

Zone 9 was perforated with 77 holes spread evenly over the interval at 1 hole/2 ft. After perforating there was strong gas flow with 650 psi developed across a ¾ inch choke. The usual breakdown treatment was carried out with 20,000 gallons of 2% KCl water pumped at 12 bbl/min. There was no evidence of ball sealer action. After some swabbing, flow increased rapidly to 2 MMcfd and then declined to 1 MMcfd within 2 days. Pressure buildup measurements gave the results plotted in Figure 7.14 with kh = 2.0 md-ft, more than adequate to justify a massive frac.

Figure 7.14: Zone 9 PBU After Breakdown

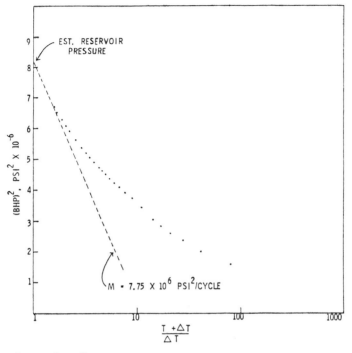

Source: Symposium II

Figure 7.15: PCU 31-13 Production Behavior

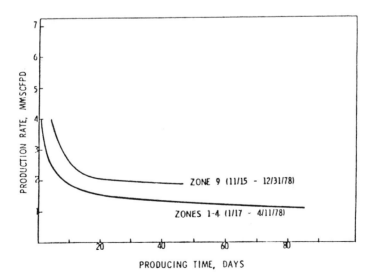

Source: Symposium II

Since zone 9 was considered to have more ideal fracturing properties a high rate thin fluid treatment design was used. This design is considered to be most appropriate for massive fracturing except in zones where pressure increase is a severe problem. The design called for 700,000 lb of 20-40 sand to be carried in 365,000 gallons of 40 lb/Mgal modified guar gum gel pumped at 60 bbl/min. The treatment was carried out as planned with only modest pressure increase during the last fourth of the treatment. The ISIP increased from an initial 500 psi to a final 800 psi. Radioactive tracer was included in all of the sand. A gamma log run the following day defined the vertical fracture extent to be limited to little more than the perforated interval.

After fracturing flow was started without swabbing and increased rapidly. Within two days the gas rate was 4.0 MMcfd, after which it began to decline slowly. Following cleanup, production from zone 9 was flowed into a sales line through December 31, 1978, for evaluation. Figure 7.15 shows the production rate during this period in comparison with the long term test of zones 1 to 4 during the previous winter. This comparison shows zone 9 to be the most productive interval in the well.

Zone Evaluation Procedure

Pressure build-up tests were performed on each zone of interest (with the exception of zone 2). In 1977, zones that were fractured (No. 1 and 3) were tested both before and after fracturing. The prefrac tests appeared to give the most reliable results. Therefore, in the zones (No. 5 through 9) completed in 1978, only prefrac PBU tests were performed.

The technique used to analyze these data was a standard Horner plot. In most cases, the data were still curving upward at the time the test was terminated. (Zone 1 prefrac is possibly the one exception.) It is recognized that the actual kh of a zone is not determined from such a test. However, by connecting the last measured pressure value to a known or reasonable value for reservoir pressure on the Horner plot, a maximum value for kh can be obtained. Following that, a judgement must be made as to the degree of correction required to adjust this maximum kh to obtain true kh.

Results for zones 5 through 9 are reported herein. It is believed the values obtained using this procedure are reasonably accurate. Additional confidence in this method is provided through analysis of short term flow data measured from each of the various zones before and/or after the PBU tests. Reasonable matches of actual production data were obtained for zones 1 through 4 using a computer model (3) and kh approximations obtained from PBU tests. For the zones fractured, reasonable fracture lengths based on quantities pumped were used. The best match obtained for each zone is illustrated in Figures 7.16 through 7.20.

Zone 1: Flow rates taken after the PBU following fracturing were matched using the model. A small amount of cleanup may have still been taking place during this measurement period. Formation kh determined from the prefrac PBU was used.

Zone 2: No PBU test was performed. Assuming darcy radial flow for the brief (4 day) flow period, a kh estimate of 0.3 md-ft was obtained. This was used to match the one measured flow rate of 137 Mcf/day after four days of swabbing and flowing following breakdown.

Zone 3: There were several flow periods following fracturing of this zone. The one used for the purpose of doing a history match followed the PBU test. While frac and workover fluids were still being produced at this time, the effects of cleanup should have been reduced compared to the other flow periods. Because they still existed, however, a match at the end of the flow period should be more valid than a match at the beginning.

Zones 4A and 4B: While tested separately, the behavior and treatment of each of these zones was quite similar. Following breakdown, they flowed back easily without swabbing, declining quickly (over about 3 days) from 500+ Mcf/day to a little more than 100 Mcf/day. PBU tests on the zones indicated kh too low to justify fracturing. The PBU tests indicated a significant negative skin in each of the zones, perhaps due to natural fracturing. The best production matches were obtained by using short fractures along with kh as approximated from the PBU tests.

Figure 7.16: Zone 1 Flow After Frac

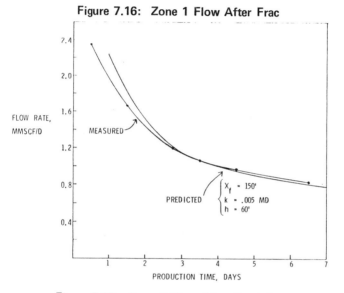

Figure 7.17: Zone 2 Flow After Breakdown

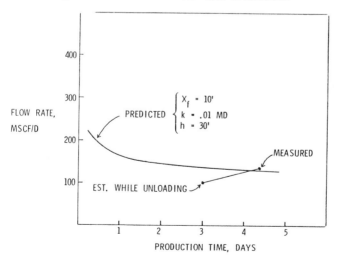

Source: Symposium II

Figure 7.18: Zone 3 Flow After Frac

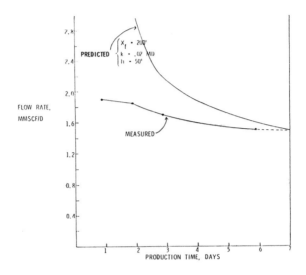

Source: Symposium II

Figure 7.19: Zone 4A Flow After Breakdown

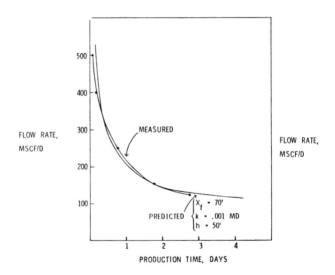

Source: Symposium II

Figure 7.20: Zone 4B Flow After Breakdown

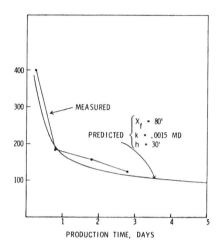

Source: Symposium II

Composite of All Zones: Of the individual zone production matches, some esthetically are more pleasing than others. These are, however, considered the best matches. They were then extended over a longer period (80 days) of time, and a composite of all zones producing at once was determined. This composite was then compared to actual commingled production from all zones over a period from January 17 to April 11, 1978. This comparison is shown in Figure 7.21.

Figure 7.21: Composite Flow From Zones 1-4 January 17-April 11, 1978

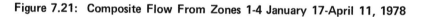

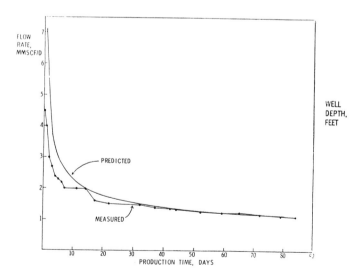

Source: Symposium II

In addition, the percent contribution of each zone to the total was determined from the composite prediction. This is plotted in Figure 7.22 along with a flow profile determined from a borehole flowmeter survey. The profile plotted is an average of three surveys run by taking stations above each of the completed intervals.

Figure 7.22: Production Profiles

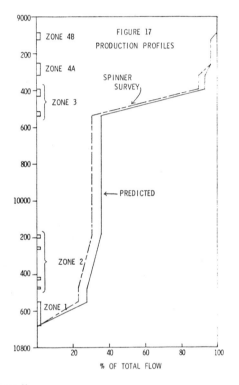

Source: Symposium II

Conclusion: The match of the composite prediction with production data for a three month period is excellent. Therefore, the approximations of kh from the Horner analyses of PBU data must be reasonably accurate.

References

(1) Fitch, J.L., "Demonstration of Massive Hydraulic Fracturing, Mesaverde Formation, Piceance Basin, Colorado," Third Annual ERDA Symposium on Enhanced Oil and Gas Recovery and Improved Drilling Methods, Tulsa, OK, (August 30-Sept. 1, 1977).

(2) Appledorn, C.R. and Mann, R.L., "Massive Hydraulic Fracturing Gas Stimulation Project," Third Annual ERDA Symposium on Enhanced Oil and Gas Recovery and Improved Drilling Methods, Tulsa, OK, (August 30-September 1, 1977).

(3) Strubhar, M.K., Fitch, J.L., and Uhri, D.C., "Simplified Performance Prediction Method for Vertically Fractured Gas Wells," SPE 6021, Deep Drilling and Production Symposium of SPE, Amarillo, TX, (April 17-19, 1977).

MASSIVE HYDRAULIC FRACTURING THE COTTON VALLEY LIME MATRIX

The information in this section is based on "A Case History for Massive Hydraulic Fracturing the Cotton Valley Lime Matrix, Fallon and North Personville Fields, Limestone County, Texas" by H.G. Kozik and B.G. Bailey of Mitchell Energy Corporation and S.A. Holditch of Texas A&M University (Symposium II).

History and Geology

The Jurassic Cotton Valley group on the west flank of the East Texas Basin generally consists of a succession of about 1,000 ft of terrestrial and marine sands and shales, 800 ft of dark shale (Bossier shale) and 300 to 500 ft of limestone known as the Cotton Valley Lime or Haynesville, at the bottom. The group covers over a quarter of a million square miles in the East Texas Basin and adjacent parts of Louisiana and Southeast Arkansas.

Mitchell Energy has drilled and completed nine wells in the Cotton Valley Limestone reservoir in Fallon and North Personville Fields, Limestone County, Texas. The Cotton Valley Limestone formation is generally gray, massive, oolitic to pisolitic and finely crystalline to micritic. The usually low permeability to gas of the pay (mostly 0.01 to 0.04 md) is locally enhanced by fractures as shown by limited core information. Also, production decline curves are supportive of that fact. The matrix porosity—which is in the 2 to 12% range, with local thin zones up to 14%—appears related to zones composed mainly of oolites in the 1 mm range. These zones are erratically developed and occur mainly in the upper several hundred feet of section.

In the Fallon Field area, a very dense section, from 30 to 80 ft thick, is normally found at the top of the lime. Histograms of core porosities and air permeability are shown in Figure 7.23. The relationship of the permeability to porosity is shown in Figure 7.24. These data are based on cores from two wells with 92% recovery, the Lawrence No. 1 and the Muse-Duke No. 1.

A summary of the collected data of measurable and nonmeasurable natural fractures from the Muse-Duke No. 1 cores is shown in Figures 7.25 and 7.26. Figure 7.25 presents the percentage distribution of fracture strike orientations principally in the northwest-southeast direction. Figure 7.26 illustrates the fracture dip angles and the dip directions found in the oriented sections of the core. Histograms of the fractures indicate that the majority of fractures occur in the 85 to 90 degree range with no fracture dip angles being noted of less than 50 degrees.

The first six wells completed in the Cotton Valley Lime received either acid frac or limited water-sand fracs (under 50,000 gallons). Figure 7.27 illustrates the performances after these treatments. The best well, Burleson No. 1, after a fracture treatment of 48,000 gallons of gelled water and 178,000 lb of sand, began producing 4.4 MMcfgpd which declined to 2.5 MMcfgpd in three months and to 0.500 MMcfgpd in two years. The other five wells, after normal acid stimulation treatments began at about 1 MMcfgpd per well, which soon declined to 0.200 to 0.300 MMcfgpd.

After the initial flush production, the decline rate for the Burleson Well was 45% for about the first two years, then dropped to 9%. The average of the other five wells declined at a rate of 57% and then dropped to 15%.

The Muse No. 1 was the first Mitchell well to receive a massive hydraulic sand-fracture stimulation treatment. A total of 340,000 gallons of gelled water and 450,000 lb of 20-40 mesh sand was pumped.

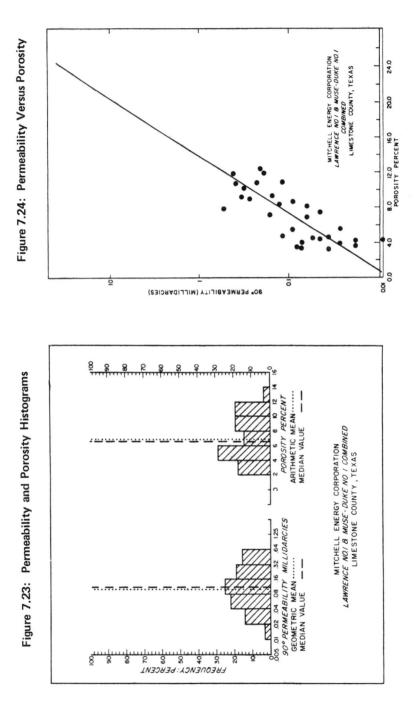

Figure 7.24: Permeability Versus Porosity

Figure 7.23: Permeability and Porosity Histograms

Source: Symposium II

Figure 7.26: Fracture Dip Angle and Direction

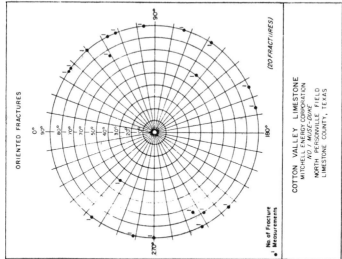

ORIENTED FRACTURES

(20 FRACTURES)

No. of Fracture
Measurements

COTTON VALLEY LIMESTONE
MITCHELL ENERGY CORPORATION
NO. I MUSE-DUKE
NORTH PERSONVILLE FIELD
LIMESTONE COUNTY, TEXAS

Figure 7.25: Comparison of Fracture Orientation
Distributions

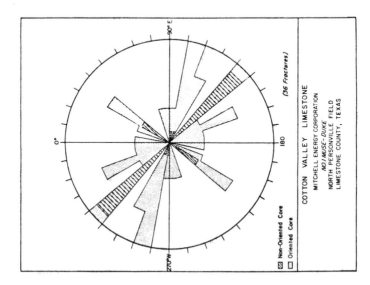

(36 Fractures)

Non-Oriented Core

Oriented Core

COTTON VALLEY LIMESTONE
MITCHELL ENERGY CORPORATION
NO./MUSE-DUKE
NORTH PERSONVILLE FIELD
LIMESTONE COUNTY, TEXAS

Source: Symposium II

Figure 7.27: Performance Comparison—Fallon and North Personville Field

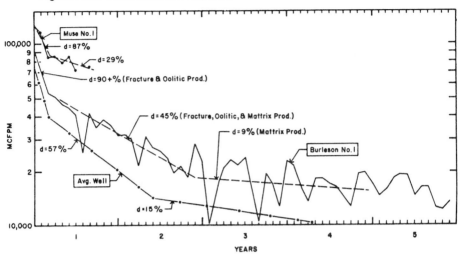

Source: Symposium II

An additional 48,000 lb of 100 mesh sand was pumped ahead of the 20-40 mesh sand to help control fluid loss. The initial after-frac rate was in excess of 4 MMcfpd and the rate has gradually declined to its present capability in excess of 1.5 MMcfpd after 20 months. After the initial flush production, its beginning decline was only 29% as compared to the others above.

The Quinn No. 1 was the second well to receive a massive hydraulic stimulation similar to the Muse No. 1. It is too soon to tell, but its after-frac production is better than the Muse No. 1 and it appears to be about the same as the following Muse-Duke No. 1.

The Muse-Duke No. 1 was the most recently drilled and completed well in the Cotton Valley Limestone reservoir. Following acid stimulation treatment, the well flowed in excess of 4 MMcfpd, dropping very quickly to 1.7 MMcfpd. After fracture stimulating with 2.8 million pounds of sand, the well flowed 6+ MMcfpd and two months after fracturing had a 5+ MMcfpd deliverability. Six months later, it was still flowing 4 MMcfgpd.

Reservoir Simulation

A computer reservoir simulator (1)(2), was used to history match the production and pressure transient data of the Muse No. 1 and the Muse-Duke No. 1. Basic formation permeability, fracture length, and fracture conductivity were varied to match the well performance and assess the sensitivity of project profit potential. Also, the study was made to optimize the development of the Cotton Valley Lime reservoir with respect to fracture length and well density. The results are shown below.

The Muse-Duke No. 1 was successfully history matched using the following reservoir parameters.

Well spacing	640 acres
Original pressure	6,750 psia

(continued)

Gas permeability	0.013 md
Gas porosity	3.6 %
Net gas pay	50 ft
Reservoir temperature	286°F
Fracture half-length	1,500 ft
Fracture conductivity	450 md-ft

The Muse-Duke No. 1 was also successfully history matched using the above parameters and a 2,500 foot fracture half-length.

Because the post-frac buildup of the Muse No. 1 was of short duration, a unique history match was not obtained. Adequate matches of the Muse No. 1 were also made using: 0.023 md with a 1,000 ft fracture half-length, and 0.0425 md with a 500 ft fracture half-length.

Even though a unique match was not obtained, the important reservoir parameters were bracketed, 0.01 to 0.04 md for permeability and 500 to 1,500 ft for fracture half-length.

For the four matches obtained during this study, the following values for a 30-year ultimate recovery were calculated assuming 640 acre drainage.

	Formation Permeability (md)	Fracture Half-Length (ft)	30-Year Recovery (Bcf)
Muse No. 1	0.013	1,500	7.244
Muse No. 1	0.023	1,000	7.850
Muse No. 1	0.0425	500	8.109
Muse-Duke No. 1	0.013	2,500	7.958

Without further pressure transient tests at constant pressure rates, it would be difficult to say what the optimum well spacing and propped fracture half-length should be. It could either be 640 acres and 2,500 ft or 320 acres and 1,500 ft, depending upon the matrix permeability to gas.

Mechanics of a Super Massive Hydraulic Frac Job

One question that is always asked during the evolutionary process of fracturing is: "How big a job can be performed?" The answer is: "Just about as big a job as you would wish. The technology is available today." The real question is: "Will it be effective?"

The objective here is to outline some of the major details in designing and executing a "super" massive hydraulic frac job for the Mitchell Energy Muse-Duke No. 1, Fallon and North Personville Fields, Limestone County, Texas.

Rock Mechanical Properties: A very important factor in the basic design of a frac job is a knowledge of the in situ stresses (3)(4)(5) of the rock; not only of the objective formation, but also the bounding formations. The in situ rock stresses will control the behavior of the induced fracture away from a well bore.

To determine these stresses, laboratory measurements were performed on core samples from three distinct zones in the Muse-Duke No. 1: Bossier Shale (11,116 to 11,207 ft); Cotton Valley Limestone (11,223 to 11,336 ft); Buckner Lime and Anhydrite formation (11,505 to 11,510 ft). The pertinent mechanical properties determined were: dynamic elastic properties including ultrasonic velocities, Poisson's ratio and elastic moduli; triaxial compression tests to evaluate static moduli, Poisson's ratio, maximum compressive strength and fracture toughness tests.

The tests were carried out under in situ conditions of effective pressure and temperature (285°F). Some tests were also performed at standard conditions to delineate the effects of in situ conditions on the measured properties.

A series of three in situ stress measurements in the wellbore were also scheduled in the Muse-Duke No. 1 Well. The first test was attempted at a depth of 11,456 feet. The packers on the testing tool failed during the first test and the program was subsequently cancelled. No in situ data were gained during the first test.

Although the attempt to measure the in situ stress did not produce the required results, the laboratory measurement program to determine the elastic properties of the various layers resulted in an over-all evaluation of the possibility of fracture containment within the Cotton Valley Limestone formation. The laboratory measurements indicate that the underlying zone (anhydrite-limestone sequence) appears to favor fracture containment; i.e., downward migration of the induced fracture may be impeded by the higher moduli.

On the other hand, the top Bossier Shale layer shows tendency not to contain upward fracture propagation. However, laboratory determined dynamic elastic moduli indicate that the top 40 ft section of the Cotton Valley Limestone formation possesses higher elastic moduli than the middle section. This moduli contrast is favorable to limiting fracture growth upward if the top 40 ft section is not perforated.

The prevention of early fracture initiation in the top limestone section would help produce a deep penetrating fracture with reduced fracture extension inside the upper Bossier Shale zone. That is, eventually, the whole limestone formation would be fractured, since fracture migration upward cannot be entirely prevented. Furthermore, the use of heavy fracturing fluids (such as gelled water based fluids) should reduce the rate of upward fracture migration.

Frac Design Data: The objective was to create a fracture in the Cotton Valley Lime to effectively drain 640 acres. Also, the fracturing fluid had to be pumped down 2⅞ inch tubing under 9,600 psi maximum through 15 perforations from 11,235 to 11,418 ft. Basically, this comes down to programming a pumping schedule to create a 210+ foot high fracture with a radius of 2,500+ ft, keep it within the section and pack it properly to keep it open.

This was accomplished by utilizing Halliburton's "Prop" computer programs. This program was used because it integrates Daneshy's (6)(7) concept of wide fracture fluids. This was more in keeping with the thought that an induced fracture in the Cotton Valley Lime would have to be wide in order to extend the fracture out any appreciable distance. This was also indicated by the measurement of mechanical properties of the rock.

The relevant well completion, formation, and treatment input data used in the program are listed in Table 7.1.

Table 7.1: Fracturing Process Input Data

Well completion data	
Depth of formation	11,200 ft
Depth of packer	11,130 ft
Tubing	2.441 i.d.
Casing	4.778 i.d.
No. perforations	15
Perforation diameter	0.38 inch
Treatment data	
Type of gel	Versagel
Gel concentration	1500–1600

(continued)

Table 7.1: (continued)

Injection rate	22.0 bbl/min
Treatment fluid	1.020 SG
N	0.5100
K (slot)	0.037000 lbf/sec N/ft^2
CW-fluid loss coefficient	0.00260 ft/sqrt (min)
Spurt volume	0 gal/ft^2
CVC-spurt loss coefficient	0.00029 ft/sqrt (min)
Spurt time	0 min
Damage ratio	1.0
Apparent viscosity @ 0.995 inch width	869 cp
Formation data	
Young's modulus	5.76 + 06 psi
Permeability	0.09 md
Porosity	8.5 %
Reservoir fluid compressibility	1.66E-04 l/psi
Reservoir fl viscosity	0.02 cp
BHTP	8607 psi
Reservoir fl pressure	6350 psi
Closure pressure	7100 psi
Gross fracture height	210 ft
Net fracture height	36 ft
Wellbore diameter	5.50 inches
Drainage radius	2640 ft
Well spacing	640 acres

Source: Symposium II

The results of the "Prop" program are listed in Tables 7.2, 7.3 and 7.4. These tables outline the calculated fracture dimensions, the pumping schedule, and the sand depositional profile. To capsulate, the program calculated that it would take 974,000 gallons of fluid and 2,965,200 lb of 20-40 mesh sand to create and prop a 2,708 foot fracture. Also, it would take 5,182 hydraulic horsepower for 18 hr of pumping at 22 bbl/min at 9,610 psi.

Table 7.2: Calculated Fracture Dimensions

	Volume		Created Length	Width Average	Prop Length	Prop Height	. . . Prop. . . . . .Total Capacity .		. . Relative Prod. . . Increase	Efficiency
Design No.	Total	Pad	Length	Average	Length	Height	.Total	Capacity.	Increase	Efficiency
	. (gal/1,000).		(ft)	(inch)	(ft)	(ft)	(Sx)	(ft)	(T)	(%)
1	974.3	200.0	3,522.2	0.995	2,708.4	199.7	29,652	5.53	4.5	94

Source: Symposium II

Table 7.3: Pumping Schedule

95,000 gal treated water prepad*
70,000 gal Versagel 1525 pad
70,000 gal Versagel 1525, 1 lb/gal Oklahoma #1
60,000 gal Versagel 1525 spacer
15,000 gal Versagel 1524, 1 lb/gal 20–40 R/A sand
15,000 gal Versagel 1524, 2 lb/gal 20–40 R/A sand
20,000 gal Versagel 1524, 3 lb/gal 20–40 R/A sand

(continued)

Table 7.3: (continued)

90,000 gal Versagel 1524, 4 lb/gal 20–40 R/A sand
180,000 gal Versagel 1622, 5 lb/gal 20–40 R/A sand
200,000 gal Versagel 1621, 5 lb/gal 20–40 R/A sand
120,000 gal Versagel 1600, 5 lb/gal 20–40 R/A sand
2,850 gal treated water displacement*

*The treated water will utilize 10 lb/gal brine water
as the base fluid.

Source: Symposium II

Table 7.4: Depositional Profiles

At the end of pumping
Carry distance	2,708.4 ft
Maximum bed height	3.9 ft
Average bed height	2.2 ft
% prop deposited	4.1 %

Distance from Well (ft)	Deposited Prop . . Bed Height (ft) . . End of Pumping	Final	. . . .Suspended Prop . . . . (height/ft)	(conc/gal)
12.1	3.6	62.1	210.0	5.0
109.3	3.6	61.9	209.3	5.0
206.5	3.8	61.9	208.7	5.0
303.6	3.7	61.8	208.1	5.0
400.8	3.6	61.7	207.4	5.1
498.0	3.4	62.6	206.4	5.1
655.9	3.3	63.0	205.4	5.1
813.7	3.1	63.1	204.4	5.1
971.6	2.9	63.3	203.3	5.1
1,129.5	2.7	63.7	202.3	5.2
1,287.4	2.5	63.1	201.4	5.2
1,433.2	2.3	63.3	200.5	5.2
1,278.9	2.1	64.2	199.6	5.3
1,724.7	1.8	61.7	198.8	5.3
1,882.5	1.5	61.6	197.9	5.3
2,040.4	1.3	51.2	194.5	4.3
2,125.5	1.1	47.5	194.0	4.3
2,210.5	1.0	48.3	193.5	4.3
2,295.5	0.8	49.3	193.0	4.4
2,380.5	0.5	50.3	192.5	4.4
2,465.5	0.4	37.9	187.7	3.3
2,502.0	0.3	34.1	187.5	3.3
2,538.4	0.3	33.7	187.2	3.3
2,574.8	0.2	27.2	180.6	2.2
2,611.3	0.1	25.5	180.3	2.2
2,647.7	0.1	15.7	170.9	1.1
2,684.1	0.0	12.4	170.5	1.1

Note: Equivalent bed — length = 2,708 ft; height = 199.7 ft; and bed concentration = 2,741 lb/1,000 ft^2.

Source: Symposium II

Discussion of Frac Design: All frac designs including the one shown for this well are set up to create fractures in a homogenous rock having constant mechanical properties. Very little consideration is given to the bounding formations above and below the section to be fractured. For short fractures, the effect of the bounding formations on the vertical propagation of an induced fracture may be ignored but for long fractures, such as the one designed for this well, it cannot be ignored. Consideration must be given to the vertical propagation (8) of the fracture into the bounding rock.

The laboratory measurements indicate that the anhydrite-limestone sequence of the Buckner formation below the Cotton Valley Lime would impede downward vertical migration of the fracture. On the other hand, the top Bossier Shale above shows tendency not to contain upward fracture propagation. Therefore, one could expect that a long fracture initiated in the Cotton Valley Lime might climb out of the section into the Bossier Shale.

Recent work (3) has demonstrated that the vertical propagation can be controlled by controlling the density of the fracturing fluid, the rates and pumping pressure. Therefore, the pumping program was modified to initiate the fracture by pumping a 10 lb/gal salt water pad and pumping the fluid down 2⅞ inch tubing at rates no greater than 24 bbl/min at 9,600 psi tubing pressure. The modifications apparently were successful since preliminary production tests after the frac job indicated that a 2,500+ foot fracture radius was created in the Cotton Valley Lime pay.

Logistics: Bringing in and arranging the equipment for this job was a bit involved, but not unwieldy. After the location was enlarged to accommodate the equipment, it was placed as shown by Figure 7.28. All the equipment was trucked in and in place in ten days as follow:

Sixty-four frac tanks were lined up for mixing the gel and water.

Twenty-two thousand, five hundred barrels of water were trucked in from neighboring municipalities.

Twelve pumpers and five intensifiers were tied together to provide the power for pumping.

The sand was brought in by eight mountain movers plus assorted other carriers.

Five blenders were lined in the flow stream for mixing the chemicals into the water and various other trucks and equipment movers were provided to move the equipment and chemicals.

One hundred twenty-five men were on location to set up and operate the equipment in shifts.

The flow of the frac fluid is shown on the schematic diagram of Figure 7.28.

Execution: The performance by Halliburton throughout the entire operation, from the installation of the equipment to the end of the pumping, was exceptional. Although a hundred percent standby pumping equipment was provided, not a single unit failed in the 18 hr of pumping at 22 bbl/min between 8,800 to 9,600 psi. As shown by the pressure and rate chart of Figure 7.29, the pumping went smoothly and without interruption except toward the end. This was because there was apparently a slight miscalculation of the fluid loss to the formation toward the end of the pumping period and the fracture began to close prematurely. This caused a pressure increase and the fracture began to "lock up" with sand and screening-out about 100,000 gallons before the designed total amount.

Figure 7.28: Equipment Layout for Frac Job-No. 1 Muse-Duke

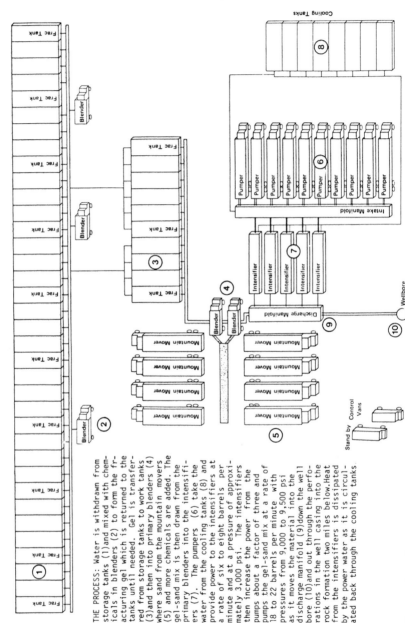

THE PROCESS: Water is withdrawn from storage tanks (1) and mixed with chemicals in blenders (2) to form the fracturing gel which is returned to the tanks until needed. Gel is transferred from storage tanks to work tanks (3) and then into primary blenders (4) where sand from the mountain movers (5) and more chemicals are added. The gel-sand mix is then drawn from the primary blenders into the intensifiers (7). The pumpers (6) take the water from the cooling tanks (8) and provide power to the intensifiers at a rate of six to eight barrels per minute and at a pressure of approximately 3,000 psi. The intensifiers then increase the power from the pumps about a factor of three and pumps the gel-sand mix at a rate of 18 to 22 barrels per minute with pressures from 9,000 to 9,500 psi as it moves the material into the discharge manifold (9) down the well bore (10) and out through the perforations in the well casing into the rock formation two miles below. Heat from the intensifiers is dissipated by the power water as it is circulated back through the cooling tanks.

Source: Symposium II

Figure 7.29:   Pressure-Rate Recordings During Frac Job #1 Muse-Duke

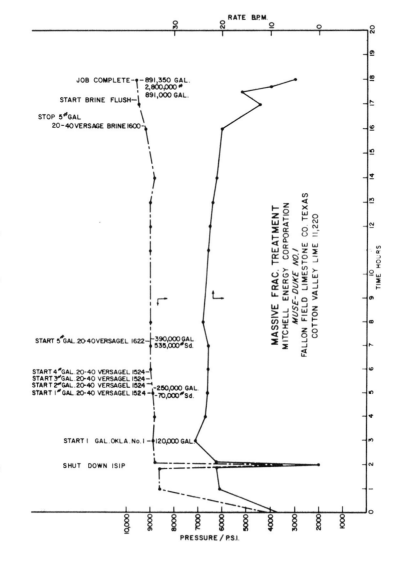

Actually, this was a minor miscalculation considering the parameters involved in designing this type job. Nevertheless, one remarkable aspect of the job was the pumping of 5 lb/gal of sand proppant for over 10 hr before locking up the fracture with sand. Furthermore, this was also a record amount of sand placed in a single fracture.

A comparison of the designed job and the actual performance is made in the Treatment Summary of Table 7.5.

Table 7.5: Treatment Summary

	Design	Actual
Hydraulic horsepower, hhp	5,182	5,085
Surface injection pressure, psi	9,610	9,100
Injection rate, bpm	22	22.8
Pipe friction, psi	5,785	–
Total Versagel, gal	840,000	793,000
Total 20-40 Ottawa sand, lb	2,965,000	2,730,000
Total Oklahoma No. 1, lb	70,000	70,000
Total water required, bbl	22,500	21,223
Propped fracture length, ft	2,708	–
Gel pumping time, hr	16.1	17

Source: Symposium II

Summary: The design and execution of a super massive hydraulic fracture job is within the bounds of practicality with today's technology. The gel chemistry and the equipment have been developed to the state that almost any of the tight gas reservoirs can be physically fractured. The question remains, "How effectively will the fracture produce the gas?"

References

(1) Holditch, S.A., Jennings, J.W. and Neuse, S.H., "The Optimization of Well Spacing and Fracture Length in Low Permeability Gas Reservoirs," SPE Paper 7496 presented at the 53rd Annual Fall Meeting of the SPE in Houston, Texas (October 1-3, 1978).

(2) Holditch, S.A. and Laufer, W.L., "The Analysis of Fractured Gas Wells Using Reservoir Stimulation," SPE Paper 7473 presented at the 53rd Annual Fall Meeting of the SPE in Houston, Texas (October 1-3, 1978).

(3) Simonson, E.R., Abou-Sayed, A.S. and Clifton, R.J., "Containment of Massive Hydraulic Fractures," Journal of Petroleum Technology, Vol. 18 #1 (February 1978).

(4) Abou-Sayed, A.S., Brechtel, C.E. and Clifton, R.J., "In Situ Stress Determination by Hydrofracturing—A Fracture Mechanics Approach," Journal of Geophysical Research, Vol. 83 #B-6 (June 10, 1978).

(5) Clifton, R.J., Simonson, E.R., Jones, A.H. and Green, S.J., "Determination of the Critical Stress-Intensity Factor K_{IC} in a Circular Ring," Experimental Mechanics, Vol. 16, pp. 233-238 (1976).

(6) Daneshy, A.A., "On the Design of Vertical Hydraulic Fractures," Journal of Petroleum Technology (January 1973).

(7) Daneshy, A.A., "Numerical Solution of Sand Transport in Hydraulic Fracturing," Journal of Petroleum Technology (January 1978).

(8) Jones, A.H., Rogers, L.A., Simonson, E.R., Green, S.J. and Clifton, R.J., "The Role of Rock Mechanics in the Design of a Massive Hydraulic Fracture," Petroleum Mechanical Engineering Conf., Tulsa, Oklahoma (September 25, 1975).

NATURAL BUTTES UNIT MASSIVE HYDRAULIC FRACTURING PROJECT

The information in this section is based on "Progress Report and Review of Natural Buttes Unit Massive Hydraulic Fracturing Project" by R.G. Merrill of Gas Producing Enterprises, Inc. (Symposium I).

A program to explore the possibilities of unlocking desperately needed energy reserves from low permeability sands located in the Uinta Basin was begun by the combination of Gas Producing Enterprises, Inc., and DOE.

The program was originally scheduled for six wells to be fractured; however, this has now been expanded to nine wells. One of the new wells involved in the program expansion was cored. This core will be subjected to extensive testing by Terra Tek Labs, out of Salt Lake City, Utah, several service company labs, U.S.G.S., etc. Six wells have been stimulated and the three remaining wells have been drilled and are being studied for completion procedures.

Program Objective

The primary objective of the program is to determine a cost-effective method of recovering gas from low permeability sands. The following parameters have been tested on the six wells treated to date:

(1) Fracture length–Various calculated frac lengths are shown in Table 7.6.

(2) Orientation of fractures–An electrical potential method of determining frac orientation has been used on Natural Buttes Units No. 14, 20 and 22. This test was also planned for CIGE-Natural Buttes Unit No. 21; however, geographical location is prohibitive for the equipment layout.

(3) In situ fracture conductivity has been tested through the use of various proppants and fluids.

(4) Fracture fluid efficiency–The fracture height appears to have been controlled by the barrier shales above and below the pay sand. No sand screen outs have been experienced with the thin fluid or the complexed fluid. Gels have broken satisfactorily in all cases. Formation damage will be measured on the core that was taken from CIGE No. 21. Three fluid systems have been used in the MHF treatments: (a) complexed low-residue guar gum gel; (b) complexed high-residue guar gum gel; and (c) lightly gelled systems 30 lb to 60 lb guar gum per 1,000 gal of 2% KCl water.

(5) The ability to treat 20 to 30 zones in one continuous treatment has been proven. Limited entry techniques have been used, but are generally limited to 10 zones or less, and the overall length of a gross interval should be kept to ±1,000 ft.

Table 7.6: Summary of MHF Treatments

							. . . . Average Frac Job Size				
Well	Zones Perf	Net Ft of Pay	Net Ft per Zone	ϕ	SW	% SD	Type of Fluid	Gallons of Gel	Pounds of Proppant	Calc Frac Length (ft)	Avg. Prod Rate (mcfd)**
Nat. Butte #18 DOE	18	224	12.5	10	48	88	Versa frac	695,000	1,480,000	882	1,200

(continued)

Table 7.6: (continued)

				. . . . . . Average Frac Job Size							
Well	Zones Perf	Net Ft of Pay	Net Ft per Zone	ϕ	SW	% SD	Type of Fluid	Gallons of Gel	Pounds of Proppant	Calc Frac Length (ft)	Avg. Prod Rate (mcfd)**
Nat. Butte #19 DOE	19	194	10.2	9.5	47	87	40% guar gum	655,000	1,237,000	950	60
Nat. Butte #14 DOE	15	271	18.0	9.9	49	65	YF4-PSD	544,000	1,082,000	879	600
Nat. Butte #20	8	65	8.1	9.9	44	88.5	YF4-PSD	309,000	826,000	1,150	800
Nat. Butte #22	24	196	8.1	12.0	45	85.0	PSD	478,758	1,091,000	—	—
Nat. Butte #9	(35)	779*	22*	***	***	***	—	—	—	—	—

*Using GR as only indication net sand which more closely equals gross sand.
**Estimated average for first year.
***Unknown.

Source: Symposium I

Program Review

Six wells have been fractured under the program. Table 7.6 contains the summarized results of these fracturing treatments. The wells are discussed separately as follows:

Natural Buttes Unit No. 18: 4½ inch casing was set at 9,140 ft. Eighteen zones were perforated over the gross interval of 6,490 ft to 8,954 ft, 4 ft/zone, 1 perf/ft. The well was blown dry with nitrogen and averaged 50 Mcfd on a prefrac production test. The well was fractured with 695,000 gal of crosslinked, low-residue guar gum fluid and 1,380,000 lb of 20-40 sand. The treatment was pumped in nine continuous stages (designed to treat two zones per stage). Each stage was spearheaded with 1,000 gal of 15% acid and the stages were diverted with perforation ball sealers. A temperature survey was run before flow-back. The well was flowed back and placed on production on September 30, 1976. The well initially produced 1,500 Mcfd and is now stabilized at 820 Mcfd.

Natural Buttes Unit No. 19: 4½ inch casing was set at 9,697 ft. Nine zones were perforated in the gross interval from 8,676 ft to 9,664 ft with 4 ft/zone, 1 perf/ft. These zones were treated with 275,000 gal of 40 lb guar gum/1,000 gal fluid, 363,000 lb of 40-60 sand and 61,000 lb of 100-mesh sand. The well was flowed back immediately at a high rate.

Ten zones were perforated in the gross interval from 7,224 ft to 8,676 ft and treated with 358,000 gal guar gum gel, 712,000 lb of 40-60 sand, and 72,000 lb of 100-mesh sand. Perforation ball sealers were used for diversion in both treatments. The well was flowed back at a high rate. A Gamma Ray Survey was run to locate radioactive sand used during the treatment. The well did not clean up and it was determined that the well was producing formation water. The water was shut off by isolating with packers and the well placed on production in February 1977. The well is currently producing about 15 Mcfd.

Natural Buttes Unit No. 14: This was an old well with 4½ inch casing set at 8,053 ft. The well had 15 zones perforated over the gross interval from 6,826 ft to 8,004 ft which had originally been acidized with 10,000 gal of 15% HCl. The well was treated with 576,000 gal of crosslinked, low-residue guar gum fluid, 1,053,000 lb of 20-40 sand and 40,000 lb of 40-60 sand.

During the treatment, Sandia Laboratories of Albuquerque, New Mexico, (Sandia) personnel took data by measuring electrical potentials to determine fracture length and orientation. The well was flowed back, cleaned up and placed on production in May 1977. This well had been producing for about two years prior to treatment and declined to a rate of 40 Mcfd. After treatment, the well produced about 790 Mcfd.

Natural Buttes Unit No. 20: $4\frac{1}{2}$" casing was set at 9,807 ft. Eight zones over the gross interval from 8,498 ft to 9,476 ft were perforated with 20 holes for limited entry. The well was stimulated on June 22, 1977, with 309,000 gal of crosslinked, low-residue guar gum fluid and 56,000 lb of 100-mesh sand, 745,000 lb of 40-60 sand and 25,000 lb of 20-40 glass beads. During treatment, Sandia personnel took data by measuring electric potentials to determine fracture length and orientation. However, the data taken is inconclusive. Before flow-back, a Gamma Ray Log was run to locate radioactive tracer sand used during the treatment. The well was placed on production in July 1977 at a rate of about 1,500 Mcfd.

Natural Buttes Unit No. 22: $4\frac{1}{2}$ inch casing was set at 8,618 ft. Twenty-four zones over the gross interval from 6,838 ft to 8,550 ft were perforated with 35 holes. The well was stimulated on November 21, 1977, with 478,000 gal of complexed, low-residue guar gum gel with a 2% KCl base, 60,000 lb 100-mesh sand, 1,066,000 lb of 40-60 mesh Ottawa Sand and 25,000 lb of glass beads. During the treatment, Sandia personnel took data by measuring electric potentials to determine fracture length and orientation.

After flowing to pit to clean up, a Gamma Survey was run to determine if positive identification of frac fluid entry could be established in each of the perforations and/or zones. Seventeen zones were recorded as having positively received some of the treating fluids. Eighteen zones did not have a radioactive deflection, which could indicate that they were not treated. A very short distance will completely mask the radioactive source from the gamma tool; and, if the radioactive source were located one foot or more from the casing wall, there would be no kick on the gamma log. Therefore, the zone could have been treated.

The well was placed on production on February 4, 1978, at an initial rate of 875 Mcfd. Average rate for March 1978 was 211 Mcfd. Gas Btu rating is 1,045 Btu.

Natural Buttes Unit No. 9: This is an old well with $4\frac{1}{2}$ inch casing set at 8,982 ft. The well had 35 zones perforated over the gross interval from 5,661 ft to 8,934 ft. The well was treated with 7,000 gal of $7\frac{1}{2}$% HCl acid, 35,000 gal 2% KCl water with 10 lb/1,000 gal guar gum added, 56,000 gal of 2% KCl water with 60 lb/1,000 gal guar gum gel, 258,000 gal of 2% KCl water with 40 lb/1,000 gal guar gum gel, 28,000 lb 100-mesh Ottawa Sand, 525,000 lb 40-60 mesh Ottawa Sand, 2,200 lb 50/50 Unibeads, $174\frac{7}{8}$ inch, 1.4 SG RCN balls and 110 tons of CO_2.

This was the first well to be treated using liquid CO_2 as part of the frac fluid. The well was turned to the pit to clean up after frac.

This well had produced for about $5\frac{1}{2}$ years prior to treatment and had declined to a rate of about 15 Mcfd. [The "rate-time" production curves for all of the abovedescribed wells are included in Symposium 1.]

Natural Buttes Unit No. 21: This well was drilled, cored and logged extensively. Approximately 575 ft of core was cut on this well, and a program submitted for evaluation of the core data.

Natural Buttes Unit No. 2 (CIGE 2-29-10-21): $4\frac{1}{2}$ inch casing was set at 9,896 ft. Eleven zones were perforated in the gross interval from 9,237 ft to 9,653 ft with a total of 22 holes. These perforations were broken down and balled out with 3,000 gal 15% MS HCl acid. The acid treatment was performed at 7.5 bbl/min at 4,000 psig, balled out at 5,000 psig and held 30 min. The well is presently dead.

The zones in the interval perforated meet the parameters used in conventional log analysis, but do not look favorable when using other types of approaches for log analysis. By treating only these zones, it will be possible to determine the best approach to log analysis with respect to water-free completions.

Natural Buttes Unit No. 23 (CIGE 23-7-10-22): 4½ inch casing was set at 9,560 ft. If the completion procedure on Natural Buttes Unit No. 2 (CIGE 2-29-10-21) proves that the conventional log analysis should be modified more heavily with mud log shows, drilling rate breaks and cross plots, then the zones perforated will be picked using the new parameters.

Results and Conclusions

The progress of the Massive Hydraulic Fracture program is nearing completion; however, until the remaining three wells have been stimulated and production tested, final results and conclusions will not be presented.

Early project data does present areas in which preliminary conclusions appear to be correct and eventually will be final conclusions.

Proppant size does not appear to be a factor in the success of the stimulation. Sand sizes of 40-60 mesh or larger do not affect the rate at which the wells are producing as long as the sand is able to withstand the crushing effects of the closing of the fracture (See Table 7.7).

Table 7.7: Effect of Sand Size on Stabilized Production Rate

				Production Rate from Sales Date		
	Sand Proppant			First	Third	End of
Well Name	lb 20-40	lb 40-60	lb 100 mesh	Month	Month	First Year
	. . . . (in thousands)			. . . . (Mcf/mo)		
NBU No. 18	1,380	100	0	27,000	21,000	18,500
NBU No. 19	0	1,075	133	580	3,600**	300
NBU No. 14	1,053	40	0	23,000	17,800	13,450
NBU No. 20	20*	745	56	36,029	27,000	25,500***
NBU No. 22	25*	1,066	60	10,500	NA	NA
NBU No. 9	0	525	28	NA	NA	NA

*Glass beads.
**This well was damaged due to extensive fishing job performed to recover coil tubing.
***Only 8 months available.

Source: Symposium I

Sand used in the MHF program does not appear to have suffered from crushing. However, in order to test out this theory, high-strength glass beads were used to tail-in with. Sustained productivity increases have not been in evidence where the glass beads were used; therefore, crushing does not seem to be a problem in this area and depth of proppant placement.

Fracture growth is generally determined to be extending above and below the sand body into barrier shales; however, fracturing out of zone has not been a problem. The primary cement job on most of the wells treated has enabled the fracturing fluids to be placed satisfactorily in the zones of interest.

Fracture length and orientation data taken to date have been inconclusive, according to Sandia personnel.

Water production has been a problem. Most of the water production seems to be associated with zones that also give up most of the gas. Due to the cost of constructing disposal systems for the large volumes of water concerned and the relatively small volumes of gas, economics and environmental impacts dictate that the water zones be shut off.

Mobilization of materials and equipment for a massive hydraulic fracturing treatment requires careful planning and cooperation of all people involved. The following is an example of what is involved:

(1) Location layouts must be considered before a well is spudded, due to the area needed to handle 20 to 50 500 bbl frac tanks, 1 to 1.5 MM lb of sand, chemicals, fuel trucks, etc. An operator could find himself limited or hindered if he built a drilling pad, drilled his well and ran pipe only to find that he did not have a big enough location to accommodate all the equipment and materials necessary for his frac. [An equipment layout is shown in Figure 7.30.] Waiting for location enlargement adds interest expense to the investment in the well already drilled. Delays could be rather long, depending on availability of dirt-moving equipment, weather or permits.

(2) Well bore tubular goods must be graded and sized in order to allow for treating pressures, to minimize friction losses, to accommodate safe and orderly frac sand clean out operations, etc.

(3) Well heads and Christmas trees have to be designed to accommodate high rates and/or volume, as well as considerations for unusual wear due to sand abrasion.

(4) Perforation programs must be carefully studied. Size of perfs and number of holes are used to regulate the volumes of frac fluids and proppants that each zone receives, especially in limited-entry type jobs.

(5) Close supervision is necessary when spotting materials and equipment in order to assure that:

Pumping equipment will be placed close enough to the water supply so that efficiency of the pumps does not suffer;

Placement of equipment must be such that chemicals and other materials can be handled and utilized efficiently;

Hauling of one million gallons of water in trucks with a capacity of only 5,000 gallons takes a considerable amount of time and can lead to great expense if long distances are involved;

Heating of water in winter months is necessary to prevent freezing as well as obtaining timely breaking of gels, etc;

Treatment of water after it has been placed in the 500 bbl tanks is necessary to remove solids from the frac water and to make sure it is compatible with the chemicals which will be added to produce the final frac fluid;

Coordination of arrivals of materials and equipment is very important to avoid excessive stand-by time or confusion in the "rig-up" process;

Figure 7.30: Equipment Layout for Natural Buttes No. 9

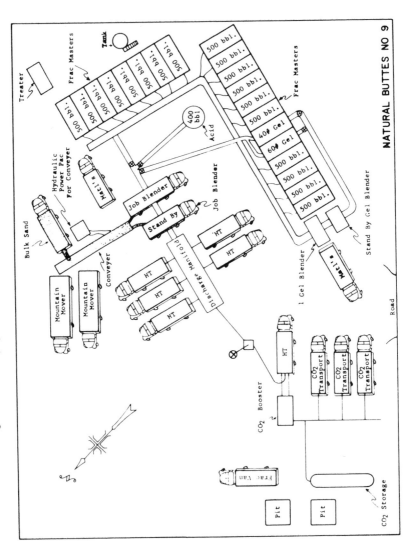

Trucking must be arranged with plenty of lead time in order to get all of the frac tanks on location in a timely fashion;

Frac tanks need to be arranged for ahead of time due to the number of tanks and availability;

CO_2 (if used) must be coordinated closely due to the nature of handling a liquid at -10°F and 300 psig in a transport. CO_2 will become dry ice at 60 psig and -70°F. Availability of CO_2 and N_2 could be a problem due to their wide variety of applications.

In summary, the massive hydraulic fracturing program in the Natural Buttes Unit has demonstrated the ability to unlock badly needed gas reserves from tight, low permeability sands. Modification of processes and designs are needed to make the program more economically attractive due to the risks involved in drilling.

FRACTURE MAPPING

The information in this section is based on "Fracture Mapping Has Become a Viable Technology" by C.L. Schuster of Sandia Laboratories (Symposium II).

Sandia Laboratories has continued to perform diagnostic experiments on massive hydraulic fracture stimulations with Amoco Production Research. These experiments were concentrated in the Wattenberg area northeast of Denver, CO.

The purpose of this experimental program was to correlate several different techniques for determining fracture orientation and to evaluate these techniques one against the other. Two of the techniques developed by Sandia were deployed on these experiments. These included the surface electrical potential system and the borehole seismic system. M.D. Wood, Inc. deployed their tiltmeter array and Amoco provided wellbore diagnostic and core correlations with Texas A&M. This intensive program has resulted in all techniques providing a measure of fracture orientation determination and excellent agreement between all these techniques.

The surface electrical potential system has been the primary means for determining fracture orientation and has been developed over the previous four years. The system has undergone a continuing development program and improvements in the data collection and analysis systems.

The electrical system is an adaptation of a four-element resistivity array where the fracture well acts as one of the current electrodes and the electric potential field is measured around the fracture well. The geometry of the fracture well changes as the pumping of conductive fluid progresses and the resulting change in the electrical potential pattern is thus measured. The orientation and asymmetry of the fracture can then be determined from the surface (1).

Extensive testing of the surface electrical potential system has been conducted over the previous four years. Although the results of the electrical system have been encouraging, the deployment of the system, data acquisition and data analysis problems tend to make this system rather cumbersome to use. The onset of the borehole seismic system has produced some encouraging results.

Seismic recording of signals associated with hydraulic fracturing was initiated on the original hydrofrac mapping experiments. The results of this were less than satisfactory as seismic signals from the fracture interval were not received with sufficient amplitude at the surface for their source location to be determined. The surface seismic efforts have been shifted to a borehole seismic package. This package can be emplaced in the fracturing wellbore during nonproppant pumping times or postfracturing time and record seismic signals associated with the fracture.

Four experiments were conducted in 1978 utilizing the borehole seismic package in the Wattenberg area, and a series of experiments were conducted at the Nevada Test Site to evaluate the system. It is significant that a considerable amount of seismic activity is occurring within the fracturing interval and can be detected during quiet periods. Analyses of these signals have resulted in determining fracture orientations away from the wellbore.

Surface Electrical Potential System

The surface electrical potential system has been evolving over the last several years by the continual testing, improving, and updating of the system. The basic system description, block diagrams, and software are given in reference (3); however, the system has undergone some minor changes and improvements since its publication.

The system is comprised of a current generator, the potential measurement system and the data collection and analyses system. The current generator, which was designed to provide the capability of bipolar current pulse of 50 amperes, is controlled by the PDP-11 computer. The computer not only controls the current pulse but collects the electrical potential data, normalizes it, analyzes it and stores it in a retrievable data system. The potential system allows for 48 sets of potential measurements to be collected simultaneously using two separate current injection schemes.

Two examples of the surface electrical potential data are given in references (4) and (5). The data collected from the shallow experiment, as presented in reference (4), was later verified by actually drilling and locating the fracture with injection flow tests. This sort of independent verification of the electrical analysis has led to an increased credibility of this technique. On deeper experiments, as given in reference (5), this direct verification cannot be obtained; however, excellent agreement in determining fracture orientations from the electrical data has been obtained using tiltmeter analysis, core analysis, and the seismic signals received by the borehole seismic system.

Borehole Seismic System

The borehole seismic system uses a three-axis geophone package that amplifies and multiplexes the three-axis signals to the surface. The configuration and systems design of this system is given in reference (6). The use of a wall clamped system with an orientation device allows for the direction of arriving seismic signals to be determined. This pointing vector provides the orientation of the seismic source that is assumed to be associated with the induced hydraulic fracture. By determining the coda of the incoming seismic wave and locating the compressional and shear wave arrival, the distance to the seismic source can also be determined.

The two experiments conducted early in 1978 resulted in the detection of a significant number of seismic events associated with the fracturing phenomena during the quiet periods or nonflowing periods in the wellbore. It was also determined that even the smallest amount of fluid flowing past the wall-clamped geophone package induced extreme seismic activity and no other signals could be detected.

During the quiet periods, though, several seismic signals were observed. These signals were observed during both shut-ins at the early breakdown stages of hydraulic fracturing as well as at the conclusion of, and for several hours after, a massive fracture treatment. These first two experiments indicated not only a large number of seismic signals received, but, in addition, that the frequency content of these signals was such that it induced oscillations in the geophone mounts and overall systems.

The second set of two experiments was conducted late in 1978 with a somewhat improved geophone mounting system. The system at this time was also upgraded to include operation in a higher temperature environment as the fracturing horizon was considerably deeper. Although the orientation device was not designed to operate at these high temperatures, sufficient film was recovered to allow the system orientation to be determined.

On both these experiments, sufficient seismic data were obtained to determine fracture orientation. The fracture orientation was determined by plotting the Lissajous pattern of the arriving wave on two orthogonal horizontal geophones. An example of this is shown in Figure 7.31.

Figure 7.31: Lissajous Pattern of Two Horizontal Seismic Signals Showing Received Vector Orientation and Time-of-Arrivals

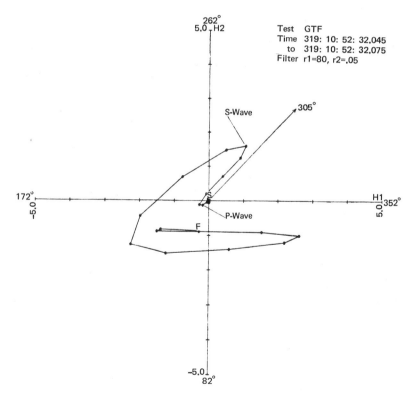

Source: Symposium II

As can be seen, the orientation to the seismic source is determined by the relative amplitude of the signal on each of the horizontal geophones and the arrival of the compressional wave and the shear wave can be readily determined. The time difference between these two arrivals, knowing the seismic velocities, can give the distance to the seismic source. The orientation determined by this method was in excellent agreement on these experiments with all other methods deployed for fracture orientation determinations. As all these techniques rely on different physical phenomena for determining fracture orientation, their agreement lends credence to the overall program for determining fracture orientations.

The borehole seismic system was also taken to the Nevada Test Site and installed in a vertical hole drilled from the floor of G-Tunnel. A triaxial geophone mount was colocated with the seismic system and grouted into place for use as a reference for receiving seismic signals. The purpose of this experiment was two-fold: to evaluate the determination of detecting source locations and for evaluating the system resonance problems associated with the borehole mounting.

The preliminary results indicate that excellent agreement is obtained between the grouted package and the borehole package for determinations of source locations; however, several system resonant problems do exist and need to be examined for methods of reducing this problem.

Borehole Stacked Hydrophone System

A new borehole tool is being designed and fabricated by Sandia to further investigate seismic signal arrivals. The stacked hydrophone array uses four detectors on ten-foot spacings. The hydrophone, being a pressure sensitive device, does not require clamping to the side of the wellbore and, being a much smaller transducer allows for the design of a system that can be installed through tubing. This system will detect the same signals as the borehole seismic system and will be able to determine the elevation from which these signals arrived. Hence, this tool can be used for determining fracture heights out away from the borehole but will not be able to determine fracture orientations.

Conclusions

The surface electrical potential system design has been carried from its inception to a fieldable system for determining fracture orientation. This system will probably continue to be deployed as a baseline for evaluating follow-on borehole systems for detecting fracture orientations. The system has been operated at a wide range of fracture depths and treatment volumes and is limited as depths become greater and treatments become smaller.

The borehole seismic package appears to be a valuable diagnostic tool and its development and evaluation will continue. This tool has the potential for being utilized on a commercial basis as it lends itself to easily being included in the fracture design. It offers the ability to obtain fracture orientation without a major fielding effort. The borehole hydrophone system development will hopefully contribute a viable way of determining fracture heights.

References

(1) Bartel, L.C., McCann, R.P. and Keck, L.J., "Use of Potential Gradients in Massive Hydraulic Fracture Mapping and Characterization," *Proceedings of the 51st Annual Fall Technical Conference and Exhibition of the Society of Petroleum Engineers of AIME, New Orleans, LA*, SPE 6090 (October 1977).
(2) Powers, D.V., Hay, R.G., Schuster, C.L. and Twombly, J., "Detection of Hydraulic Fracture Orientation and Dimensions in Cased Wells," presented at SPE 50th Annual Fall Meeting, Dallas, TX, September 28-October 1, 1975 (SPE 5626); *Journal of Petroleum Technology* (September 1976).

(3) Keck, L.J. and Seavey, R.W., *Instrumentation for Massive Hydraulic Fracture Mapping*, Sandia Laboratories Report SAND-77-0195 (April 1977).

(4) Keck, L.J. and Schuster, C.L., "Shallow Formation Hydrofracture Mapping Experiment," *Proceedings of the Energy Technology Conference and Exhibition, ASME, Houston, TX, September 18-23, 1977; Journal of Pressure Vessel Technology (February 1978).*

(5) McCann, R.P., Bartel, L.C. and Keck, L.J., *Massive Hydraulic Fracture Mapping and Characterization Program: Surface Potential Data for Wattenberg 1975-76 Experiments,* Sandia Laboratories Report, SAND-77-0396 (August 1977).

(6) Schuster, C.L. "Detection Within the Wellbore of Seismic Signals Created by Hydraulic Fracturing," *Proceedings of the 53rd Annual Technology Conference and Exhibition, AIME, Houston, TX,* SPE 7448 (October 1978).

FRACTURING FLUID INTERACTIONS WITH CORES FROM THE COTTON VALLEY LIMESTONE FORMATION

The information in this section is from "Massive Hydraulic Fracture Containment and Productivity Analyses of MEC Muse-Duke Well #1" by U. Ahmed, A.S. Abou-Sayed, L.M. Buchholdt and A.H. Jones of Terra Tek, Inc. and H.G. Kozik of Mitchell Energy Corporation (Symposium II).

Acreage in the Fallon and Personville Fields in the Limestone County, TX is being actively developed by Mitchell Energy Corporation (MEC) of Houston. In support of a successful MHF treatment in the MEC Muse-Duke Well No. 1, Terra Tek Incorporated of Salt Lake City has provided laboratory and field tests in their program.

Results to date of Massive Hydraulic Fracturing (MHF) vary from extremely successful to extremely disappointing failures (1). Studies over the years and a recent one (2) of some 500 MHF treatments suggest that the majority of the MHF failures result from not achieving the "necessary propped fracture geometry."

In an earlier study (3) by the authors it has been identified that there are three aspects which impair the achievement of "the necessary propped fracture geometry": (a) fracture deviation from the zone of gas concentration (fracture containment analysis), (b) productivity damage due to the fracturing fluid interaction with host rock face (swelling, flocculation, release of fines, embedment, etc.) and, (c) fracture conductivity damage due to the fracturing fluid interaction with the propped bed. The present status of the fracture containment analysis is well presented in the literature (1)(4)(5).

A recent study (3) elaborately documents the damaging effect of fracturing fluid on the productivity of hydraulically fractured wells. These studies (1)(3)(6) show the use of laboratory tests on cores at simulated in situ conditions (stress, temperature, saturation and fracture treatment procedure) to delineate the extent of the above mentioned three aspects. This paper presents the analyses of the three aspects in respect to the MHF treatment performed in Mitchell Energy Well Muse-Duke No. 1 in the Cotton Valley limestone formation.

Background

The Cotton Valley Limestone formation (pay zone) is mostly gray, massive, oolitic to pisolitic and finely crystalline to micritic (7). In the Muse-Duke Well No. 1 the pay formation ranges from about 11,220 ft to about 11,420 ft. Above it lies a massive layer of the Bossier Shale. Below it is the Buckner Limestone followed by alternate thin sections of limestone and anhydrite.

The pay zone permeability to gas is low (mostly 0.01 to 0.04 md) with locally enhanced permeability areas (0.03 md to 0.1 md) due to the presence of natural fractures. The matrix porosity ranges between 2 to 12% with local thin zones up to 14% (7).

The MHF treatment is intended to stimulate gas production from the pay zone by draining 640 acres of the reservoir. Almost 900,000 gal of fracturing fluid and 2.8 MM lb of sand proppant were pumped into the fracture. It was estimated that a fracture 210 ft high by 5,400 ft long (2,700 ft each wing) would be generated.

Fracture Containment Analysis

Laboratory and field measurements have been performed to evaluate the MHF containment in the Cotton Valley Limestone formation of MEC Muse-Duke No. 1 well.

Laboratory measurements were performed on samples from three distinct zones: Bossier Shale, Cotton Valley Limestone and Anhydrite formation. Pertinent mechanical properties determined were: dynamic elastic properties including ultrasonic velocities, Poisson's ratio and Young's moduli; triaxial compression tests to evaluate static moduli, Poisson's ratio, maximum compressive strength and fracture toughness tests. The tests were carried out under in situ conditions of effective pressure and temperature (285°F). Description of the experimental procedure can be found in earlier work (1)(4)(6). Some tests were also performed at standard conditions to delineate the effects of in situ conditions on the measured properties.

The effect of in situ conditions (pressure and temperature) on the measured ultrasonic velocities was very small. The static moduli were lower than the dynamic moduli: for samples tested, the difference was between 15 to 25%. A series of three field in situ stress measurements (Bossier Shale, Cotton Valley Limestone and Anydrite formation) were scheduled. The first test was attempted at a depth of 11,456 ft. It was conducted using a Lynes straddle inflatable testing and treating tool with an eight foot straddle section. Detailed description of the test is described elsewhere (8).

Failure in the packer resulted at a shut-in pressure of 7,980 psi and the bottom hole peak pressure reached 9,380 psi. Further evidence against a successful test is found in the results of the MHF treatments conducted by Mitchell Energy on Wells Muse A No. 1 and Quinn Estate No. 1 (9). Further field in situ stress measurements were subsequently cancelled.

The laboratory measurement program to determine the elastic properties of the various layers resulted in an overall evaluation of the possibility of fracture containment within the Cotton Valley Limestone formation. Figure 7.32 presents the variation of dynamic and static moduli with depth. Figure 7.33 presents the variation of the critical stress intensity factor with depth.

The laboratory measurements indicate that the underlying zone (Anydrite-Limestone sequence) appears to favor fracture containment, i.e., downward migration of the induced fracture may be impeded by the higher moduli. On the other hand, the top Bossier Shale layer shows a tendency to contain upward fracture propagation. However, laboratory determined dynamic elastic moduli indicate that the top 40 foot section of the Cotton Valley Limestone formation possesses higher elastic moduli than the middle section. This moduli contrast is favorable to limiting fracture growth upward if the top 40 foot section is not perforated.

The prevention of early fracture initiation in the top limestone section would help produce a deep penetrating fracture with reduced fracture extension inside the upper Bossier shale zone. That is, eventually, the whole limestone formation would be fractured, since fracture migration upward cannot be entirely prevented.

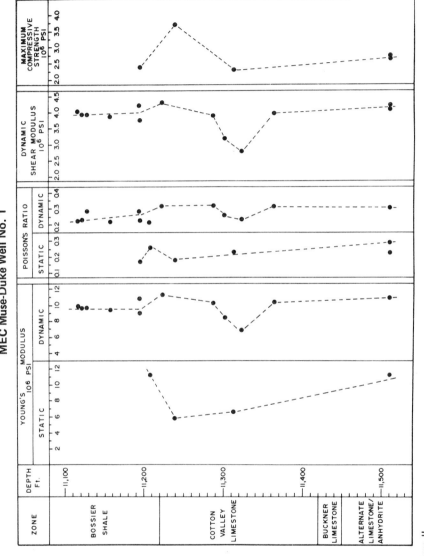

Figure 7.32: Variation of Moduli and Mechanical Properties with Depth
MEC Muse-Duke Well No. 1

Source: Symposium II

Figure 7.33:   Variation of Critical Stress Intensity Factor with
Depth MEC Muse-Duke Well No. 1

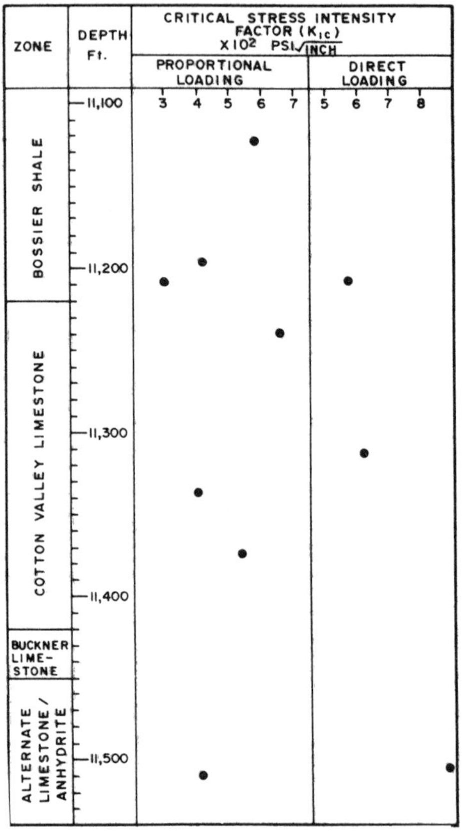

Source: Symposium II

Furthermore, the use of heavy fracturing fluids (such as jelled water based fluids and high sand concentration) would reduce the rate of upward fracture migration. Therefore, the use of foams in treating the Cotton Valley formation in Muse-Duke No. 1 well would not work.

Fracturing Fluid Damage

From the fracture containment analysis it appears that a massive hydraulic fracture will vertically extend to cover the entire Cotton Valley Limestone formation and that it will be contained downward by the limestone/anhydrite sequence and upward by the upper portion of limestone formation and the Bossier Shale. Even if the fracture with right orientation and dimension is created the ultimate productivity of the fractured well will depend on the flow of gas into the fracture and the conductivity of the fracture to transport the gas to the wellbore. Fracturing fluid can considerably affect the resulting conductivity.

Three fracturing fluids were used to determine their damaging nature to the pay zone fracture face permeability and the propped fractured conductivity. The three fracturing fluids are basically the same—crosslinked refined guar gum water gelled. They were supplied by Halliburton, Dowell and Western Companies and are hereby referenced (respectively) as Water-gelled I, Water-gelled II and Water-gelled III.

The system used to measure permeability and propped fracture conductivity of the tight reservoir rocks at varying stress levels and in situ temperature conditions is shown schematically in Figure 7.34. A detailed description of the testing procedure can be found in Reference (3).

Figure 7.34: Schematic of the Laboratory System

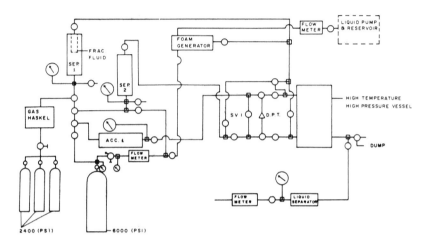

Source: Symposium II

The transient technique for permeability determination (10)(11) is used because of the very low permeability of the cores (μd range). Horizontal permeability at bench conditions for core at in situ saturation was first determined. After the samples have been subjected to in situ effective pressure (stress) and temperature conditions, permeability was again determined. The in situ effective pressure used is the difference between the confining pressure and pore pressure of the formation when the well is produced after the MHF job. While the confining pressure, pore pressure and temperature were maintained constant, the fracturing fluid was then flowed across one face of the sample for four hours at an appropriate injection pressure (the particular fracture-fluid flow time and pressure used simulated the field fracture job conditions).

Shut-in time of 12 hr was allowed (to closely simulate field shut-in) to assure the fracturing fluid was completely broken before backflow. On termination of shut-in, permeability measurements were taken to assess the amount of damage. Clean up was simulated by introducing nitrogen at pressure equivalent to the pore pressure at the sample back face and reducing pressure at the fracture face. The "pressure reduction" is essentially the differential pressure available in the well. Tests were stopped upon reaching the specific "reduction pressure" of the well even if 100% damage was not recovered. At the termination of tests, samples were visually inspected to determine whether or not the fracturing fluid had completely penetrated the core.

Fracture conductivity was measured by using fluid flow rate low enough to avoid non-Darcy flow inside the fracture or by extrapolating measurements to zero flow rate. The core samples were saw cut and propped with 20-40 mesh Ottawa sand. Proppant concentrations used were similar to field concentrations. Initially the cores were subjected to confining pressures of 0.7 MPa (100 psi) for the proppants to settle in place. By flowing dry nitrogen gas through the propped channel, conductivity measurements were taken. The change in conductivity with effective pressure was determined by keeping the pore pressure constant and varying the confining pressure.

This approach provided proof of the dependence of fracture conductivity on the effective stress and eliminates the Klinkenberg effect from the results. To simulate fracturing time, fracturing fluids were subsequently flowed through the propped fracture for 4 hr. Afterward, nitrogen gas was flowed at in situ stress conditions to simulate opening of the well and conductivity measurements were taken until it stabilized. After each sequence of tests ended, the fracture faces were examined with an optical microscope to assess the degree of sand embedment, sand crushing, fracture face chipping, clay flocculation and chemical reaction at the fracture face.

The permeability and the clean-up data on the cores from the pay zone are presented in Table 7.8 and Figure 7.35.

Table 7.8: Fracturing Fluid/Rock Interaction
MEC: Muse-Duke Well No. 1 11,313 ft

Fracturing Fluid Water-Gelled	Permeability* Before Fracturing Fluid Flow K_i, µd	Permeability After Fracturing Fluid Flow K_d, µd	Time of Nitrogen Backflow ...... (hr) ...... Δp of				Final Permeability K_f, µd	Estimated 100% Recovery Time (hr)
			500	1000	1500	2000		
I	31	0.3	1.5	—	—	—	25	1.7
II	23	0.2	2	—	—	—	22	2.2
III	83	0.1	3	0.5	—	0.5	—	Inconclusive

*At *in situ* conditions.

The initial in situ permeabilities of the samples interacted by Water-gelled I and II were in the neighborhood of 25 µd, whereas, the sample interacted by Water-gelled III was much higher (83 µd). The damaged permeability of all the samples at initiation of clean up were in the tenths of microdarcy range, irrespective of the initial permeability of the sample and fluid used. There was no evidence of the fracturing fluid flowing totally through the sample.

A differential pressure of 500 psi was adequate for the clean-up of the samples damaged by Water-gelled I and II. The sample damaged by Water-gelled III cleaned up about 50% on a 500 psi differential pressure; permeability appeared to stabilize at this level even after 3 hr of backflow. Upon increasing the differential up to 2,000 psi there was a 10% increase in damage recovery.

The various levels of permeability reduction are attributed to the fact that different fracturing fluids leave different types of solid and liquid residue in the sample pores (3). Nitrogen backflow helps remove some residues. However, varying amounts of damage remain even after backflowing for unrealistically high differential pressures.

Figure 7.35: Changes in Permeability During Damage
and Clean Up MEC Muse-Duke Well No. 1

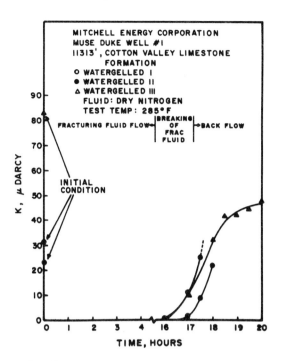

Source: Symposium II

Figure 7.36 illustrates the effect of in situ conditions (stress and temperature) and three different water-gelled fracturing fluids (Water-gelled I, Water-
gelled II and Water-gelled III) on the fracture conductivity of samples from Cotton
Valley Limestone formation. The fractures were fully packed (4 lb/ft^2) with
20-40 mesh Ottawa sand. The curve for the virgin samples is the fracture conductivity to nitrogen-effective stress relationship at in situ temperature. Point
A represents fracture conductivity at effective in situ stress. The gentle slope of
this curve indicates that the reduction in fracture conductivity with stress is
principally due to the packing of the proppants. The other curves represented
by individual reacting fracturing fluids have two points at in situ effective stress
and temperature.

Points B', C' and D' represent the lowest fracture conductivity values at
maximum damage. The stabilized values of fracture conductivity upon nitrogen
flow clean-up procedure are shown as points B, C and D. These values represent
the expected fracture conductivity in the field after the well has been opened and
flowed to clean-up. The remainder of the curves show the conductivity-effective
stress relationship at virgin condition. After interaction, the three different fracturing fluids damaged the fracture conductivity in different amounts. At in situ
conditions of stress and temperature the fracture conductivity reduction ranges
between one and two orders of magnitude depending on the fracturing fluid.

Figure 7.36: Fracture Conductivity-Effective Stress Relationship

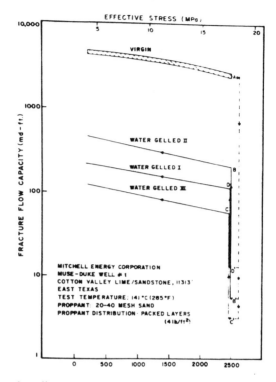

Source: Symposium II

Examination of the samples after each test under an optical microscope and scanning electron microscope (SEM) revealed that clogging of channels due to the residue from fracturing fluids is the predominant reason for the reduction in fracture conductivity. Other pertinent reasons identified were the packing of proppants, proppant embedment due to the softening of the fracture face, proppant crushing and release of fines.

Productivity Assessment

Based on the fracture containment analysis, a massive hydraulic fracture initiated vertically at the middle of the Cotton Valley Limestone formation is expected to be contained within the formation. The lower limestone/anhydrite sequence will act as the lower barrier and the upper portion of the Cotton Valley Limestone formation and Bossier Shale will act as the upper barrier. Because of a less strong upper barrier the use of heavy fracturing fluids (such as the water-gelled fracturing fluids) would reduce the rate of upward fracture migration. The experimental data concerning the heavy fracturing fluid show considerable reduction in fracture conductivity due to the residue left behind by the fracturing fluid and the damages to the fracture face (softening, swelling, release of fines, etc.) by the fracturing fluid.

The reduced fracture conductivity is the expected conductivity in the field and it needs to be incorporated in the evaluation of the particular MHF stimulation treatment. It is seen that there is considerable damage to the formation permeability by the fracturing fluid. However, numerical study (12) has shown that 95% permeability damage up to a depth of six inches is a prerequisite for any significant reduction in well productivity. In calculating the actual stimulation ratio the lab measured damaged in situ fracture conductivity should be used along with in situ formation permeability, predicted fracture dimensions (fracture length and width) based on the volume of sand and fracturing fluid to be pumped, well data and spacing.

Table 7.9 presents the relevant reservoir and laboratory measured properties used in the following productivity assessment study.

Table 7.9: MEC Muse-Duke Well No. 1
Reservoir and Lab Measured Properties

Reservoir pressure	6,400 psi
Effective stress	2,400 psi
Reservoir temperature	285°F
Gas porosity	0.036
Well spacing	640 acres
Wellbore radius	0.25 ft
Fracture geometry	210 ft*
Range of undamaged formation permeability	23–90 μd
Range of damaged formation permeability	20–48 μd
Undamaged fracture conductivity	2,500 md-ft
Range of damaged fracture conductivity	59–280 md-ft

*In height with 800 to 2,700 ft of each wing in the Cotton Valley Limestone formation.

Source: Symposium II

The field MHF job in which pretreatment laboratory analysis has been performed is presented here. The benefits of using the lab-derived actual data in assessing the post fracture production of the well is shown.

Here posttreatment production is determined. It is predicted based on both damaged and undamaged fracture conductivities and compared with predictions that, at present, are widely used in the industry. The field job consisted of pumping fracturing fluid (refined guar gum crosslinked Water-gelled I) and 20-40 mesh Ottawa sand through tubing perforations at 11,220 to 11,430 ft. The laboratory data presented earlier concerns the core samples from the same formation depths.

The formation has an overall matrix permeability of 90 μdarcys. Figure 7.36 shows that the fracture conductivity at in situ conditions of stress and temperature of the Cotton Valley formation core material (using same proppant and proppant concentration as used in the field job) at undamaged condition is 2,500 md-ft (Point A) and 59 md-ft (Point C) after clean-up of the Water-gelled I fracturing fluid. Figure 7.37 shows a comparison of fracture conductivity-effective stress relationship of the sample before and after the interaction with Water-gelled I at in situ contitions. Also included in the figure are some results presented by the Service Company (13). Their work dealt with proppant embedment in formations characterized as soft to hard.

Figure 7.37: Fracture Conductivity-Effective Stress Relationship for Four Different Prediction Methods

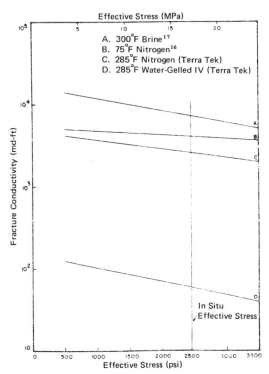

Source: Symposium II

Data from work by Cooke (14) are shown in the figure as well. The data deals with interaction of brine with soft to hard formations at in situ stress and temperature condition. Table 7.10 presents a list of data used to calculate stimulation ratio using the equation presented by McGuire and Sikora (15) for vertically fractured wells for the four different approaches: Cooke (14), Service Company (13), Terra Tek (Undamaged data), Terra Tek (damaged data).

Table 7.10: Stimulation Ratio Calculation

Data Similar to All Prediction Methods					Variable Data Source	Variable Data			Stimulation (15) Ratio	Representative Curve on Figure 7.37
r_w (ft)	r_e (ft)	r_f (ft)	K_o (μd)	W (ft)		K_d (μd)	K_w (md-ft)	r_d (ft)		
0.25	2,979	2,700	90	0.04	Cooke (14)	90	7,200	—	15.2	A
0.25	2,979	2,700	90	0.04	Service Co (13)	90	4,000	—	13.9	B
0.25	2,979	2,700	90	0.04	Terra Tek*	90	2,500	—	13.2	C
0.25	2,979	2,700	90	0.04	Terra Tek**	22	59	0.08	3.4	D

Note: r_w = wellbore radius; r_e = drainage radius; r_f = created fracture radius; K_o = undamaged formation permeability; W = created fracture width; K_d = damaged formation permeability; K_w = fracture conductivity; r_d = radius of formation penetration of fracturing fluid.

*Undamaged.
**Damaged.

Source: Symposium II

Actual well spacing and well data, formation permeability and fracture dimensions are used. Pretreatment production trend is used with the calculated stimulation ratio to predict the post fracturing production. Figure 7.38 presents the actual field production data of the well before and after the fracturing stimulation.

**Figure 7.38: Production Decline Curve Prediction
MEC Muse-Duke Well No. 1**

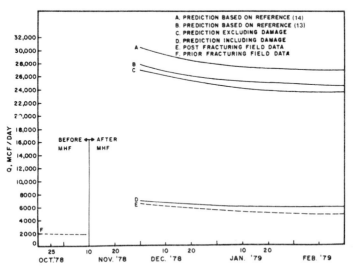

Source: Symposium II

Post stimulation predicted production data are also presented for all four stimulation ratios. Curve D represents the anticipated production based on the laboratory data for damaged permeability and fracture conductivity. There is a close agreement with the actual field data with a deviation of only about 6%. Curves A, B and C are substantially different from the actual field data since damage was not considered. Again curves B [Service Company (13)] and C (Terra Tek, undamaged tests) are very close to each other due to the similarity of fracture conductivity measurement by flowing nitrogen only. One needs to note that the relationship presented by McGuire and Sikora (15) is applied to semisteady state situation. For this system, the reservoir reaches semisteady state after 69 days and Figure 7.38 presents prediction beyond 69 days.

Figure 7.39 shows the dependency of stimulation ratio on fracture radius for various fracture conductivity values for this particular reservoir well spacing, wellbore radius and formation permeability values (damaged and undamaged). When the actual fracture conductivity value is in the 10 to 100 md-ft range, fracture longer than 0.3 to 0.4 of the drainage radius does not significantly affect the stimulation ratio. Holditch (12) has shown similar results in his numerical study.

It is important to note that for the measured fracture conductivity value (59 md-ft) of this particular well, production associated with a 2,700 ft fracture is essentially identical to a 600 to 800 ft fracture.

Figure 7.39: Relationship Between Stimulation Ratio and Fracture Length for Different Fracture Conductivity

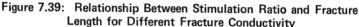
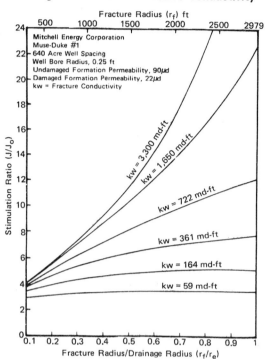

Source: Symposium II

When estimated fracture flow capacity values are in thousands of md-ft range, fracture radius does have a very significant impact on the stimulation ratio. This further strengthens the issue that knowledge of realistic fracture conductivity can economize the MHF job size.

Conclusions

The experimental and field results presented here give insight into aspects of the massive hydraulic fracture containment and productivity analysis of MEC Muse-Duke Well No. 1. Laboratory measured values of elastic moduli are important in delineating a fracture containment and the measured values of permeability and fracture conductivity indicate that the amount of formation damage is dependent on the interaction of fracturing fluid, the host rock and proppant bed design.

In general, the following conclusion can be drawn from results of the reported work.

It appears that a MHF treatment will vertically extend to cover the entire Cotton Valley Limestone formation and that it will be contained downward by the limestone/anhydrite sequence and upward by the upper portion of the Limestone formation and the Bossier Shale.

Different types of fracturing fluid used in this study can cause as much as 100% matrix permeability damage. The damage clean-up is dependent not only on high differential pressure but also on the fracturing fluid causing the damage.

Fracture conductivity reduction as much as 89 to 90% for fully packed proppant beds following fracturing fluid flow is seen.

The pronounced reason for fracture conductivity reduction at in situ condition has been the residue left behind by the fracturing fluid. Other pertinent reasons were proppant packing, proppant embedment due to softening of fracture face, and release of fines from the fracture face.

The use of damaged fracture face and fracture conductivity (measured at in situ condition) in the stimulation ratio calculation provided a better method to predict post fracturing production realistically.

Knowledge of the fracturing fluid damaging nature can be useful for careful planning of fracture design optimization. Depending on the fracture conductivity there exists an upper bound on fracture length beyond which productivity does not increase significantly. For the well in question, the bound appears to be 800 ft.

The laboratory measured values show that the revenue from the incremental gas sales has a potential of defraying the stimulation cost in about six months.

References

(1) Simonson, E.R., Abou-Sayed, A.S. and Jones, A.H., "Containment of Massive Hydraulic Fractures," *Society of Petroleum Engineers Journal,* Vol. 14, No. 1, pp. 27-32 (Feb. 1978).

(2) Murphy, D.L. and Carney, M.J., "Massive FRAC-A Second Look," *Proceedings of the Massive Hydraulic Fracturing Symposium,* University of Oklahoma, pp. 57-75 (Feb. 1977).

(3) Ahmed, U., Abou-Sayed, A.S. and Jones, A.H. "Experimental Evaluation of Fracturing Fluid Interaction with Tight Reservoir Rocks and Propped Fractures," paper SPE 7922, *Proceedings of the 1979 SPE Symposium on Low Permeability Gas Reservoirs,* Denver, Colorado, pp. 109-126 (May 20-22, 1979).

(4) Brechtel, C.E., Abou-Sayed, A.S. and Jones, A.H., *An Evaluation of Fracture Containment within the Castlegate Sand,* Terra Tek Report TR 77-24 (March, 1977).

(5) Brechtel, C.E., Abou-Sayed, A.S. and Jones, A.H., "Fracture Containment Analysis Conducted on the Benson Pay Zone in Columbia Well 20538-T," Second Eastern Gas Shales Symposium, Vol. 1, U.S. Department of Energy (METC), Morgantown, West Virginia, pp. 264-272 (October, 1978).

(6) Abou-Sayed, A.S., Brechtel, C.B. and Clifton, R.J., "*In Situ* Stress Determination of Hydrofracturing—A Fracture Mechanics Approach," *Journal of Geophysical Research,* Vol. 83, No. B6 (June, 1978).

(7) Kozik, H.G., Bailey, B.G. and Holditch, S.A., "A Case History for Massive Hydraulic Fracturing the Cotton Valley Lime Matrix Fallon and Personville Fields—Limestone County, Texas," paper SPE 7911, *Proceedings of the 1979 SPE Symposium on Low-Permeability Gas Reservoirs,* Denver, Colorado, pp. 15-32. (May 20-22, 1979).

(8) Brechtel, C.E., Ahmed, U. and Abou-Sayed, A.S., *Laboratory and Field Measurements in Support of Fracture Treatment in the Cotton Valley Limestone Formation in Muse-Duke Well No. 1,* Terra Tek Report, TR 78-6, (November, 1978).

(9) Field data, Mitchell Energy Corporation, Houston, Texas.

(10) Brace, W.F., Walsh, J.B. and Frangos, W.T., "Permeability of Granite Under High Pressure" *Journal of Geophysical Research,* Vol. 69, pp. 259-273 (1968).

(11) Yamada, S.E. and Jones, A.H., *Analysis of Pulse Technique for Permeability Measurement,* Terra Tek Report, TR 79-8 (January, 1979).

(12) Holditch, S.A., "Factors Affecting Water Blocking and Gas Flow from Hydraulically Fractured Gas Wells," paper SPE 7561 presented at SPE 53rd Annual Fall Meeting, Houston, Texas (October 1-3, 1978).

(13) Halliburton, *The Fracbook-Design/Data Manual for Hydraulic Fracturing,* Halliburton Services, Duncan, Oklahoma, p. 92 (1971).

(14) Cooke, C.E., Jr., "Conductivity of Fracture Proppants in Multiple Layers," *Journal of Petroleum Technology,* pp. 1101-1106 (September, 1973).

(15) McGuire, W.J. and Sikora, V.J., "The Effect of Vertical Fractures on Well Productivity," *Trans. AIME,* 219, pp. 401-403 (1960).

REPORT OF FOAM FRACTURE

Mitchell Energy and Development Corp. has attempted the massive foam fracture stimulation of No. 1 Stone Well, Limestone County, Texas ("Foam fracturing opens the way to more gas," *Chemical Week,* March 26, 1980). If successful, the fracturing was expected to increase the well's production from the present 50 to 75 Mcfd to 2.5 MMcfd.

Mitchell Energy and Development considers massive foam fracturing to be a step beyond massive hydraulic fracturing as a viable technique to improve productivity from tight, depleted reservoirs.

Although the No. 1 Stone fracture, the largest foam fracture undertaken to date, was aborted after a mechanical failure in a sand concentrator, use of foam fracturing in other parts of the area is planned.

Technology Studies—Devonian Shale

The information in this chapter is based on *Fourth DOE Symposium on Enhanced Oil and Gas Recovery and Improved Drilling Methods, Volume 2—Gas and Drilling, Tulsa, Oklahoma, August 29-31, 1978,* DOE CONF-780825-P3, edited by B. Linville of Bartlesville Energy Technology Center. This symposium will be referred to as Symposium I. References in this chapter are at the end of each section.

INVESTIGATION OF HYDRAULIC FRACTURING TECHNOLOGY

The information in this section is based on "Investigation of Hydraulic Fracturing Technology in Tight Gas Reservoirs" by R.M. Forrest and S.F. McKetta of Columbia Gas System Service Corporation (Symposium I).

In two jointly sponsored DOE programs, Columbia has been investigating hydraulic fracturing techniques to enhance gas recovery from impermeable, low pressure reservoirs in the eastern United States. The scope of investigation encompasses varying the size of hydraulic fracturing treatments from conventional to MHF-type treatments in several sandstone reservoirs and the Devonian shale.

Types of Hydraulic Fracturing Utilized

There are basically two components that can be varied in the hydraulic stimulation process; fluid and sand. Table 8.1 illustrates the types of components that were varied in the two Columbia/DOE research programs. The various types of fluids that were used to create the fractures are foam, cryogens (low temperature liquids) and gelled water. The use of foam as a fracturing fluid came about after reviewing various frac proposals prepared by the service companies from both a technical and economic viewpoint.

In view of the potential clean-up problems in the low pressured reservoirs under investigation, it was decided to try "foam-fracturing" because of its beneficial energy-assist mechanism with gaseous nitrogen. The same reasoning was used to determine the application of cryogens as fracturing fluids. In foam-fracturing, gaseous nitrogen is directly injected into the fluids at the surface, while in the case of a "fizz frac", the cryogen, carbon dioxide, is injected as a liquid. Upon reaching reservoir conditions (at proper temperature and pressure), this cryogen flashes to a gas and thus helps to promote fluid recovery after treatment.

To test extension capabilities, the size was varied from conventional to massive (MHF) type treatments. The MHF generally utilizes about 6 to 10 times more of the components than the conventional treatment. In general, a conventional-sized foam, fizz and gelled water treatment or an MHF-sized foam, fizz and gelled water treatment can be utilized to create a fracture.

Lastly, two types of sand were used for different purposes. Generally, the 80/100 mesh, which has the smaller grain size of the two, is used to prevent leak-off into small natural fractures that the artificially created fracture encounters. This technique allows more of the fluid pressure to be maintained inside the fracture to permit further extension. This is followed up with 20/40 mesh sand to pack the fracture to maintain it for gas flow into the wellbore.

Table 8.1: Types of Components Utilized

Type of Fluid	Size Variation	Type of Sand
Foam	conventional	80/100 mesh
Cryogen	MHF	20/40 mesh
Gelled water	—	—

Source: Symposium I

Many authors (1)(2)(3) have reported on the benefits of the use of foam as a fracturing fluid and on its physical properties (1)(3)(4). The main reason for using the foam process is its ability, with the gaseous nitrogen, to return the frac fluid. Also, it has a low liquid content, for use with shales where water is sensitive to clay migration or swelling. Although foam, as a fracturing medium has the capability of getting the sand into the extremities of the created fracture, it is limited in transporting a high sand concentration at any instance. This arises because of surface limitations.

In the case of the 80% quality foam mixture, the gaseous nitrogen, which is injected directly at the wellhead, is incapable of sand transport. Therefore, a sand concentration of 10 pounds per gallon at the blender is transported in only 20% of the total mixture. Once this sand is mixed with the gaseous nitrogen, its concentration becomes only 2 pounds per gallon which translates to the sand's downhole conditions.

The other type of treatment that was utilized is the fizz frac. This fizz frac consists mainly of gelled water but also contains a substantial amount of liquid carbon dioxide. The quantities illustrated in Table 8.2 are somewhat arbitrary. The main factor controlling these quantities is one of logistics. The available liquid carbon dioxide transporting units are generally limited, by their pump capacity and manifolds, to deliver sufficient quantities of the liquid carbon dioxide to the frac pumping equipment to increase it appreciably above 25%.

Once again, the main reason for the use of this cryogenic fluid is the ability to return the frac fluids. After the liquid carbon dioxide flashes to a gas under reservoir conditions, it becomes very miscible with the injected water and, therefore, has a higher potential of tying up the water than gaseous nitrogen, which has a tendency to "slug-out" of solution. Also, since it is injected as a liquid, it is much easier to handle than gaseous nitrogen; and since it is the smaller component of the mixture, a much higher sand concentration can be injected in it than with a foam frac. In these experiments, the sand concentration was increased to as high as 6 pounds per gallon.

A third major type of hydraulic fracturing treatment being investigated uses gelled water. Its obvious advantages in use are: ease of handling, lower cost, lower cost per unit volume and high sand carrying capability (up to 10 pounds per gallon). Some of the gelled water experiments have utilized gaseous nitrogen to some extent (about 100 scf/bbl) for energy-assist purposes to help remove the frac fluids used in the low pressured reservoirs being investigated.

Table 8.2: Types of Technology Utilized and Defined

	Constituents	Quantities	Benefits
Foam	gaseous nitrogen	80%	Low fluid loss; high sand-carrying
Frac	water with surfactant	20%	capability; low liquid content;
	sand	1-2*	faster recovery.
Fizz	liquid carbon dioxide	25%	Prevents water blocks; reduces
Frac	gelled water	75%	liquids used; faster recovery;
	sand	1-6*	more sand can be injected.
Gelled water	gelled water	100%	Easy to handle; lower cost/unit
Frac	gaseous nitrogen	as assist	volume; more sand can be
	sand	1-10*	injected.

*Pounds per gallon.

Source: Symposium I

An offshoot of the gelled water process, not illustrated in Table 8.2, is the patented Kiel-frac or dendritic frac. This technique, which uses a cyclic injection and flow-back process, has major advantages of producing branchlike (dendritic) fractures and of having a low cost per unit volume. This process utilizes carbon dioxide rather than nitrogen as an energy-assist medium.

Devonian Shale Research

The Devonian shale, which underlies much of the eastern United States, is one of the target reservoirs under research investigation. It can be characterized as an impermeable, low pressured reservoir requiring special effort to enhance gas recovery. To delineate, characterize and study the effects of hydraulic fracturing in the Devonian shale, Columbia Gas and the U.S. DOE are jointly sponsoring a three-well research program in West Virginia and a four-well research program (as part of a separate program with thirteen wells) in Kentucky, Virginia and Ohio (5)(6)(7). Figure 8.1 illustrates the locations of these wells with respect to the historical area of shale production. Basically, the objective of this research is to determine which type of hydraulic fracturing treatment is the most effective and efficient in the Devonian shale.

In Table 8.3, some of the variables employed are listed on the Devonian shale experiments. In the three-well program, two wells (Nos. 20401 and 20403) have been stimulated with eight massive type treatments. The treatment types utilized were foam, gelled water and a modified foam treatment. The foam and gelled water type treatments were previously described. The modified foam type treatment (well 20401) consists of a combination of foam and gelled water injected in sequential order. Its use came about because of an extensive clean-up period of about 7 months following the first MHF gelled water treatment in this well. Table 8.3 also illustrates the average amount of constituents used (fluid and sand), the average cost per treatment, and the average open flows, both before and after the treatments.

Table 8.3: Statistics on Devonian Shale Experiments

	20401 Lincoln Co., WV		20403 Lincoln Co., WV	20336 Martin Co., KY	20337 Martin Co., KY	20338 Wise Co., VA	11236 Trumbull Co., OH
Number of Treatments	3	1	4	2	2	2	2
Treatment Type	MHF Modified Foam	MHF Gelled Water	MHF Foam	Fizz	Fizz	Fizz	Fizz
Average amount of fluids							
Gelled water, gal	75,600	475,000	64,400	55,800	34,800	87,000	82,000
Foam, gal	61,000	–	257,800	–	–	–	–
Liquid carbon dioxide, gal	–	–	–	12,600	8,000	25,600	21,600
Gaseous nitrogen, scf	750,000	120,000	2,700,000	–	–	–	–
Average amount of sand, lb	339,000	930,000	371,000	115,000	90,000	203,000	258,200
Average cost, $	70,000	149,000	79,400	33,100*	25,000*	63,000	NA**
Average initial open flow before treatment, Mcfd	. . . .show		145***	show	show	show	show
Average initial open flow after treatment, Mcfd	. . . . 81		–	328	NA**	134	NA**

*Estimated.

**Not available at publication of paper.

***Result of possible vertical communication of frac treatments.

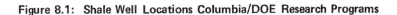

Source: Symposium I

Figure 8.1: Shale Well Locations Columbia/DOE Research Programs

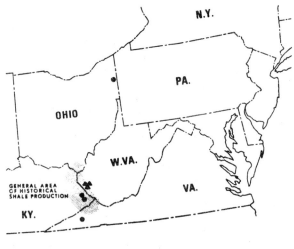

● – 13 WELL PROGRAM- SHALE WELLS ONLY
▲ – 3 WELL SHALE PROGRAM (MHF)

Source: Symposium I

In well No. 20403, the high average initial open flow before treatment is attributed to communication (fracture migration in a vertical plane) in each successive treatment. This can better be seen in Figure 8.2, which presents the open flows before and after each treatment and the intervals in the shales that were treated. For instance, after treating the bottom zone and moving up the hole and perforating, an anomalous open flow of 95 Mcfd was encountered.

After pressure testing and gas composition analyses were conducted on this zone, this communication became evident. Subsequent testing and analyses proved that the entire column was in communication after the four treatments were performed. This phenomenon did not appear to be the case in the second well No. 20401.

The treatments utilized in the other four shale wells were fizz fracs. Table 8.3 shows the data on these wells. To give a perspective view of where the zones of interest were for these treatments, the geological sections are shown in Figure 8.3.

Figure 8.2: Open Flow Wells No. 20403 and 20401 (MHF)

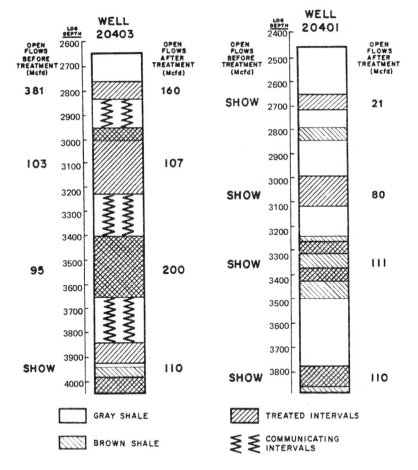

Source: Symposium I

The darker colored areas in the column refer to high gamma radiation areas (brown shales) and the lighter areas refer to gray shales. As can be seen from the graphical presentation, the research efforts in these wells have focused on the darker, richer brown or black shales.

Figure 8.3: Summary of Geological Findings, Shale Wells—13 Well Program

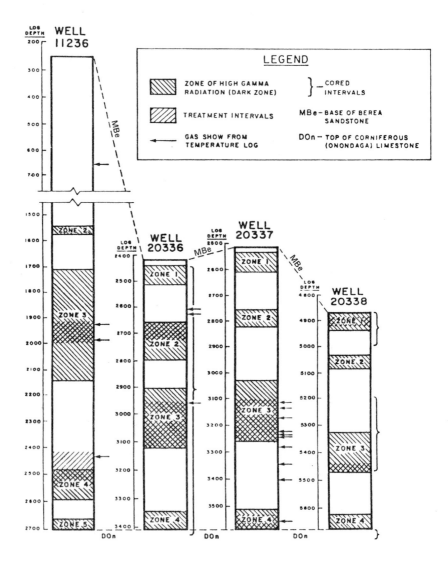

Source: Symposium I

Status of Shale Research

As far as gas production resulting from the treatments in the Devonian shale, only one well is currently producing into a pipeline. This well (No. 20403 in Lincoln County, West Virginia) is producing about 145 mcfd against a line pressure of 48 psig (as of May 1, 1978). The other research well in Lincoln County, No. 20401, is also hooked up to a line. However, full clean-up of the large amounts of water used in the four treatments (refer to Table 8.3) has not been achieved. The hookup to a pipeline of the four other research shale wells has been hampered by the inclement winter weather experienced in the area.

One research shale well (No. 20402, also in Lincoln County, West Virginia) remains to be treated. Currently, analyses of reservoir characteristics and treatment results of the other wells are being studied to treat this last well by early fall.

References

(1) Blauer, R.E. and Kohlhaas, C.A., "Formation Fracturing with Foam", SPE Paper 5003, 49th Annual Fall Meeting, SPE, Houston, Texas, October 6-9 1974.

(2) Abbott, W.A. and Vaughn, H.F., "Foam Frac Completions for Tight Gas Formations", *Petroleum Engineer*, (April, 1976).

(3) Albrecht, R.A. and Marsden, S.S., Jr., "Foams as Blocking Agents in Porous Media", *Society of Petroleum Engineers Journal*, (March, 1970).

(4) David, A. and Marsden, S.S., Jr., "The Rheology of Foam", SPE Paper 2544, 44th Annual Fall Meeting, SPE, Denver, Colorado, September 28-October 1, 1969.

(5) Ranostaj, E.J., "Massive Hydraulic Fracturing the Eastern Devonian Shales", *Second Enhanced Oil and Gas Recovery Symposium,* Volume 2, Tulsa, Oklahoma, September 9-10, 1976, U.S. ERDA.

(6) McKetta, S.F., "Massive Hydraulic Fracturing of the Devonian Shale in Lincoln County, West Virginia", *Proc. of Third Enhanced Oil and Gas Recovery Symposium,* Volume 2, Tulsa, Oklahoma, August 30-September 1, 1977, U.S. ERDA.

(7) Cremean, S.P., "Novel Fracturing Treatments in the Devonian Shale", *Proc. of First Eastern Gas Shales Symposium*, Morgantown, West Virginia, October 17-19, 1977, Morgantown Energy Research Center.

CHEMICAL EXPLOSIVE FRACTURING

The information in this section is based on "Chemical Explosive Fracturing of Devonian Shale Gas Wells", by S.J. LaRocca and A.M. Spencer of Petroleum Technology Corporation (Symposium I).

The Technique

PTC's chemical explosive fracturing process, Astro Frac, has been described in previous documents and publications (1)(2). Briefly, the technique involves the remote pumping and downhole mixing of a fuel and oxidizer to form a high energy liquid explosive. The explosive is displaced down fiber glass tubing into the open hole section and formation where it is detonated. Fractures created by the detonation are propagated out into formation by expanding gases generated by the explosion. The intersection of these fractures with natural fractures in the reservoir gives rise to increased gas production. Comparisons of two PTC explosive formulations to nitroglycerin are given in Table 8.4. PTC-4 is the explosive formulation used for this test program.

Table 8.4: PTC Liquid Explosive Characteristics as Compared to Liquid Nitroglycerin

Property	Nitroglycerin	PTC-4	PTC-8
Density, g/cc	1.47	1.35	1.32
Viscosity, cp	7–35*	8–10*	8–10*
Detonation velocity, m/sec	8,000	8,250	8,100
Detonation pressure, kbar	210	260	250
Heat of detonation, cal/g	1,290	940	890
Gas generated, cc/g	730	1,046	1,056
Sensitivity to:**			
Impact, J/sec	3.4×10^4	$>2.4 \times 10^5$	$>2.4 \times 10^5$
Friction, N/m^2 at m/sec	0.4×10^8 @1.2	$>5.4 \times 10^8$ @2.4	$>5.4 \times 10^8$
Electrostatic discharge, J	0.024	>5	>5
Cap sensitive, initiation	yes	yes	no
Propagates in:			
Thin cracks, <1/64 inch	yes	yes	no
Porous media, 20/40 mesh sand	yes	yes	no
Thermal application limit, °F	<200	>250	>250

*Variable.
**Tests conducted by Allegany Ballistics Laboratory, Hercules Inc. In these tests on PTC-4, there was no reaction at the maximum limits of the test equipment. Values for PTC-4 are the maximum limits of the equipment.

Source: Symposium I

Test Program

This research field test program involved locating and drilling three Devonian shale wells which were treated using PTC's liquid chemical explosive. Gas production was measured before and after treatment in each case in order to gain some indication of the effectiveness of the treatment. In each case, the stimulation effort resulted in increased gas production and suggested changes in the downhole hardware which improved the overall process.

Site and Well Location: The site of this three well test program is a 1,285 acre tract in Lincoln County, West Virginia. The test site is approximately 30 miles south of Huntington, West Virginia. A second test site consists of some 600 acres in Pike County, Kentucky, where PTC is cooperating jointly with DOE on a similar Devonian shale test program.

Selection of each well location was based on a detailed interpretation of U-2 and Landsat imagery, surface joint strikes and offset production data. The dominant strike of the surface joints was found to be N 20° by 30° W. In each case the sites of the three wells were selected at the intersection of two to three lineaments, or surface joints. The strategy associated with imagery analysis and joint strikes is aimed at locating the wells such that the wellbores will intersect the Devonian shale in highly fractured regions. This rationale is believed to offer the optimum opportunity to effectively stimulate the shale with explosives (3).

The average thickness of the Devonian shale underlying this area is 1,300 feet, which includes the commonly referred to grey, middle brown, and lower brown or Marcellus sections. The top of the shale lies between 2,100 and 2,400 feet below the surface. The three wells were drilled through the shale using an air rotary rig, and cased with 7 inch casing, which is required to run the necessary downhole equipment used in the PTC process. Figure 8.4 shows the well diagrams and depths of various strata penetrated by the wellbores.

Treatment and Results: Figure 8.5 illustrates the downhole equipment configuration employed to treat the three wells. As is generally the case, experience gained in applying the technique precipitated design changes in the downhole equipment by the second and third wells, which not only improved the technical efficacy of the overall process, but also reduced the hardware costs. Well No. 686-1 was treated in February, 1977, while wells No. 686-2 and 686-3 were treated in December, 1977 and January, 1978, respectively. Between 20,000 and 30,000 pounds of explosive were used in each well.

Well No. 686-1 had a 7 inch casing set through the Berea sand into the top of the Devonian shale at 2,505 feet. A permanent type packer was set inside the casing at 2,486 feet with a fiber glass tubing tail pipe section extending to 3,788 feet or 7 feet above the bottom of the wellbore. A secondary or back-up detonator was lowered through the tubing and released at the bottom of the borehole, using a wireline, just prior to treatment of the well. The primary detonator, which is designed to detonate the in situ explosive immediately upon displacement into the wellbore and formation, was lowered into the well on the bottom of the tail pipe.

The actual explosive treatment involved pumping sufficient oxidizer and fuel into the well to make 29,400 pounds, or roughly 2,700 gallons, of PTC-4 explosive. The explosive was manufactured and displaced into the wellbore and formation at a rate of 55 gpm and zero surface pressure.

The top wiper plug, which not only separates the explosive and the displacing fluid, but also triggers the primary detonator as it seats in the baffle at 3,633 feet, was displaced with treated fresh water. It failed to reach and seat in the baffle, however, because of an apparent split in the wall of the fiber glass tubing plus debris accumulation on top of the baffle. The explosive was subsequently detonated when a wireline sinker bar was lowered into the tubing to locate the plug and inadvertently pushed the plug into the baffle which caused the primary detonator to fire. The tubing was plugged by the wireline tools and fiber glass debris, which prevented release of the detonation gases.

After removal of the wireline and tubing, it was discovered that the 7 inch casing had partially collapsed or buckled at approximately 2,100 feet. Nevertheless, the well began producing gas during cleanout. After venting the detonation gas, which is noncombustible, the well flowed at the rate of 265 Mcfd. The well exhibited only a trace of gas prior to stimulation. Subsequent cleanout efforts resulted in two gas-air explosions and partial loss of the drill string and a decline in gas production to 133 Mcfd. The well was completed for production in April, 1977. Producing through the downhole junk and debris, the well has averaged 25 Mcfd, against a sales line pressure of 40 to 60 psi.

The second well, No. 686-2, was completed with the 7 inch casing set higher than in well No. 686-1, or slightly below the Injun sand at 1,748 feet. The first attempt to treat the well was aborted when wax and strands of fiber glass from inside the fiber glass tubing prevented proper placement of the string in the well. Approximately 650 gallons of kerosene remained in the wellbore.

The second attempt to run the equipment was successful, and the packer was set in the Berea sand at 2,236 feet, using another 650 gallons of kerosene. Aside from the casing and packer changes, noteworthy improvements in the downhole hardware are the introduction of an entirely new pumpdown secondary detonator and a more compact primary detonator. The pumpdown detonator eliminates the need for a wireline as it is displaced down the tubing by the oxidizer and explosive.

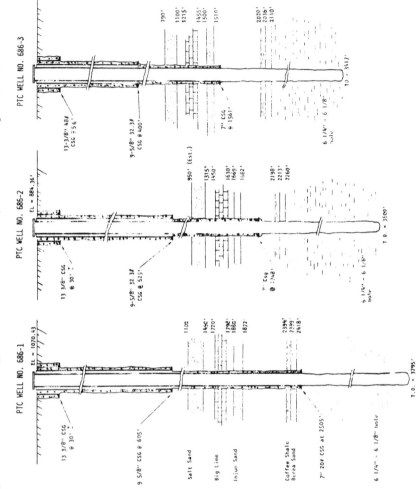

Figure 8.4: Wellbore Diagrams Prior to Chemical Explosive Fracturing Treatment

Source: Symposium I

Other changes in the hardware included the use of an aluminum tubing section to cool the explosive, which normally exits the mixer around 200°F, and a combination sand and water tamp. The second diagram of Figure 8.5 illustrates these changes.

The manufacturing and displacing phases of the treatment occurred without incident, and 27,900 pounds of PTC-4 were detonated by the primary detonator as the top plug landed in the top baffle.

After removal of the tubing above the tamp, the sand tamp was cleaned out and the packer removed. The top of the shot rubble was encountered at about 3,200 feet, indicating the 27,900 pound treatment was concentrated over the bottom 400 feet of the shale. Based on 6¼ inch hole diameter, approximately 22,000 pounds, or 1,900 gallons, were displaced into the formation. Only 60 feet of the 250 foot brown shale, the entire white shale and the Marcellus, or lower brown shale, sections were stimulated. The explosive did not reach the upper 1,000 feet of the shale section.

Lack of adequate air supply prevented complete cleanout of the shot zone. Also, the 1,300 gallons of kerosene, which apparently were driven back into the formation by the liquid explosive and subsequent detonation, began to flow back into the wellbore with formation gas. The kerosene mixed with the rubblized shale and created a gummy, gumbolike substance which further defied cleanout with the air rotary rig. Final efforts to clean out the hole with cable tools only resulted in reaching 3,280 feet.

Prior to treatment the well flowed only a trace of gas but would build up to approximately 500 psi bottomhole pressure over several weeks. After treatment, gas flow was measured on several occasions during cleanout operations. Following venting of the detonation gases, the maximum flow rate was 220 Mcfd, then dropped to 134 Mcfd. The well had a final open flow potential of 77 to 80 Mcfd. After connection to the sales line in April, 1978, the well has produced at an average rate of 40 Mcfd.

The third and final well treated under this joint DOE/industry contract was well No. 686-3. Additional modifications were made in the downhole equipment as a result of the test on well No. 686-2, and are illustrated in the third wellbore diagram of Figure 8.5. Use of a conventional metal cement basket with 3,000 pounds of sand tamp greatly simplified placement of the equipment in the well, as well as prevented the introduction of extraneous fluids into the open hole section which could possibly damage the shale. Compared to the permanent type packer and the inflatable packer used in the first two wells, use of the cement basket represented a substantial reduction in costs, not only for materials, but in rig time and specialized service personnel.

Unusually cold weather, with high winds driving the chill factor down to a –30°F, prevented treatment of the well with the planned 30,000 pounds of explosive. Approximately 22,000 pounds of PTC-4 explosive were ultimately manufactured at the mixer, 155 feet below the surface. Treated water was used to displace the top plug and explosive to the baffle at 3,224 feet. The plug seated only 7 gallons short of the calculated displacement volume, and the primary detonator immediately initiated the displaced explosive.

Unlike the two previous treatments, detonated gases vented through the treatment tubing string one hour after detonation. Sixteen hours later, Devonian shale gas flowed from the well at a rate of 300 Mcfd. The well exhibited only a trace of gas which was too small to measure using a 1 inch Pitot tube prior to explosive fracturing.

Figure 8.5: Wellbore Diagram During Chemical Explosive Treatment

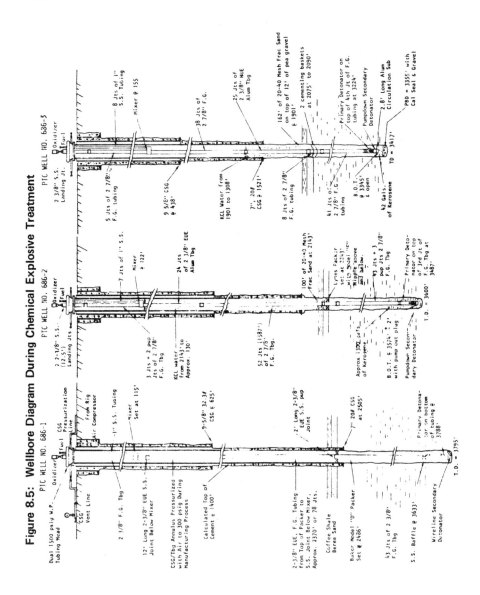

Source: Symposium I

After setting a drillable tubing plug at 2,000 feet, the fiber glass tubing was cut using a string shot (P-cord) at 1,865 feet, or roughly 35 feet below the sand tamp. After removing the tubing strings and downhole mixer, the water above the tamp was blown out and the sand cleaned out to approximately 2,000 feet. At this depth the tubing plug blew out and traveled up the annulus, through the return line and arrived at the far side of the pit in short order. After flowing 24 hours, the flow rate was measured and found to be 311 Mcfd. Further clean-out efforts were aborted due to an insufficient air supply to prevent a possible downhole gas and air explosion while drilling the metal cement baskets.

Further chemical explosive fracturing tests will be conducted using fiber glass bridging baskets if current qualification tests prove successful, thereby eliminating the potential gas-air fire and/or explosion, such as experienced at well No. 686-1.

The well was connected to the 40 to 60 psi sales line in March, 1978. Daily production has averaged 81 Mcfd through May, 1978. Comparison of this well's initial potential with historical data from nearby offset shale wells and data from a shale well treated with PTC-4 explosive three years ago, indicate that gas deliverability to the sales line should be higher. These data suggest that gas deliverability should be on the order of 135 to 150 Mcfd, or approximately 55 percent of the initial open flow potential.

The possibility exists that the tubing section which extends through the sand tamp and down into the shot zone could have been plugged with fiber glass debris as the well was produced. Consideration is being given to reentering the wellbore and drilling out the sand tamp, then either fishing out the basket or pushing it down to the top of the rubble, and thus removing any possible restriction to the gas flow.

References

(1) A Proposal to the Department of the Interior, Bureau of Mines, "Chemical Explosive Fracturing Field Demonstration, Volume I-Technical Proposal", March 11, 1974, in response to RFP No. H0242036.

(2) LaRocca, S.J. and Spencer, A.M., "Chemical Explosive Fracturing of Eight Tight Gas Wells", a Paper submitted to ERDA, October, 1977.

(3) Komar, C.A., *Development of a Rationale for Stimulation Design in the Devonian Shale,* SPE Paper 7166, Omaha, Nebraska, June, 1978.

CHARACTERIZATION AND RESOURCE ASSESSMENT FROM CORE ANALYSIS

The information in this section is based on "Characterization and Resource Assessment of the Devonian Shales in the Appalachian and Illinois Basins" by R.E. Zielinski and S.W. Nance of Monsanto Research Corporation.

The national goal of increasing self-sufficiency in energy resources is resulting in the investigation of less conventional sources for hydrocarbons. One such potentially major source is the late Devonian-early Mississippian organic-rich shales from the Eastern interior basins (Figure 8.6). These dark (black, brown and gray) shales underlie hundreds of thousands of square miles near major energy-consuming population and industrial centers. Previous production from these low-permeability (0.001 to 0.0001 millidarcy) rocks has been restricted by marginal exploration economics (1). The DOE through the Morgantown Energy Research Center (MERC) has begun the Eastern Gas Shale Project (EGSP). The purpose of this project is to assess the hydrocarbon-producing potential (natural and induced) of the dark (gas) shales of the Eastern interior basins.

Mound Facility has initiated a detailed investigation of the chemical and physical characteristics of the dark shales in the Illinois and Appalachian basins (2)(3)(4)(5)(6). This information will provide the basis for characterizing the shale, identifying the source, assessing the resource, and locating the most promising reservoirs. The purpose of this section is to summarize selected results from the project. Emphasis will be placed on recent information obtained from Appalachian basin shales.

Data and Techniques

Shale samples were obtained from cores of eleven wells located in the Illinois and Appalachian basins (Figure 8.6 and Table 8.5). In most instances, the entire Devonian shale interval was cored and representative samples were selected at approximately 30 foot intervals. The cores were placed in airtight containers at the drilling site and shipped to the laboratory. The gases in the containers released by the shales and the shale samples were subjected to detailed geochemical and fuel yield analyses (Figures 8.7 and 8.8). The results of the analyses indicated by the cross-hatched areas of Figures 8.7 and 8.8, along with biostratigraphic studies of the shales, are discussed herein.

General details for the procedures are given in the following sections. Specific details of analytical procedures were not described here but have been previously reported (2)(3)(4)(5).

Geologic Setting

Basin outlines and major structural elements are shown in Figure 8.6. The outline of the Appalachian basin corresponds to the outermost occurrence of Silurian rocks. The outline for the Illinois basin is delineated as the approximate outermost occurrence of Pennsylvanian-age sediments.

The Appalachian basin extends east to the Blue Ridge province and is bounded on the west and northwest by the Cincinnati and Findlay arches, respectively. These arches became effective barriers by Ordovician time. The Appalachian highlands were the prominent source of clastics for the basin. However, the Cincinnati arch may have acted as an intermittent source during the Devonian period. As the basin matured, successive formations were generally thinner and the basin depocenter progressed westward as the sea retreated to the northeast (7).

The Illinois basin is bounded by a series of low-relief arches and domes which were emergent features by Ordovician time (Figure 8.6). The effectiveness of the Pascola arch as a southern barrier during the entire Paleozoic era is questionable (8). Unlike the basin, there was no major source of clastics near the Illinois basin margin. This in part accounts for only 400 feet of Devonian dark shales in the Illinois basin as compared to 1,500 feet in the Appalachian basin.

Results and Discussion

Biostratigraphy: The ages and environments of basin deposits were determined by palynological studies of acritarchs and spores in samples from wells

VA-1, KY-2 and I-2 and I-3. Rock samples were treated with specific mineral dissolving acids to remove most of the silicate and carbonate matrices. The remaining organic material was studied by an experienced palynologist.

Figure 8.6: Eastern Interior Basins

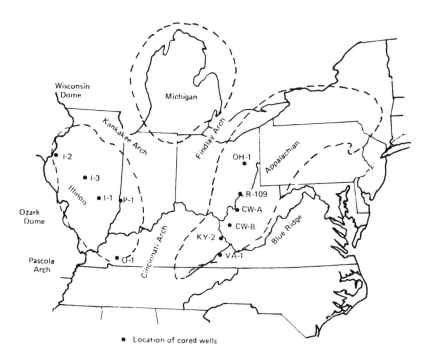

Source: Symposium I

Table 8.5: Location of Wells

Illinois Basin Wells		Appalachian Basin Wells	
I-2	Henderson County, Illinois	R-109	Washington County, Ohio
I-3	Tazewell County, Illinois	CW-A	"Cottageville Wells" Jackson County, West Virginia
P-1	Sullivan County, Indiana		
I-1	Effingham County, Illinois	CW-B	"Cottageville Wells" Lincoln County, West Virginia
O-1	Christian County, Kentucky		
		KY-2	Martin County, Kentucky
		VA-1	Wise County, Virginia
		WV-5	Mason County, West Virginia
		OH-1	Carroll County, Ohio

Source: Symposium I

Polymorphs of spores and acritarchs are known to be indicative of geologic age and environment of deposition. The shales from wells I-2 and I-3, VA-1 and

KY-2 are all Frasnian or Famennian in age (5)(6)(7)(8)(9). The predominant environment of deposition for the Illinois basin shales was initially marine, then restricted marine, with finally a greater influence of nonmarine deposition (Figure 8.9) (10).

Figure 8.7: Flow Diagram for Organic Geochemical Analysis

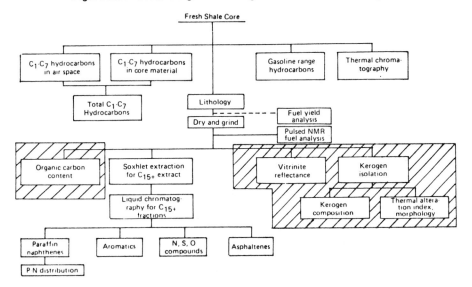

Source: Symposium I

Figure 8.8: Flow Diagram for Fuel Yield Analysis

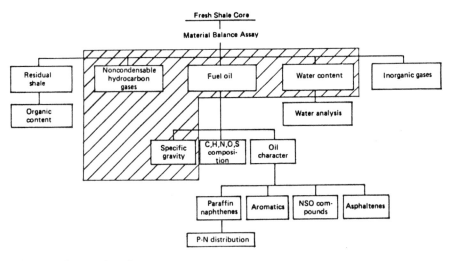

Source: Symposium I

Figure 8.9: Biofacies Results

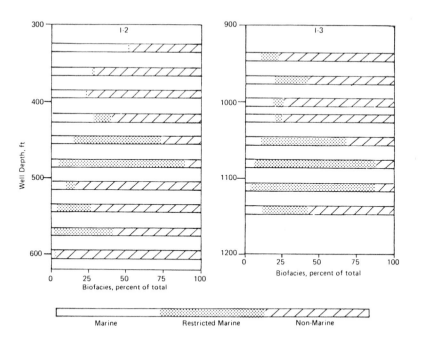

In contrast, shales at well VA-1 were almost entirely deposited under nonmarine conditions, an obvious effect of the nearby Appalachian highlands. The more basinward KY-2 site was a basin hinge-line setting. Environments fluctuated rapidly from among marine, restricted marine, and nonmarine as a function of variations in the sediment-influx and basin-subsidence (9)(10).

Organic Geochemistry: The organic material of the dark shales is primarily composed of the detritus from marine fauna and flora, with terrestrially derived organic debris from the Appalachian uplift and occasionally the Cincinnati arch. Diagenesis of the organic debris proceeds through microbial degradation and thermal alteration to produce kerogen. Kerogen, a complex heterogeneous material, is the precursor of oil and gas.

The biological precursor of kerogen can be determined by visual inspection (11)(12). Kerogen is primarily composed of the remnants of the following: algal, amorphous (sapropel), herbaceous (spores, pollen, cuticles, and membrane debris), woody (structured), and coaly (inertinite) materials. This list proceeds for kerogen characterization from marine to nonmarine environments. The analytical results from the Appalachian shale samples indicate that the marinelike kerogen, for the entire well profile, increases from wells VA-1 to CW-A, CW-B and R-109 (Figure 8.10). This is an indication of the basin environments during deposition. The influence of woody-coaly (nonmarine) detritus supplied from the Appalachian highlands decreased in the wells located farther north. The wide range of kerogen types for samples from the KY-2 well (Figure 8.10) is consistent with the previous explanation of that basin position being affected by frequent depositional environment changes. The absence of a prominent bordering highland obviously resulted in the decreased presence of woody and coaly material in the Illinois basin shales from wells I-2 and I-3 (Figure 8.10). The hydrocarbon potential of the organic debris in the dark shales will depend on the degree of thermal maturation, type of kerogen debris, and volume of organic material.

Figure 8.10: Kerogen Description Zones

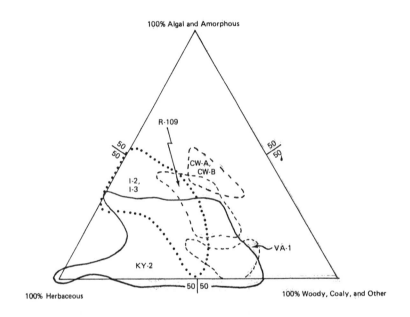

100% Algal and Amorphous

R-109

CW-A, CW-B

I-2, I-3

VA-1

KY-2

50 | 50

100% Herbaceous

100% Woody, Coaly, and Other

Source: Symposium I

One method of determining maturation is kerogen coloration (12). The kerogen coloration of plant cuticle and spore pollen debris is measured in transmitted light. The measured degree of thermal alteration is compared to one of five stages of coloration. Light greenish-yellow kerogen corresponds to stage 1 of the Thermal Alteration Index (TAI). Stage 1 is mildly altered kerogen. The sequence of color designations proceeds with increasing thermal effect through orange and brown to black (stage 5). Stages 2 to 3 represent the degree of thermal maturation necessary for oil generation.

TAI values for all available samples (except from VA-1) from both the Illinois and Appalachian basins were very uniform (6). The kerogen coloration was yellow to orange-brown. This gives a TAI of 1 to 2, indicative of the early stages of petroleum generation. A more detailed evaluation was made using vitrinite reflectance. This method has higher resolution and less subjectivity. Vitrinite reflectance (R_0) is the measure of the reflectivity of small organic grains exposed on a polished section of core. R_0 values of 0.2 to 0.6 indicate insufficient thermal alteration for efficient oil generation. An R_0 of 0.6 to 1.2 is the range for maximum petroleum generation. An R_0 of 1.2 to 3.0 indicates wet gas and methane production. R_0 values greater than 3.0 indicate exhaustion of hydrocarbon-generating capacity (13).

The method used was a modification of the technique described by Hacquebard and Donaldson (14). The Appalachian basin results are given in Figure 8.11. Both the mean and the range of R_0 values are plotted versus depth. The mean values of R_0 for the entire well profiles of KY-2, OH-1, R-109 and VA-1 are 0.52, 0.71, 0.58 and 1.02, respectively.

Figure 8.11: Vitrinite Reflectance Results

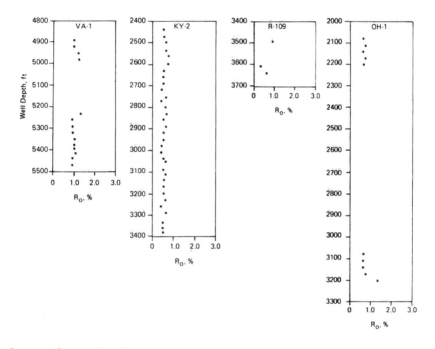

Source: Symposium I

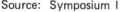

These values indicate adequate thermal alteration for petroleum generation. There is also a trend of increasing maturation of the shale in a southeasterly direction. This may be an effect of the presence of older highly altered organic detritus from the Appalachian highlands being reworked and supplied to the southeast regions of the basin.

Vitrinite reflectance measurements of Illinois basin shales (wells P-1, I-1, and O-1) yield mean R_0 values between 0.45 and 0.50 (6). This suggests that the dark shales from this basin have not experienced adequate thermal maturation to become a significant hydrocarbon source. Larger values with depth would be anticipated as a result of the temperature increase with greater depth of burial. Well KY-2 was sampled over an interval of 900 feet. No trend is apparent in Figure 8.11. This may be the result of a very small increase in R_o values resulting from the variation in the source of organic material. Current analysis of well cores (sampled on a smaller interval) from the Devonian basin center will provide results for thicker sections with more uniform organic material. Perhaps then an R_o-versus-depth relationship may be discerned.

The tendency for the dark shales to yield oil or gas is also controlled by the source of organic material. Algal and amorphous detritus tend to yield oil-prone hydrocarbons; woody and coaly materials, gas-prone hydrocarbons. The yield from herbaceous material is intermediate in nature (15)(16). Thus hydrocarbon yields tend from oil to gas from environments characterized as more marine to more nonmarine, respectively. Elemental analysis of kerogen, expressed as ratios of H/C versus O/C, indicates source and potential of the sediments (15). Nonmarine environments tend to accumulate material rich in aromatics. This gives high O/C ratio and low H/C ratio (curve C in Figure 8.12).

The kerogen capable of producing oil is rich in aliphatic structures and subsequently results in high H/C and low O/C ratios. These sediments indicate a marine environment of deposition (curve A in Figure 8.12).

Figure 8.12: Atomic Ratio Results

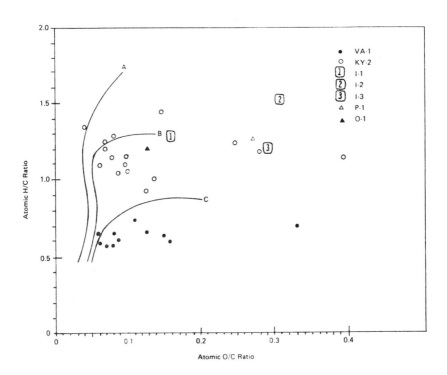

Source: Symposium I

Curve B represents the evolution path of organic material intermediate in nature. The first effects of increased thermal maturation are indicated by the elimination of oxygen, resulting in a decreased O/C ratio. Subsequently the H/C ratio is reduced by hydrocarbon generation (15). These results are indicated by the arrow on curve C.

The H/C versus O/C results from the Appalachian KY-2 and VA-1 wells are given in Table 8.6 and plotted on Figure 8.12. It is immediately evident that samples from VA-1 are gas-prone and thermally mature. Samples from KY-2 suggest hydrocarbon yields will be of a nature intermediate between oil and gas. The scatter may be the result of the various organic sources evident at this basin location. In general, the sediment appears to be thermally mature. No correlation was evident between depth of sample and thermal maturation indicated by O/C versus H/C results.

Table 8.6: Appalachian Basin Atomic Ratio Values

VA-1

Depth	Atomic H/C	Atomic O/C
4890	0.69	0.33
4920	0.62	0.15
4980	0.56	0.07
5260	0.56	0.08
5290	0.59	0.16
5320	0.57	0.06
5350	0.60	0.09
5380	0.66	0.13
5390	0.74	0.11
5440	0.63	0.08
5470	0.63	0.06

KY-2

Depth	Atomic H/C	Atomic O/C
2440	1.07	0.10
2470	1.00	0.14
2540	0.91	0.13
2660	1.17	0.28
2690	1.13	0.39
2720	1.13	0.08
2770	1.04	0.10
2860	1.19	0.07
2920	1.14	0.10
3010	1.22	0.07
3040	1.34	0.04
3090	1.44	0.15
3110	1.22	0.25
3200	1.03	0.09
3230	1.28	0.08
3360	1.06	0.06

Source: Symposium I

The results for Illinois basin wells were expressed as mean values of O/C and H/C for the entire vertical well profile (Table 8.7). The results in Figure

8.12 suggest these sediments will yield hydrocarbons intermediate in nature. The samples from this basin also indicate a relatively mild degree of thermal alteration.

Table 8.7: Illinois Basin Atomic Mean Ratio Values

Well	Atomic H/C	Atomic O/C
I-1	1.26	0.16
I-2	1.47	0.31
I-3	1.17	0.29
P-1	1.25	0.27
O-1	1.17	0.13

Source: Symposium I

The hydrocarbon potential of the shales is also related to the organic carbon content. The values of the organic carbon content in weight percent versus depth, are given in Figure 8.13 for the Appalachian well profiles. The mean values for wells VA-1, KY-2, R-109 and OH-1 are 1.96, 2.04, 1.35 and 1.86, respectively. Comparatively, the averages for the Illinois basin wells I-1, O-1 and P-1 are 6.77, 8.75 and 4.73, respectively (6). The mean values all exceed the 0.4% minimum organic carbon content found in productive basins (17).

The range of values is large. For example, values at KY-2 vary from 0.01 to 7.75%. This indicates the rapid changes which can occur in a well profile and subsequent need for a less than 30 foot sampling interval. Part of this variation at KY-2 is due to its basinal position. As confirmed by the biostratigraphic study, frequent changes of environmental setting in the source area influence the volume and preservation of organic matter in shales at this site.

Hydrocarbon Yields: Estimates of total hydrocarbon potential in the dark shales were derived from the results of organic carbon content and material balance assay (MBA). The indigenous gas from shales can be estimated by using average organic carbon content, the Mott factor (1,350 ft^3 of gas per ton of organic matter), and assuming a shale density of 162 lb/ft^3 and the weight of organic matter to be 1.25 times the weight of organic carbon (6). Results indicate total gas reserves in MMcf per unit of reservoir rock (1 mile x 1 mile x 1,000 feet) for wells KY-2, VA-1, OH-1 and R-109 are 77,900, 74,800, 71,200 and 51,600, respectively (Table 8.8). This reflects the large total reserves of the Appalachian basin Devonian shales as has been discussed (18). Results for the Illinois basin wells, per unit reservoir, were substantially higher. The indigenous gas reserves per unit of reservoir rock from wells I-1 and O-1 were 237,000 and 306,000, respectively (6).

This approach for estimating reserves does not take into account the type of organic matter in the shales or their state of thermal maturation. Analysis of the volume of gas released from the shale core to the sample canisters and that retained within the shales more accurately indicates the actual volume of gas within the formation (Table 8.8). These results give a minimum value because of the unknown volume of gas released during coring and canning of the samples. The average gas contents (expressed in MMcf/unit of reservoir rock) from wells VA-1, KY-2, I-1 and O-1 are 96,000, 9,000, 12,000 and 9,000 (6).

Figure 8.13: Organic Carbon Content Results

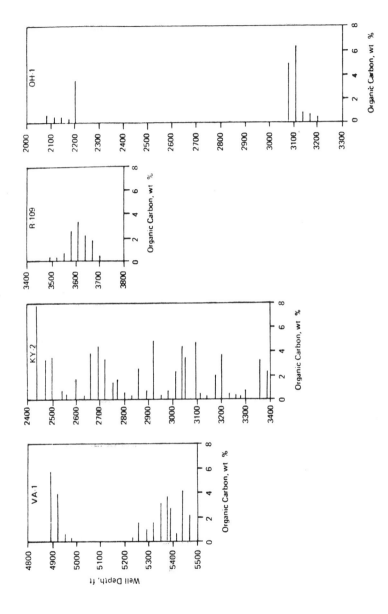

Comparison of these results with those predicted by using the organic carbon content indicates not only are actual gas yields likely to be lower, but also the yields depend on several variables rather than solely on organic carbon content. This interpretation assumes that the volume of gas released before sample canning is proportional to that measured.

Table 8.8: Estimated Indigenous Gas Content

. . . . .Predicted Gas Content.			
. . . . .Well			
VA-1	KY-2	R-109	OH-1
Organic carbon content (wt %) 1.96	2.04	1.35	1.86
Gas content (MMcf/unit) 74,800	77,900	51,600	71,200

. . . . Measured Gas Content.			
. . . . .Well			
VA-1	KY-2	O-1	I-1
Mean C_{1-4} hydrocarbon content (cm³/g) 0.710	0.147	0.135	0.165
Gas content (ft³/gas/ton sediment) 22.86	4.71	4.32	5.30
Gas content (MMcf/unit) 45,700	9,400	8,600	10,600

Source: Symposium I

Material balance assays were performed on selected samples from KY-2, VA-1, OH-1 and R-109. The MBA studies of shale samples determined the yields of oil, gas and water, as well as the API gravity of the oil. MBAs are done on approximately 100 gram samples of 4 to 8 mesh shale material. Distillation of the sample is carried out in an inert helium or nitrogen atmosphere to 500°C. Results of the fluid yields are given in Table 8.9 for wells KY-2, VA-1, OH-1 and R-109. Values are expressed as average MBA product yields. Average oil yields, in gal/ton, for KY-2, VA-1, OH-1 and R-109 were 1.63, <0.5, 1.73 and 1.17, respectively.

Table 8.9: Material Balance Assay Results

	. . . . Well.			
	OH-1	R-109	VA-1	KY-2
Oil yield (gal/ton)	1.73	1.17	<0.5	1.63
°API, oil gravity	34.1	36.8	*	32
Water yield (gal/ton)	3.56	2.75	2.53	4.13
pH	8.9	9.2	9.2	9.3
Hydrocarbon gas yield (ft³/ton)	31.13	35.14	16.56	31.50

*Parameter not determined.

Source: Symposium I

The variation in yields was large both within vertical well profiles and throughout the basin. For example, well KY-2 gave oil yields from 0.1 to 4.6

gal/ton. There was no identifiable trend of hydrocarbon yields. The results from VA-1 suggest that the shales are relatively gas-prone. KY-2 yields a significant amount of oil hydrocarbons relative to VA-1. These results are consistent with the H/C and O/C kerogen studies. However, total gas yields in KY-2 are also higher. This is expected because of the lower hydrocarbon yields from the nonmarine organic debris (13).

Oil yields from Illinois basin shales from I-1 and O-1 were 9.04 and 12.75, respectively (6). MBA analysis of Illinois and Appalachian basin shales indicates oil yields for these shales are less than the Western oil shales (Table 8.10). This secondary source of hydrocarbons will give added economic benefits during recovery from Eastern shale gas wells.

Table 8.10: Fuel Oil Yield (Adapted from Zielinski) (2)

	Gal/Ton
Yield based on organic carbon content	
Colorado	16.9
Wyoming (GRB)	19.4
Wyoming (WB)	17.1
Utah	19.7
Kentucky	8.9
Oil yield from MBA analysis (mean value for all basin samples)	
Appalachian basin	1.4
Illinois basin	9.6

Source: Symposium I

Conclusions

The deposition of upper Devonian dark shales of the Appalachian basin was greatly influenced by the presence of the Appalachian highlands. The amount of terrestrially derived organic debris increases in a southeasterly direction. Burial with a moderate degree of thermal alteration throughout the basin has resulted in low-yield gas-prone shales in the southern regions with higher yield oil- and gas-prone sediments in the northern areas.

The dark shales of the Illinois basin experienced a depositional setting similar to that of the northern Appalachian basin shales. Thermal maturation is less in the Illinois basin, but the hydrocarbon yields of these oil- and gas-prone shales are relatively greater.

References

(1) Negus de Wys, J. and Schumaker, R.C., *Pilot Study of Gas Production Analysis Method Applied to Cottageville Field,* West Virginia University, Morgantown, West Virginia, 1978.

(2) Zielinski, R.E., *Physical and Chemical Characterization of Devonian Gas Shale,* MLM-ML-77-43-0002, Miamisburg, Ohio, 1977.

(3) Zielinski, R.E., *Physical and Chemical Characterization of Devonian Gas Shale,* MLM-ML-77-46-0001, Miamisburg, Ohio, 1977.

(4) Zielinski, R.E., *Physical and Chemical Characterization of Devonian Gas Shale,* MLM-ML-78-41-0002, Miamisburg, Ohio, 1978.

(5) Zielinski, R.E., *Physical and Chemical Characterization of Devonian Gas Shale,* MLM-ML-78-44-0001, Miamisburg, Ohio, 1978.

(6) Zielinski, R.E., "Geochemical Characterization of Devonian Gas Shale", First Eastern Gas Shale Symposium, Morgantown, West Virginia, 1977.

(7) Woodward, H.P., "Emplacement of Oil and Gas in the Appalachian Basin", from *Habitat of Oil,* L.G. Weeks (ed.), American Association of Petroleum Geologists, Tulsa, Oklahoma, 1958.

(8) Swann, D.H. and Bell, A.H., "Habitat of Oil in the Illinois Basin", in *Habitat of Oil,* L.G. Weeks (ed.)., American Association of Petroleum Geologists, Tulsa, Oklahoma, 1958.

(9) Martin, S.J. and Zielinski, R.E., "A Biostratigraphic Analysis of Core Samples from Wells Drilled in the Devonian Shale Interval of the Appalachian and Illinois Basins", 1978.

(10) McIver, R.D., Zielinski, R.E., Overby, W.K. and Martin, S.J., "Geochemical Analysis of the Devonian Shales of the Eastern Interior Basins", American Association of Petroleum Geologists, Annual Meeting, April 9-12, 1978, Oklahoma City, Oklahoma.

(11) Staphlin, F.L., "Sedimentary Organic Matter, Organic Metamorphism, and Oil and Gas Occurrence", *Bull. Pet. Geol., 17,* (1969).

(12) Burgess, J.D., "Microscopic Examination of Kerogen (Dispersed Organic Matter) in Petroleum Exploration", Geologic Society of America, Special Paper 153, 1974.

(13) Dow, W.G., "Kerogen Studies and Geological Interpretations", *J. Geochem. Expl., 7,* (1977).

(14) Hacquebard, P.A. and Donaldson, J.R., "Coal Metamorphism and Hydrocarbon Potential in the Upper Paleozoic of the Atlantic Provinces, Canada", *Can. J. Earth Sci., 7,* (1970).

(15) Tissot, B., Durand, B., Espitolie, J. and Combaz, A., "Influence of Nature and Diagenesis of Organic Matter in Formation of Petroleum", *Am. Assoc. Pet. Geologists Bull., 58,* (1974).

(16) McIver, R.D., "Composition of Kerogen—Clue to its Role in the Origin of Petroleum", *Proc. 7th World Pet. Congr., 2.,* (1967).

(17) Ronov, A.B., "Organic Carbon in Sedimentary Rocks (in Relation to the Presence of Petroleum)", *Translation in Geochemistry, 5,* (1958).

(18) Foster, J.C., "A 'New' Gas Supply—Devonian Shales", *Oil and Gas J., 75:1,* (1977).

EVALUATION OF STIMULATION TECHNOLOGIES

The information in this section is based on "Evaluation of Stimulation Technologies in the Eastern Gas Shales Project" by C. Young of Science Applications, Inc. (Symposium I).

The Devonian shales, underlying extensive areas of the eastern United States, are estimated to contain trillions of cubic feet of natural gas but only a small fraction of this is economically recoverable using existing well completion and stimulation technologies. The Morgantown Energy Research Center (MERC) of the Department of Energy initiated in 1976 the Eastern Gas Shales Project (EGSP) which had the development, evaluation and field demonstration of new recovery and stimulation technologies as one of its key components. The individual projects within the EGSP encompass almost every conceivable stimulation technology, from massive hydraulic fracturing to in-formation high explosive detonation; studies of these processes are being conducted by universities, private industry, and the national laboratories and MERC.

With the exception of directional drilling, all the improved recovery techniques depend upon rock fracturing as the means to improve permeabilities and/or bore-hole communication with the formation. Many existing and several new fracturing techniques are being tested in the EGSP. The projects range from massive hydraulic fracturing experiments and the detonation of large quantities of high explosives injected into the formation to analytic studies supported by small scale experiments. Although all the fracture stimulation projects have the demonstration of a specific method as their objective, the complicated nature of the many features influencing rock fracture and their interactions make quantitative comparisons between the various proposed methods difficult.

Science Applications, Inc. has completed a technical review of the stimulation technologies being applied to or considered for the Eastern Gas Shales. This review included a consideration of the production characteristics of the shales amenable to stimulation as well as consideration of the physical processes associated with each stimulation method. The review efforts culminated in a workshop held in Morgantown in May, 1978. In this workshop, specific consideration was given to the known production characteristics of Devonian shales, the geologic and tectonic control of Devonian shale gas production, reservoir and production simulation methods applicable to the Devonian shale, the modeling and analysis of the various stimulation methods being considered and, finally, instrumental needs for the field evaluation of stimulation effects and the consequent changes in production characteristics.

This section summarizes the results and primary conclusions of this workshop and the related efforts conducted by Science Applications during the two months preceding the workshop.

Stimulation Methods

Due to the complicated and often poorly understood nature of the physical processes controlling the various stimulation methods, no single system is universally applicable for classifying the stimulation methods being used or considered in the EGSP. Scharff and McKay (1) suggest a classification method based on the controllable directionality of the stimulation method, with directional drilling offering the highest directional control and conventional hydraulic fracturing the least control. Komar, et al (2) employ a classification system based upon whether the method uses pumped hydraulic fluids or chemical explosives as the primary energy source.

Massive hydraulic fracturing (MHF), tailored pulsed loading with propellants and novel fracturing whether by hydraulic or explosive means are considered as subsets of these major classifications. Other classification schemes could be based upon whether the process was predominantly static or dynamic in nature, the quantities of fracturing fluid or chemical explosives utilized, etc.

For purposes of this review, it is most useful to classify stimulation methods according to the current degree of understanding of the physical processes involved. Thus, conventional methods would include those stimulation techniques wherein the interactions between the stimulation process and the rock production characteristics are well known and a significant quantity of experience and production data is available. In contrast the novel methods would include stimulation techniques whose physical effects are poorly understood and with which little experience has been gained to date.

Because the effects and benefits of the conventional methods are well documented in the literature, only minor consideration will be given to them in the following paragraphs. More extensive consideration will be given to the novel stimulation methods, the characterization and analysis of their physical effects, and their potential applicability to the Devonian shales.

Conventional Techniques

With the exception of the dendritic fracturing technique (3) and the tailored explosive pulse concept, all of the hydraulic or fluid fracture methods can be considered as conventional. The basic principles, effects and limitations of hydraulic fracturing are well summarized by Howard and Fast (4). More recently, specific consideration has been given to the behavior of fracture propagation in layered formations (5) and the containment of the induced fractures to the producing formation in massive hydraulic fracture operations (6). Modifications to hydraulic or fluid fracturing treatments, wherein special fluids are employed, do not significantly alter the rock fracturing process and thus can be considered as conventional.

Use of water (with a foaming agent) and nitrogen as the fracturing fluid in foam fracturing treatments yield a highly non-Newtonian fracturing fluid with lower fluid leak-off rates and consequently improved proppant transport and cleanup characteristics. Similarly, a cryogenic treatment employing liquid CO_2 with water and/or methanol, offers improved cleanup characteristics, but only minor modifications to the basic rock fracturing processes.

The hydraulic fracturing method, including most modifications to it, are probably most limited in that the orientation of the induced fracture will be dictated by the in situ stress field. As will be discussed in greater detail in later sections, any tendency for the Devonian shale to have an anisotropic fracture permeability will be closely related to the present in situ stress field and consequently induced hydraulic fractures would most likely parallel the naturally existing anisotropic permeability rather than transecting the fractures causing this anisotropy. It is clear that a knowledge of the formation production characteristics, including the degree of permeability anisotropy, and the orientation and magnitude of the in situ stress field must be known before the probable effects and benefits of conventional hydraulic treatments can be predicted.

Although the effects and benefits of explosive bore-hole shooting are not as well documented as those of hydraulic fracturing, the extensive use of bore-hole shooting in Devonian shale gas wells dictates that this method be considered conventional in nature. The extensive study of rock fracture and fragmentation by bore-hole explosive loading in mining (7), does provide a good physical basis for the rock fracture effected by bore-hole shooting in well stimulation. The very rapid loading rates and high induced stresses associated with bore-hole shooting certainly result in the multidirectional fracturing of the rock independent of the magnitude and anisotropy of the in situ stresses.

The high induced stresses, however, do cause extensive deformation and crushing of the rock immediately adjacent to the well bore with irreparable well bore damage being a possible result. The relatively short duration of the explosively induced stresses also limits the extent to which the induced fractures can propagate into the formation. Consequently, explosive bore-hole shooting can be considered as a multidirectional bore enlargement technique with limited radial distance effectiveness.

Novel Stimulation Techniques

For purposes of this review the novel stimulation methods are defined as those for which the controlling physical processes and their interactions with the production characteristics of the reservoir rock are not well known. Some of the novel stimulation methods were proposed and/or developed with a specific production stimulation effect in mind. Others have come into existence as modifications of existing techniques and specific physical reasons for why they might be effective are still in the process of being identified. Four stimulation methods which have been specifically identified as novel stimulation techniques within the EGSP are listed in Table 8.11. The current understanding of the physical processes and concordant production benefits associated with each method are briefly summarized in the following paragraphs.

Table 8.11: Novel Well Stimulation Techniques and Their Principal Physical Effects

Shaped Charge Penetration

Orientation independent of in situ stress
Traverse natural fractures
Offset explosive or hydraulic fractures

Dendritic Fracturing (Kiel Process)

In situ stress modification giving secondary fracturing
Spall propping of fractures

Tailored-Pulse Loading

(Dynafrac and SRI/Thiokol propellant)
Multiple well bore fractures
Fractures independent of in situ stress
Fluid or gas fracture extension

Emplaced High Explosive

Secondary (detonation) fracturing
Dynamic fracture extension
In situ stress modification giving secondary fracturing

Shaped Charge Penetration: This technique offers a potentially attractive means for communicating with the formation in directions perpendicular to the well bore and/or perpendicular to natural or hydraulically induced fractures. Efforts to optimize explosively driven shaped charge devices for well stimulation applications have been carried out at the Los Alamos Scientific Laboratory (LASL) and are summarized by Carter (8). The capability of shaped charge devices to penetrate hard dense rock to depths many times greater than the charge diameter offers a means for establishing communication channels in preselected directions from a well bore.

Inasmuch as the highly damaged rock around a conduit created by shaped charge penetration will be highly impermeable, two methods have been proposed for establishing communication between the shaped charge conduit and the surrounding rock. Liquid high explosives could be emplaced in the shaped charge channels and, when detonated, would succeed in extensively fracturing the rock around these channels with a concordant increase in permeability. Alternatively, the portion of the well bore where shaped charge entry has been effected could be packed off and the shaped charge channel hydraulically pressurized until fracture occurs. Fractures so formed would probably be parallel to fractures which would have been formed at the well bore without shaped charge penetration but these fractures would be offset into the formation at distances controlled by the depth of shaped charge penetration and the capability to have fracture initiation occur near the tip of the shaped charge channel.

While the penetration characteristics of shaped charges are well known, (8) explosive or hydraulic fracturing methods for improving communication between shaped charge conduits and the formation remain to be developed and evaluated. The greatest limitation to the development of shaped charge methods for well stimulation probably derives from the fact that shaped charge penetration can only be ten to twenty times the diameter of the well bore in which the device is placed. As communication distances of hundreds of feet in preferred directions would be required for successful well stimulation, shaped charge devices may only see application in cases where limited penetration in very specific directions and at many stratigraphic levels is beneficial.

Dendritic Fracturing: This is basically a modification of conventional hydraulic fracturing wherein multiple pressurization, shut-in and pressure relief cycles are utilized to achieve certain objectives (3). The potential success of the Kiel method probably derives from two separate and distinct processes occuring during fracturing operations. The first process involves the development of secondary fractures, preferably orthogonal to the primary fracture, due to local perturbations to the in situ stress state effected by the fracturing operation.

As illustrated in Figure 8.14, a modification of the in situ stress state is realized by sanding out the primary fracture thus causing the fluid pressure acting within the fracture to rise to pressures higher than those that would be required for propagation of the primary fracture if the fluid pressure were active over a larger portion of the fracture surface. If the perturbation to the in situ stress field can be greater than the natural in situ stress differences, then secondary fractures inclined to the primary fracture could be formed.

A second process involves the spalling of material from the fracture faces which, after relative displacement and misalignment, can serve as a propping agent. This spall is effected by suddenly dropping the frac-fluid injection pressure such that stresses acting parallel to the face of the newly formed fracture (possibly assisted by residual fluid pressure within the rock) form extension type fractures parallel to the primary fracture. Intersection of such extension fractures with the primary fracture surface could form relatively large spalls.

Both the fluid pressure variations leading to spall and the transport of fine propping sand and fines produced during spall to the fracture tip are effected by cycles of pumping, shut-in and pressure relieving.

Figure 8.14: Schematic Representation of In Situ Stress Modification by the Over-Pressurization of a Sanded-Out Hydraulic Fracture

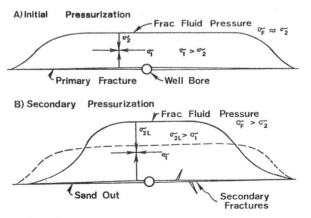

Source: Symposium I

Changes in both fracture breakdown pressure and instantaneous shut-in pressure on successive cycles are indicative of a change in the local in situ stress state and the possible propagation of new fractures in other directions. The recovery of relatively large (>¼" diameter) fragments of the formation rock indicate that spalling is occurring. Unfortunately, data which support the possibility of the two processes believed to be important in the dendritic fracturing process are scanty and not well substantiated.

Considerable laboratory and field data coupled with appropriate analytical and numerical modeling efforts will be required before any valid assessment of dendritic fracturing can be made. The required laboratory data would be related to the evaluation of potential spall behavior by means of extensional fracture formation and/or fluid pressure induced fracturing during rapid variations in confining pressure. The required field data should include well documented variations in fracture breakdown pressures and instantaneous shut-in pressures under conditions where such changes cannot be attributed to other causes such as variations in gel strength or fluid viscosity.

Modeling efforts should be devoted to both the evaluation of spall formation by extensional fracturing and pore pressure effects and modifications in the in situ stress field by pressurization of sanded-out fractures to pressures higher than initial breakdown or shut-in pressures. Additional data must be obtained and critical evaluations performed before significant field demonstrations with the dendritic-fracturing process are attempted.

Tailored-Pulse Loading: This represents a stimulation technique which has been developed towards a specific objective or goal. The initial efforts on tailored pulse loading were carried out by Physics International and have resulted in their Dynafrac process (9). The initial objective sought with tailored pulsed loading (and the one realized with Physics International's Dynafrac process) was to create multiple fractures emanating from a well bore without the well bore damage typically associated with bore-hole shooting.

As indicated schematically in Figure 8.15, the formation of multiple fractures around the well bore is dependent upon the rise time of the loading pulse. Moore, et al (9) calculate that the bore-hole stress loading rate must be greater than twice the tensile strength times the seismic wave speed in the formation rock divided by the bore hole circumference. Using typical values for rock, this criterion requires that the stress loading rate be on the order of 10^7 psi per second.

As only detonating high explosives are capable of supplying such loading rates down hole, Physics International developed a centralized explosive charge which, with appropriate fluid coupling to the formation, could supply the necessary loading rates and yet restrict peak pressures to those below those which would cause bore-hole damage by crushing or shear deformation of the rock. The centralized charge technique has its greatest limitation in that the pressure pulse is of extremely short duration and fractures of only a few well bore diameters could be expected to be created.

Physics International has proposed to enhance fracture extension by the pressurization of the buffer fluid (between the centralized charge and the well bore wall) with a conventional solid rocket motor propellant placed adjacent to the region containing the centralized charge (this is known as augmented Dynafrac). Ignition of this propellant at the appropriate time could provide the sustained pulse required for multiple fracture propagation as indicated schematically in Figure 8.15.

More recently, SRI International and Thiokol Corporation have suggested using liquid propellants as the buffer fluid between the explosive charges (de-

signed for rapid well bore loading) and the well bore wall. The controlled de-
flagration of this propellant following detonation of the primary explosive would
provide a long duration pulse similar to that which could be realized by the augmented Dynafrac process. To the extent that the explosive and propellant reaction products could flow into the multiple fractures formed around the well
bore without undue quenching, then propagation of the multiple fractures could
be enhanced by their internal pressurization. SRI International has been conducting an analytical and experimental program to evaluate the potential for
multiple well bore fracture of Devonian shale by tailored pulse loading.

**Figure 8.15: Formation and Propagation of Multiple Well Bore Fractures by
Rapid Well Bore Pressurization**

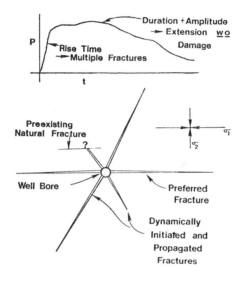

Source: Symposium I

    As illustrated in Figure 8.15 there are several aspects of tailored pulse
loading methods which are not yet well understood nor characterized. Further
efforts must be devoted to the determination of the potential for forming multiple well bore fractures in rocks of various types using practical explosive con-
figurations. Once such fractures are formed, it then remains to be demonstrated
that they may be propagated to significant distances within the formation either
by means of the radially divergent stresses created by well bore pressurization
or by direct pressurization of the fractures themselves. It must be determined
whether externally pressurized fluids (as in the augmented Dynafrac process) or
explosive and propellant reaction products (as in the SRI/Thiokol process) are
more effective for the pressure propagation of multiple fractures.
    Finally, it must be established to what extent multiple fractures can be
made to overcome the preferred fractured directions dictated by an anisotropic
in situ stress field or by the existence of preexisting fractures as indicated in
Figure 8.15. Because of the dynamic nature of many of the processes involved

in tailored pulse loading, many of these fundamental questions can be answered with careful numerical modeling employing finite-difference calculational techniques. Parameter sensitivity studies, such as those being initiated by SRI International, will be valuable in determining the relative and absolute importance of various rock properties related to multiple fracture initiation and propagation.

Emplaced High Explosive Stimulation: This probably represents one of the stimulation methods which was developed and put into practice before specific physical effects and production benefits were identified. Now that several emplaced explosive stimulation treatments are being conducted within the EGSP, consideration should be given to the probable effects of such a stimulation method. No particular claims have been made by the Petroleum Technology Corporation as to physical processes that may make the emplaced explosive stimulation method attractive. There are three conceivable processes, however, which could cause emplaced explosive stimulation to yield improved production.

As illustrated schematically in Figure 8.16, the first potentially beneficial process involves the formation of a secondary fracture system inclined to the primary fracture in which the explosive is detonated. This fracture system would be formed by the high transient stresses generated during explosive detonation. It is not felt that these fractures could be expected to extend significant distances into the formation and so these fractures would be of limited value to increased production.

Figure 8.16: Modification of In Situ Stress Field and Formation of Secondary Fractures by the Dynamic Explosive Pressurization of a Primary Fracture

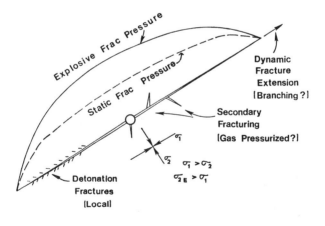

Source: Symposium I

The second process of potential value involves rapid extension of the primary fracture due to the rapid pressurization of the explosively filled primary fracture. This rapid fracture propagation could cause the fracture to traverse geologic variations or discontinuities that a slowly moving fracture would not traverse and possibly could cause more spalling and concordant propping due to fracture branching associated with rapid fracture propagation.

The third process by which emplaced explosive detonation might provide beneficial results derives from a modification of the in situ stresses (similar to that effected by fracture sand-out in the dendritic fracturing process) which could allow secondary fractures to be formed and propagated into the formation laterally, away from the primary, explosively loaded fracture.

Due to the dynamic nature of the three processes discussed above, their potential merits and contributions could be best evaluated using numerical modeling employing explicit finite-difference methods. Two-dimensional calculations could be used in parameter-sensitivity type studies to evaluate transient stress loading and concordant secondary fracture formation during detonation of an explosive in a primary fracture. Similar two-dimensional calculations could be utilized to evaluate the potential for modifying the in situ stress field and the propagation of secondary fractures by their pressurization with explosive reaction products.

Evaluation of the second potentially beneficial process, the rapid pressurization and propagation of the primary fracture, cannot be evaluated without additional data on dynamic fracture propagation. The difficulty of obtaining and interpreting such data on rock formations of interest precludes extensive and quantitative analysis of dynamic fracture propagation at this time.

Production Characteristics

The Eastern gas shales, or more precisely the Devonian shales, contain a typical reservoir rocks having production characteristics which could well require unique and highly specialized well completion and stimulation methods. Consequently, any program to develop, evaluate and optimize stimulation methods must include the careful characterization of the controlling production characteristics and their interactions with the stimulation processes.

There are at least two aspects of gas production from the Devonian shales which could be significantly different from typical, isotropic-permeability reservoirs. The first aspect relates to the potential for a dual or even triple porosity reservoir in which gas production characteristics and release times are uniquely different. As proposed by de Wys and Shumaker (10), gas production from fractured Devonian shale reservoirs could have three distinct origins.

Initial production would be dominated by the free gas contained in the larger interconnected fracture system encountered by the well or artificially induced fractures radiating therefrom. As soon as the formation pressure in this fractured reservoir begins to drop then adsorbed gas contained on the fracture surfaces would be released and could contribute to production. Further reductions in the reservoir pressure would result in the release of chemically absorbed gas contained in the capillary water, heavier organic compounds, and hydrous minerals comprising the rock. At the 500 to 1,000 psi formation pressure characteristic of Devonian gas wells, adsorbed and chemically absorbed gas could be several times greater than the gas which could be contained in a chemically inert sand reservoir of comparable porosity.

A detailed evaluation of production declined curves by de Wys and Shumaker (10) reveals production characteristics which require a complex reservoir gas evolution model to be examined. The common observation that the composition and heating value change significantly during the production life of Devonian gas wells further supports the thesis that production is from a dual or triple porosity type of reservoir.

The second aspect of importance to production characteristics in the Devonian shales relates to the anisotropy and/or fracture interconnectedness of the naturally fractured reservoir rock. If tectonic history and/or present day residual in situ stresses have resulted in a highly anisotropic fracture distribution or fracture conductivity, then production from wells in these reservoirs will not be uniformly distributed azimuthally around the well bore. Significantly anisotropic production characteristics could strongly control well interference effects, dictate optimum well spacing with respect to the high conductivity fracture orientation and could dictate the development of specialized stimulation techniques specifically designed to overcome this feature of reservoir production.

The interconnectedness of naturally fractured reservoirs in the vertical direction as controlled by stratigraphic variations could be as important as the interconnectedness in horizontal directions controlled by fracture conductivity anisotropy. Any lack of reservoir interconnectedness in the vertical or stratigraphic direction could require that specialized well completion and stimulation techniques be considered for development.

The potentially unique production characteristics of Devonian shale reservoirs dictate that these production characteristics be carefully identified and quantitatively evaluated. Because these production characteristics could have an important bearing upon the selection, application and interpretation of both conventional and novel stimulation methods, it is imperative that efforts to better delimit Devonian shale production characteristics be carried on simultaneously with any efforts to apply and evaluate stimulation methods.

A proper definition of Devonian shale production characteristics will require extensive laboratory and field data. The data requirements can be roughly characterized into one of three groups: laboratory core testing, field well testing and geologic controls and correlations.

The laboratory core measurements should be carried out on material obtained by carefully supervised pressure coring operations. As the core canisters are systematically depressurized the daily gas production rates should be monitored (11). Variations in gas composition during this depressurization and gas evolution history should be carefully monitored. Finally, the cores should be carefully logged and evaluated for macrofracture intersections, microfracture orientations and densities, water sensitivity (in terms of both expansive clays and capillary water retention), porosity and gas diffusivity.

The field well test data should include the careful evaluation of individual wells followed by several multiple well experiments. Individual well test data should be directed towards determining the well production characteristics prior to stimulation including stratigraphic variations in production, correlations between production intervals and naturally fractured sections of the formation, and adequate production-time data so that reservoir simulation models may be developed.

Poststimulation production tests should include a redetermination of the stratigraphic distribution in production (including correlations with naturally and artificially fractured sections of the formation), initial production tests for several months followed by a one to two week pressure buildup test and finally additional pressure buildup tests after the well has been in production for one or two years.

The multiple well tests should be designed to yield production and pressure interference data which can answer some of the basic questions on the interconnectedness of the natural fracture system both horizontally and vertically, the degree of anisotropy in the natural fracture system, the aerial extent of the fractured reservoirs contributing to the production of each well and, finally, the data base required to develop and evaluate the importance of a dual porosity reservoir.

The determination of geologic controls and correlations will require the definition of regions and subregions with specific geologic characteristics related to Devonian shale reservoir production characteristics. Determination of these controls and correlations will allow for the selection of areas most amenable to modeling and detailed stimulation evaluation and could eventually provide a basis for recommending optimum stimulation approaches for specific geologic areas. Finally, it must be reemphasized that all of the efforts devoted to reservoir modeling, the derivation of Devonian shale production characteristics and the correlation of these characteristics with geologic variations in the Appalachian area must all be developed closely with the efforts to develop and evaluate stimulation technologies.

Some effort should be given to the development and application of specialized instrumentation packages. Evaluation of the production characteristics of the Eastern gas shales and how these characteristics are modified by various stimulation treatments would be greatly facilitated by field measurements of in situ amplitudes and directions, in situ fracture densities and orientations, and gas production from discrete fractures and stratigraphic horizons.

Field data on the effects or performance of stimulation treatments, such as fracture propagation direction and distance in hydraulic treatments and explosive distribution and detonation in explosive treatments, would be especially valuable. Considerably more effort will be required for the deployment and interpretation of field measurements using existing equipment than for the development of new instrumentation.

Conclusions

The quantitative evaluation of the many well stimulation techniques being utilized in or proposed for the EGSP will require a more substantial basis than exists. During the course of this technical review it became apparent that two approaches for evaluation of stimulation technologies should be explored. The physical bases for each stimulation method must be identified and analyzed. The production benefits obtained by each stimulation method must be comparatively evaluated. The successful implementation of each of these two approaches will require a quantitative description of the pertinent reservoir production characteristics.

The physical processes which have been suggested or identified as giving each stimulation method its particular advantages need to be physically verified and quantitatively described. For explosive and propellant stimulation the rate of bore-hole loading is the critical parameter. The SRI International program will answer some key questions on the effects and benefits of loading rate but additional studies will be required on such features as fracture pressurization and fracture propagation to significant distances from the explosively loaded region.

As summarized in the previous discussion on novel stimulation methods, many aspects of the dynamic (explosively driven) stimulation methods can be analyzed using explicit finite-difference modeling techniques. The greatest restriction to the successful application of numerical modeling lies with the lack of reliable constitutive equation data for input to the models and the appropriate interpretation of the necessarily simplistic models in terms of complex geologic variations which cannot be explicitly included in the models.

For conventional hydraulic fracturing it is the further determination and quantitative description of parameters controlling the direction and vertical migration of fracture propagation which are most in need of study. For the dendritic fracturing process the potential to effect a significant perturbation in the natural in situ stress field needs to be critically analyzed. For evaluation of the quasi-static stimulation methods, traditional finite element methods can be readily employed. As for the dynamic analyses, a critical need exists for representative constitutive equations describing rock behavior. It is recommended that all future review and consideration of well stimulation technologies in the EGSP be directed toward the identification and quantitative description of the beneficial and detrimental physical processes acting in each of the methods.

In addition to critical physical evaluation, the conventional and novel stimulation technologies must also be subjected to a comparative evaluation. It is difficult to conduct such evaluations as each method is tested under field and geologic conditions which are nearly always uniquely different from those of other tests. The high costs of field tests probably precludes the execution of enough experiments to average out the geologic and field variations and provide a data base which could be meaningfully subjected to statistical analyses.

A physical description of the various stimulation processes can be combined with the reservoir production characteristics in parameter sensitivity evaluations directed towards determining the relative and absolute importance of the various geologic, rock property, stimulation and production parameters. Once the importance of and interactions between the various parameters have been defined, then test matrices can be derived for statistically determining the relative merit of various stimulation technologies. The significance of and additional requirements for statistical analyses might be further evaluated by the application of Monte Carlo parameter perturbation techniques such as have been utilized recently in geotechnical engineering design problems (12).

Certainly, the need to make statistical comparisons between the relative benefits of the various stimulation technologies requires that significant efforts be devoted to the establishment of bases for making such comparisons.

The evaluation of stimulation technologies in the EGSP cannot be carried out without a quantitative description of the controlling production characteristics. Both the naturally occuring production characteristics and how these characteristics are modified by the various stimulation methods need to be studied in detail. While many of the existing stimulation technologies could be evaluated with a minimum of information on production characteristics, the development, application and evaluation of completely new and untried stimulation methods will require a detailed knowledge of Devonian shale production characteristics. Only if the natural production characteristics are well known can the stimulation effects desired for optimum production be defined.

References

(1) Scharff, M.F. and McKay, M.W., *A Perspective on Stimulation of Eastern Gas Shales by Fracturing with Chemical Explosives,* Report for the U.S. Energy Research and Development Administration, Morgantown Energy Research Center, Contract No. EY-77-C-21-8113, 1977.

(2) Komar, C.A., Frohne, K.H. and Yost, A.B., *Stimulation Technology Development in the Eastern Gas Shales Project*, Morgantown Energy Research Center (1978).

(3) Kiel, M., *Hydraulic Fracturing Process Using Reverse Flow,* U.S. Patent 3,933,205 (January 20, 1976).

(4) Howard, G.C. and Fast, C.R., *Hydraulic Fracturing,* SPE-AIME Monograph, Vol. 2 (1970).

(5) Daneshy, A.A., "Hydraulic Fracture Propagation in Layered Formations." SPE-AIME 51st Annual Fall Technical Conference and Exhibition (1976).

(6) Simonson, E.R., Abou-Sayed, A.S. and Clifton, R.J., "Containment of Massive Hydraulic Fractures." SPE-AIME 51st Annual Technical Conference and Exhibition (1976).

(7) Kutter, H.K. and Fairhurst, C., "On the Fracture Process in Blasting." *International Journal of Rock Mechanics and Mining Sciences,* Vol. 8, No. 3, p. 181-202 (1971).

(8) Carter, W.J., "New Developments in Shaped Charge Technology." *Petroleum Engineer,* (April, 1978).

(9) Moore, E.T., Mumma, D.M. and Seifert, K.D., *Dynafrac—Application of a Novel Rock-Fracturing Method to Oil and Gas Recovery.* U.S. Energy Research and Development Administration, Washington, D.C. (1977).

(10) de Wys, J.N. and Shumaker, R.C., *A Pilot Study of Gas Production Analysis Methods Applied to Cottageville Field.* Report for the Department of Energy, Morgantown Energy Research Center, Contract No. EY-76-C-05-5194, 1978.

(11) Chase, R.W., "Natural Gas Production from Coal Seams." Presented at the Eastern Regional Meeting of the Society of Petroleum Engineers of AIME, SPE 6629, October 27-28, 1977.

(12) Kim, H.S., Major, G. and Ross-Brown, D.M., "Application of Monte Carlo Techniques to Slope Stability." 19th U.S. Symposium on Rock Mechanics, 1978.

Sources Utilized

DOE BETC/IC-79/3

Review of Fracture Fluid-Reservoir Interactions in Tight Gas Formations, prepared by B.A. Baker and H.B. Carroll, Jr. of Bartlesville Energy Technology Center for the U.S. Department of Energy, June 1979.

DOE CONF-780825-P3

Fourth DOE Symposium on Enhanced Oil and Gas Recovery and Improved Drilling Methods, Volume 2—Gas and Drilling, Tulsa, Oklahoma, August 29-31, 1978, edited by B. Linville of Bartlesville Energy Technology Center.

DOE CONF-790805-P3

Fifth DOE Symposium on Enhanced Oil and Gas Recovery and Improved Drilling Technology, Volume 3—Gas and Drilling, Tulsa, Oklahoma, August 22-24, 1979, edited by B. Linville of Bartlesville Energy Technology Center.

DOE EF-77-C-01-2705

Enhanced Recovery of Unconventional Gas: The Program—Volume II (of 3 Volumes), prepared by V.O. Kuuskraa and V.P. Brashear of Lewin and Associates, Inc., T.M. Doscher of University of Southern California and L.E. Elkins for the U.S. Department of Energy, October 1978.

NTIS PB80-128929

Help for Declining Natural Gas Production Seen in the Unconventional Sources of Natural Gas, report to the Congress by the Comptroller General of the United States, prepared by the General Accounting Office, Energy and Materials Division, January 1980.

OTA-E-57

Status Report on the Gas Potential from Devonian Shales of the Appalachian Basin, prepared by the staff of the Office of Technology Assessment for the Congress of the United States, November 1977.

DOE/FERC-0010

National Gas Survey Report to the Federal Energy Regulatory Commission by the Supply-Technical Advisory Task Force on Nonconventional Natural

Gas Resources, "Gas in Tight Formations," prepared by Sub-Task Force IV, F. Stead of the U.S. Geological Survey, T. Jennings of National Gas Survey, P. Brown of Columbia Gas Transmission Corp., J.M. Dennison of University of North Carolina, W. deWitt, Jr. of the U.S. Geological Survey, W. K. Overby, Jr. of Energy Research and Development Administration and A.B. Waters of Halliburton Services for the Department of Energy, June 1978.

DOE/FERC-0029

National Gas Survey Report to the Federal Energy Regulatory Commission by the Supply-Technical Advisory Task Force on Nonconventional Natural Gas Resources, Sub-Task Force I—Gas Dissolved in Water, prepared by J.W. Harbaugh of Stanford University, Task Force Chairman and Sub-Task Force I, P. Jones of Louisiana State University, T. Jennings of National Gas Survey, P.A. Dennie of Shell Oil Company, C.R. Hocott of University of Texas, P.E. LaMoreaux of P.E. LaMoreaux and Associates, D. Lombard of U.S. Department of Energy and R.H.Wallace, Jr. of U.S. Geological Survey for the Department of Energy, March 1979.